Analysis of Variance
in
Complex
Experimental
Designs

Analysis of Variance
in
Complex
Experimental
Designs

Harold R. Lindman

Indiana University

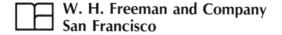 **W. H. Freeman and Company**
San Francisco

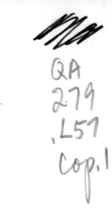
Library of Congress Cataloging in Publication Data

Lindman, Harold R
    Analysis of variance in complex
experimental designs.

    1. Analysis of variance. 2. Experimental
design. I. Title.
QA279.L57   519.5′35   74–11211
ISBN–0–7167–0774–8

Printed in the United States of America

10 9 8 7 6 5 4 3 2 1

Contents

Appendix   313

Preface

This book is intended primarily for graduate students who have already had a solid, noncalculus course in statistics. My own students cover most of volume one of Hays and Winkler's *Statistics* before studying ANOVA.

The emphasis in this book is on applied problems, although I believe one cannot use ANOVA intelligently without a solid grounding in theory. I have ignored or treated only lightly some traditional techniques of ANOVA that, in my experience, have proved to be of little or no use in practice. Instead, I present other, newer, techniques to help solve problems that arise in wrestling with real data.

The validity of some of these new techniques has not as yet been demonstrated. For example, the approximate formula for planned comparisons with unequal population variances (Chapter 4) is still being investigated. (Preliminary results indicate that it is valid, and I expect to establish its validity in a published journal article). Similarly, the criticisms of the Neuman-Keuls and Duncan tests (Chapter 4) are the result of an investigation still underway.

The emphasis throughout is on planned comparisons and combinations of planned comparisons. There are three reasons for this. First, I believe that students can more readily understand the nature of the standard tests if the tests are viewed as "combined" planned comparisons. Second, planned comparisons enable the experimenter to ask specific, thoughtful questions of data, instead of taking the traditional "shotgun" approach of the overall F test. Third, my own bias is toward Bayesian statistics (although the text is definitely not Bayesian), and I feel that planned comparisons are more in line with the spirit of the Bayesian approach than are overall F tests.

July, 1974   Harold R. Lindman

Analysis of Variance
in
Complex
Experimental
Designs

Review of Statistical Concepts

This text is written for students who have already had a relatively advanced, noncalculus course in statistics. In the first two chapters we shall review certain basic concepts and cover some fine points that may have been overlooked in earlier study. These chapters will also introduce the special notation used in the book, as well as my own statistical biases.

Statistics is in fact, a rather controversial subject. Although there is wide agreement on the principles of statistics, there is considerable disagreement on the application of these principles. Since my concern in this book is with knowledgable application of the principles of statistics, certain controversial subjects will necessarily be taken up. An explanation of my own biases may help clarify my stand in these controversies.

As to the special notation, unfortunately, much statistical notation is not standardized. Although the notation for new concepts will be introduced with the concepts, it seems simplest to present the basic statistical notation used in this book at an early point. (But see the symbol list at the end of the book.)

This text is concerned exclusively with statistical inference, in which probability theory plays a basic role.

PROBABILITY THEORY

Mathematical Concepts of Probability

Mathematically, probability theory is concerned with assigning numbers to events in such a way that those numbers represent, in some sense, how likely each event is to occur. To do this plausibly, the numbers must have certain properties. The basic properties are: that the probability (Pr) of an impossible event is zero; that no probability is greater than one; and that if two events, A and B, are mutually exclusive (if they cannot *both* occur), then Pr $(A$ or $B) =$ Pr $(A) +$ Pr (B).

From these basic properties, a number of other important properties are derived. Some of the main ones are:

$$\text{Pr (not } A) = 1 - \text{Pr } (A);$$
$$\text{Pr } (A \text{ or } B) = \text{Pr } (A) + \text{Pr } (B) - \text{Pr } (A \text{ and } B);$$
$$\text{Pr (not both } A \text{ and } B) = \text{Pr (not } A \text{ or not } B).$$

An important concept in statistics is *conditional probability,* defined generally as the probability that an event will occur on condition that another event will also occur, and defined mathematically as:

$$\text{Pr } (A|B) = \text{Pr } (A \text{ and } B)/\text{Pr } (B), \text{ Pr } (B) \neq 0.$$

Conditional probabilities play an important role in hypothesis testing. Acceptance or rejection of a null hypothesis (H_0) depends on the experimental outcome, and the experimental outcome is conditional on H_0.

APPLIED PROBABILITY THEORY

In the basic mathematical theory, neither events nor probabilities are given any practical definition. That task is left to applied probability theory and statistics.

The most commonly used definition limits the use of probabilities to the description of the outcomes of *experiments.* An experiment is defined technically as a set of acts that result in one of a group of possible events, and that can in principle be repeated an infinite number of times under identical circumstances. By this definition, the outcomes of flipping a coin have probabilities associated with them because the coin can be flipped over and over again under circumstances that, for all practical purposes at least, are identical. The outcomes of a boxing match, on the other hand, do not have probabilities associated with them because there is no way in which a boxing match can be repeated under even remotely identical circumstances.

For this definition, usually called the *relative frequency* definition, the probability of an event is the limiting relative frequency of the event as the number of repetitions of the experiment approaches infinity. All of the basic theory of statistics in this text was originally developed with the relative frequency interpretation of probability.

A different applied definition, considered disreputable by statisticians until recently but now growing rapidly in popularity, is the *subjective* (or *personalistic*) definition. It holds that a probability need not be tied to any particular relative frequency of occurrence, but that it is a measure of an individual's belief about the likelihood of the occurrence of an event. According to this view, insofar as a person has a belief about the likelihood of a particular outcome of a boxing match, the outcome of the boxing match has a probability. The probability of an event may therefore be different for different people, since people can differ in their beliefs about the outcomes of boxing matches. Moreover, the probability need not be related in any rigid way to any relative frequency, even if one exists, although the mathematics of probability theory asserts that under certain circumstances beliefs are influenced by relative frequencies.

Detailed discussion (including the philosophical bases) of each view of probability are not within the scope of this text. Nevertheless, these opposing views reflect differences in the general approach to statistics. Those who hold to relative frequency have generally had a rather rigid approach to statistical inference, setting up specific rules that must be followed exactly to obtain valid statistical results. For example, the probability that a null hypothesis is either true or false is not allowable. Since a null hypothesis is not the outcome of an infinitely repeatable experiment, it is considered improper to assign it a probability. Another example is the insistence on the unqualified acceptance or rejection of a null hypothesis on the basis of a predetermined significance level.

Personalists, on the other hand, freely talk about the probability that a null hypothesis is either true or false, but refuse to unqualifiedly accept or reject it on the basis of inconclusive evidence. As a general rule, the personalist uses statistics more freely, feeling that exact interpretation is more important than exact procedure. I am a personalist, and although the mathematics in this text have all been developed within the framework of relative frequency theory, much of the discussion on interpretation and proper procedures is colored by my bias for freer use of the procedures with intelligent interpretation of the results.

There is a third view, the *necessary view*. The necessary view asserts that probabilities are objectively defined, rather than being a matter of personal opinion, but that probabilities need not be defined as limiting relative frequencies. In the practical application of statistics, holders of the necessary view usually differ little from personalists.

DISTRIBUTION THEORY

The basic theory of statistics is built around probability distributions on random variables. Earlier I stated that probabilities are defined on outcomes. When we attach numbers to these outcomes so that outcomes on different trials of the experiment can be added, subtracted, etc., then the set of possible numbers is a random variable, and the numbers with their associated probabilities form a probability distribution.

Distributions may be either *discrete,* like the outcomes of tossing dice, or *continuous,* like heights of people. In a discrete distribution, every possible number has a definite probability; in a continuous distribution, single numbers have probability zero, but the probability of the random variable falling in some interval is nonzero. If, by six feet tall, we mean exactly six feet no matter how accurately we measure (i.e., up to an infinity of decimals), then the probability of being six feet tall is zero. However, the probability of being between five feet eleven inches and six feet one inch is greater than zero. Most commonly used distributions in statistics are continuous because they are mathematically simpler to use; the most important distribution in statistics, the normal distribution, is continuous.

Joint Distributions

In some cases each event may have two numbers of interest paired with it. In sampling an individual, for example, we may be interested in both his chronological and mental age. We then talk about the *joint distribution* of the two or more random variables involved. In a joint distribution of two random variables, probabilities are assigned to pairs of numbers, one from each random variable, instead of to single numbers. In a joint distribution of three random variables, probabilities would be assigned to triplets of numbers.

The probability distribution of any single random variable in such a set, or of any given subset of the random variables, is then called the *marginal distribution* of that random variable or subset of random variables. The *conditional distribution* of a random variable, X, conditional on Y, is the distribution that X has when Y is limited to only one of its possible values. The random variables X and Y are assumed to be independent if and only if the conditional distribution of X is the same as its marginal distribution. If X and Y are independent, then the conditional distribution of Y will also be the same as its marginal distribution.

Summary Values

It is common practice to characterize distributions by single numbers that represent certain important properties of the distributions. The most commonly used such numbers are measures of central tendency: the mean (or expected value), the median, and the mode. Other important measures are the measures of dispersion, usually the range, interquartile range, mean absolute deviation, and the standard deviation, or its square, the variance. The variance is most commonly used by statisticians because of its useful mathematical properties. It can be defined as the mean, or expected value, of the squared deviations about the mean of the distribution. If we let X_i be the value taken on by a random variable, and $\Pr(X_i)$ be the probability that it takes on that value, then the mean of a discrete distribution is

$$\mu = E(X) = \Sigma_i \, X_i \, \Pr \, (X_i),$$

and the variance is

$$\sigma^2 = V(X) = E[(X - \mu)^2] = \Sigma_i \, (X_i - \mu)^2 \, \Pr \, (X_i).$$

The mean and variance of a continuous distribution are found by replacing the summations with integrals.

Two other summary measures that are important in this text are *skewness* and *kurtosis*. They are best defined in terms of the third and fourth *moments about the mean*. For any value of r, the rth moment about the mean is defined as the expected value of the deviations from the mean, raised to the rth power:

$$\mu_r{}^* = E[(X - \mu)^r] = \Sigma_i \, (X_i - \mu)^r \, \Pr\,(X_i).$$

The first moment about the mean is zero, and the second is the variance.

To find the skewness, we divide the third moment about the mean by the cube of the standard deviation:

$$\mathrm{Sk} = \mu_3{}^*/\sigma^3.$$

The skewness of a symmetric distribution is zero. If the skewness is positive, the distribution tends to have a long tail to the right when graphed; if it is negative, the long tail is to the left. Figure 1-1 shows distributions with different amounts of skewness.

To find the kurtosis, we take the fourth moment about the mean, divide it by the fourth power of the standard deviation, and subtract three from the result:

$$\mathrm{Ku} = [\mu_4{}^*/\sigma^4] - 3.$$

The 3 plays an important role and cannot be neglected. It serves to standardize the kurtosis measure so that the kurtosis of the normal distribution is zero.

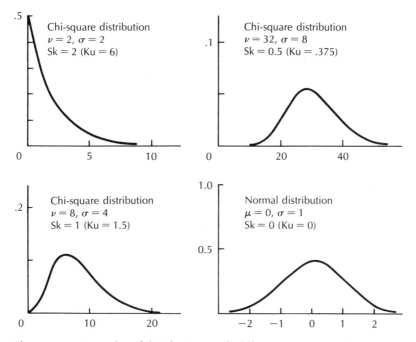

Figure 1-1. Examples of distributions with different amounts of skewness.

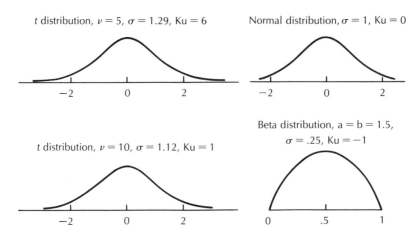

Figure 1-2. Examples of distributions with different amounts of kurtosis. Each has a skewness of 0.

The kurtosis is the peakedness of the distribution. If the kurtosis is positive, the distribution tends to have a very high sharp peak near the center and long thin tails at both ends. If it is negative, it tends to be relatively flat in the center and to drop off rapidly to zero near the ends. Kurtosis differs from skewness primarily in that kurtosis measures the overall tendency toward long tails at both ends; skewness measures the degree to which this tendency differs at the two ends. The skewness can take on any value between plus and minus infinity, although it will seldom be greater than plus or minus two. The kurtosis can only range from minus two to plus infinity. The rectangular distribution has a kurtosis of −1.2. Figure 1-2 shows sample distributions having different amounts of kurtosis.

Covariance and Correlation

An important summary value, the *covariance*, is analogous to the variance except that it applies only to joint distributions. In a discrete joint distribution on X and Y, the covariance is defined to be:

$$C(X,Y) = E[(X - \mu_x)(Y - \mu_y)] = \Sigma_i \ (X_i - \mu_x)(Y_i - \mu_y) \ \text{Pr} \ (X_i, Y_i),$$

where μ_x and μ_y are the means of X and Y, and Pr (X_i, Y_i) is the joint probability of X_i and Y_i. The similarity between this formula and the previous formula for calculating a variance should be apparent; the two formulae differ only in that one contains a sum of squares and the other a sum of cross-products. In effect, a variance is the covariance of the random variable with itself.

The covariance is a kind of measure of dependence of two random variables. If the random variables are independent, the covariance is zero; however, it may be zero even when the random variables are not independent. The covariance is a measure of the *linear dependence* of X and

Y. It measures the accuracy with which we can predict X from a knowledge of Y, using the equation

$$X = aY + b,$$

where a and b are constants. If we add to this equation the term ϵ, representing the error in predicting X from Y, we have

$$X = aY + b + \epsilon.$$

ϵ is a random variable with a probability distribution that can be found if the joint distribution of X and Y is known. If we choose a and b so as to minimize the variance of ϵ, then it can be shown that

$$a = C(X,Y)/V(Y)$$
$$b = E(X) - aE(Y),$$
$$E(\epsilon) = 0$$
$$V(\epsilon) = C(X,\epsilon) = V(X) - aC(X,Y)$$
$$C(Y,\epsilon) = 0$$

A more standardized measure of linear dependence is the *correlation coefficient*,

$$r_{xy} = C(X,Y)/(\sigma_X \sigma_Y).$$

The correlation coefficient is zero if there is no *linear* dependence between X and Y; it is plus or minus one if X is perfectly predictable from Y and vice versa; it is positive if a is positive, and negative if a is negative.

Finally, if we take the sum of two random variables, the result will be a random variable whose mean and variance are

$$E(X + Y) = E(X) + E(Y),$$
$$V(X + Y) = V(X) + V(Y) + 2C(X,Y).$$

If the covariance is zero, then the variance of the sum is just the sum of the variances.

SAMPLING

The basic task of inferential statistics is to use a sample from a distribution to make inferences about the distribution. Usually, the sample values are assumed to be randomly and independently sampled. The two basic problems of inference discussed at greatest length in this book are estimation and hypothesis testing.

Point Estimation

The basic problem of point estimation is to find a sample value that will be a good estimator of a similar value of the underlying distribution, or *population*. For example, the mean of the sample is frequently used to estimate the mean of the population.

Such an estimator is usually called a *statistic,* and the value being estimated is called a *parameter.* The statistic itself can be regarded as a random variable, with the distribution being that which we would obtain if we took a very large number of independent samples and calculated the value of the statistic for each sample. In this text, when we refer to a "best estimate" of some parameter, we mean that the statistic, regarded as a random variable, has two properties: it is unbiased and it has minimum variance.

By "unbiased" we mean that the expected value of the statistic is equal to the parameter; by "minimum variance" we mean that the variance of the statistic is smaller than that of any other unbiased estimator of that parameter. Frequently, however, both assertions will be strictly true only if the underlying population is normal.

The two most commonly estimated parameters are the mean and variance. Usually the mean of the population is estimated by the mean, $\overline{X}$, of the sample, and the variance of the population is estimated by

$$\hat{\sigma}^2 = ns^2/(n-1), \tag{1-1}$$

where n is the sample size, and s^2 is the variance of the sample:

$$s^2 = \frac{1}{n}\sum_{i=1}^{n}(X_i - \overline{X})^2.$$

Both of these estimators are unbiased, no matter what the underlying population mean and variance may be. They are "best" in that when the population distribution is normal, they are unbiased *and* have minimum variance; if the population is not normal, they may not have minimum variance.

Hypothesis Testing

In hypothesis testing, instead of trying to estimate the value of a parameter or set of parameters, we begin with the assertion that the parameters have specific values (the *null hypothesis*). We then use the sample to test that assertion.

We make a one-tailed test by first calculating a statistic that has a known distribution if the null hypothesis is true. We then determine how far out in the tail of that distribution the calculated value for our sample is, and the probability that it would be that far into the tail if the null hypothesis were true.

Technically, the proper way to proceed is to choose in advance some minimum probability, or *significance level,* α. If the actual probability is below α, then the null hypothesis is said to be *rejected* at the α level of significance. More and more, however, experimenters are simply reporting the obtained probability, rather than some previously chosen value such as the α level. Technically, this is incorrect within the relative frequency approach to testing null hypotheses; but it is in the spirit of the personalistic approach. It uses the α level as an index of how confident the data allows one to be when rejecting the null hypothesis. *In this text I will follow the procedure of always reporting the obtained probability as* α. By convention, I will usually regard something as significant if $\alpha \leq .05$, although I would be unwilling to impose either practice on anyone else.

In two-tailed tests we recognize that the data could be in either tail. The probability computed is the probability that the statistic would be as far as it is into either tail. The two-tail probability is usually about twice as large as the one-tail probability. Just as in a one-tailed test, the obtained probability can be used as the alpha level.

Note that statistical significance does not mean the same as "meaning-fulness" or "importance." If the statistical test is very sensitive to small differences, a difference too small to be of interest may be significant. Conversely, if the statistical test is not sensitive, a very large and potentially important difference may not be significant. Significance is a strictly statistical quality, based on the probability that the results could have been obtained by chance.

Power

The power (β) of a test is technically the probability of rejecting the null hypothesis when it is false. It is a function of the type of test used, the size of the sample, the significance level (α), and the difference between the true value of the parameter and that specified by the null hypothesis.

In general, however, when comparing two types of tests, I will talk about relative power without specifying any of the above. When I say that one type of test is more powerful than another, I will mean that it is more powerful for any (or almost any) values of n, α, or true value of the parameter.

SUMMATION

This text necessarily makes use of relatively complicated equations involving summation. It is important, therefore, to have a solid understanding of the definition and important properties of the summation sign.

The summation sign itself has already been used earlier. It is simply a short way of representing the addition of a list of numbers:

$$\Sigma_i X_i = X_1 + X_2 + X_3 + X_4 + \ldots$$

Usually the values of i over which the sum is to be taken are assumed to be obvious from the nature of the equation. If they cannot be assumed to be obvious, the limits of the summations are stated explicitly below and above the summation sign:

$$\sum_{i=3}^{5} X_i = X_3 + X_4 + X_5,$$

or simply,

$$\sum_{3}^{5} X_i = X_3 + X_4 + X_5,$$

it being implicitly understood that the summation is to be over values of the subscript i.

Some important properties of summations are

$$\Sigma_i (X_i \pm Y_i) = \Sigma_i X_i \pm \Sigma_i Y_i,$$

$$\sum_{i=1}^{n} (aX_i + b) = a \sum_{i=1}^{n} X_i + nb,$$

$$\Sigma_i \Sigma_j X_{ij} = \Sigma_j \Sigma_i X_{ij},$$

$$\sum_{i=1}^{n} \sum_{j=1}^{i} X_{ij} = \sum_{j=1}^{n} \sum_{i=j}^{n} X_{ij}.$$

These properties are all directly derivable from the simple properties inherent in addition of numbers.

NOTATION

Most of the mathematical terms in this book follow standard mathematical notation. However, there are some differences, and not all students are familiar with mathematical terms, so let us review the basic system.

First, both Greek and Roman letters are used for symbols. With two exceptions, Greek letters are used exclusively to represent properties (parameters) of populations. In general, then, Greek letters represent values that can be estimated but not known. The two exceptions are the letters α and β, used to represent the significance level and the probability of a Type II error, respectively. The following Greek letters are used in the text:

α (alpha)	μ (mu)
β (beta)	ν (nu)
γ (gamma)	ϕ (phi)
δ (delta)	ψ (psi)
ϵ (epsilon)	ω (omega)
λ (lambda)	

Roman letters are used to represent sample values, i.e., statistics. For example, s^2 stands for the variance of a sample, while σ^2 represents the corresponding population variance. There is a significant exception to this practice, when statistics are used to estimate population values. An estimate is represented by the symbol for the particular population parameter, but the symbol will wear a "hat." For example, the mean of a population is symbolized by the Greek letter μ, whereas the estimate of a population mean is symbolized by $\hat{\mu}$. Similarly, $\hat{\sigma}^2$ stands for the unbiased estimate of the population variance. Specifically, if the sample size is n, then,

$$\hat{\sigma}^2 = \frac{ns^2}{n-1}.$$

Subscripts $i, j, k, l, \ldots$, are used to denote different values of quantities of the same type. Suppose, for example, that we have data from 25 subjects in each of 4 schools in each of 3 towns. Then X_{ijk} might represent the observation on the kth subject in the jth school in the ith town. In such a case the number of possible values of each subscript is indicated by the corresponding capital letter. Thus, in the example just given, I would be the total number of towns studied ($I = 3$), and J would be the number of schools in each town ($J = 4$).

For the number of subjects, however, we will use a different convention. Subjects are usually randomly sampled, and n traditionally represents the number of subjects in a single random sample. We will usually use n to designate the size of such a random sample (in our example, $n = 25$). In the experiments discussed in this text, several different samples will be taken from different groups. In the above study, for example, we have taken 12 samples (4 schools in each of 3 towns) of 25 each, for a grand total of $12 \times 25 = 300$ observations. We will let N represent the total number of observations in the entire experiment (in the example, $N = 300$).

Since sample means are very important, we will have a special method of designating them. A bar over a symbol will indicate that a mean is intended, and the subscript over which we have averaged will be replaced by a dot. Thus, $\overline{X}_{ij\bullet}$ will represent the average, over the 25 subjects (subscript k, replaced by a dot), for the jth school in the ith town. Or,

$$\overline{X}_{ij\bullet} = \frac{1}{n} \sum_{k=1}^{n} X_{ijk}.$$

If we next averaged over the schools, we would have, for the ith town,

$$\overline{X}_{i\bullet\bullet} = \frac{1}{J} \sum_{j=1}^{J} \overline{X}_{ij\bullet} = \frac{1}{nJ} \sum_{j=1}^{J} \sum_{k=1}^{n} X_{ijk}.$$

Finally, the grand mean of all the observations would be:

$$\overline{X}_{\bullet\bullet\bullet} = \frac{1}{I} \sum_{i=1}^{I} \overline{X}_{i\bullet\bullet} = \frac{1}{N} \sum_{i=1}^{I} \sum_{j=1}^{J} \sum_{k=1}^{n} X_{ijk}.$$

For computational purposes, we will sometimes work with totals instead of means. The total, or sum, of a set of observations will be represented by t with appropriate subscripts. Thus,

$$t_{ij\bullet} = \sum_{k=1}^{n} X_{ijk} = n\overline{X}_{ij\bullet},$$

$$t_{i\bullet\bullet} = \sum_{j=1}^{J} t_{ij\bullet} = \sum_{j=1}^{J} \sum_{k=1}^{n} X_{ijk} = nJ\overline{X}_{i\bullet\bullet},$$

$$t_{\bullet\bullet\bullet} = \sum_{i=1}^{I} t_{i\bullet\bullet} = \sum_{i=1}^{I} \sum_{j=1}^{J} \sum_{k=1}^{n} X_{ijk} = N\overline{X}_{\bullet\bullet\bullet}.$$

2

Important Distributions

This book requires an understanding of several different distributions. Although the purpose of this chapter is simply to review these distributions, it also presents the special distributional notation used throughout the book, as well as some facts you may not already know.

THE NORMAL DISTRIBUTION

The normal distribution is basic to most statistical inference. It is the familiar bell-shaped curve seen in Figures 1-1 and 1-2. The skewness and kurtosis of the normal distribution are both zero, and a number of statistical tests assume that the population distribution is the normal distribution.

Areas of the normal distribution are usually tabled in *standard form*. The normal distribution in standard form, or the *standard normal distribution,* is the form that has a mean of zero and a variance of one.

According to the *central limit theorem,* the main reason the normal distribution is important is that it is the limiting distribution of the sample mean as the sample size approaches infinity. Regardless of the population from which we draw (within certain rather broad limits), the distribution of the sample mean will be approximately normal if the sample size is large. This principle has a broader application than one may at first suppose.

The sample variance, for example, is a kind of mean (the mean of the squared deviations from the expected value), so that the distribution of the sample variance approaches the normal as the sample size increases.

Several other important properties are unique to the normal distribution. When sampling from a normal population, the sample mean and sample variance are statistically independent. This property plays an important role in relation to the other distributions discussed here. Two other important properties concern the joint distribution of two normally distributed random variables. First, any statistical dependence that may exist between two such random variables must be linear. This means that if two normally distributed random variables are uncorrelated (i.e., their correlation is zero), they are statistically independent. Second, the sum of any two normally distributed random variables is also normally distributed.

DISTRIBUTIONAL NOTATION

We will find it convenient to use a distributional notation that is not usually found in basic statistics texts, one that is useful for defining and describing all of the distributions in this text. We will use the notation

$$X \sim N_{(\mu, \sigma^2)}$$

to indicate that X has the normal distribution with a mean of μ and a variance of σ^2. If X has the standard normal distribution, for example, then

$$X \sim N_{(0,1)}.$$

The $\sim$ sign in this case has a somewhat different meaning than an equality sign. It means "is distributed as" instead of "is equal to."

Although the $\sim$ here does not mean the same thing as $=$, many similar operations can be performed with it. For example, we know that if we add a constant to a standard normal random variable, we leave the variance the same but change the mean to the value of the constant. That is, if

$$X \sim N_{(0,1)},$$

then

$$X + a \sim N_{(a,1)},$$

which we can symbolize in a single equation by

$$N_{(0,1)} + a \sim N_{(a,1)}. \tag{2-1}$$

If we now multiply $(X + a)$ by a constant, b, the new mean will be ab and the new variance will be b^2. We can symbolize these changes with the equation

$$b[N_{(0,1)} + a] \sim N_{(ab,b^2)}.$$

We can also form functions of two or more independent random variables by this method. If, for example, we add two independent random variables, each with the standard normal distribution, their sum is normally distributed with a mean of zero and a variance of two:

$$N_{(0,1)} + N_{(0,1)} \sim N_{(0,2)}. \tag{2-2}$$

Whenever notations for two distributions are both found on the same side of the $\sim$, they will be assumed in this text to be independent.

The above equation points up some limitations of this notation. If the $\sim$ were used like $=$, we could subtract $N_{(0,1)}$ from both sides, giving

$$N_{(0,1)} \sim N_{(0,2)} - N_{(0,1)}.$$

Since the random variables on the same side of the equal sign are assumed to be independent, however, the above equation is not legitimate. Instead,

$$N_{(0,3)} \sim N_{(0,2)} - N_{(0,1)}.$$

Similarly, with Equation 2-2, we might be tempted to add the two identical terms on the left, giving

$$2N_{(0,1)} \sim N_{(0,2)}.$$

This is also illegitimate, however, since

$$2N_{(0,1)} \sim N_{(0,4)}.$$

In general, once an equation has been written, we can perform any operation we wish on the constants that appear in it, but we cannot operate in any way on the symbols representing the random variable. In Equation 2-1, for example, we can subtract a to obtain

$$N_{(0,1)} \sim N_{(a,1)} - a,$$

a legitimate equation. However, we cannot subtract $N_{(0,1)}$ to obtain

$$a \sim N_{(a,1)} - N_{(0,1)},$$

which implies that the difference between two independent random variables is a constant.

CHI-SQUARE

We can put the new notation to work immediately with the definition of the *chi-square distribution*. The chi-square distribution with ν degrees of freedom is defined to be the distribution of the sum of ν independent squared random variables, each random variable having the standard normal distribution. In our notation, we write

$$\chi_{(\nu)}^2 \sim \sum_{i=1}^{\nu} [N_{(0,1)}]^2.$$

The chi-square distribution is *not* to be confused with the popularly used chi-square test. The chi-square test was so named because the statistic that is computed when making that test has, *approximately*, the chi-square distribution. The derivation of the distribution, however, has nothing to do with the test.

The mean of the chi-square distribution is ν, and its variance is 2ν. It has both positive skewness and positive kurtosis, with both decreasing toward zero as ν approaches infinity. By the central limit theorem, the chi-square distribution is approximately normal for large ν. The distributions in Figure 1-1 are examples of the chi-square distribution.

The chi-square distribution is important primarily because, when the population is normally distributed, it gives us the distribution of the sample variance, s^2. Specifically,

$$ns^2/\sigma^2 \sim \chi^2_{(n-1)},$$

or

$$ns^2 \sim \sigma^2\chi^2_{(n-1)}.$$

If we write this in terms of the best estimate of the population variance (Eq. 1-1) we have

$$(n-1)\,\hat{\sigma}^2/\sigma^2 \sim \chi^2_{(n-1)}$$

$$\hat{\sigma}^2 \sim \sigma^2\chi^2_{(n-1)}/(n-1). \tag{2-3}$$

On the right side of this equation, we have a chi-square variable divided by its degrees of freedom, a variable that is important in defining the t and F distributions.

Another important property of the chi-square distribution is that the sum of *independent* chi-square distributions is also chi-square. The degrees of freedom of the sum is equal to the sum of the degrees of freedom of the individual chi-squares:

$$\Sigma_i\,\chi^2_{(\nu_i)} \sim \chi^2_{(\Sigma_i\nu_i)}.$$

NONCENTRAL CHI-SQUARE

A related distribution, less commonly discussed in statistics texts but important for some parts of this text, is the *noncentral chi-square* distribution. The noncentral chi-square distribution with ν degrees of freedom and noncentrality parameter ϕ is the sum of ν independent normally distributed random variables, each of which has a variance of 1, but a mean of $\phi \neq 0$. It thus differs from the chi-square distribution in the means of the variables that are squared and summed. In our special notation:

$$\chi^2_{(\nu,\phi)} \sim \sum_{i=1}^{\nu}\,[N_{(\phi,1)}]^2.$$

The noncentral chi-square distribution, and the noncentral F derived from it, are important in estimating the power of the statistical tests described in later chapters.

t DISTRIBUTION

The t distribution, with ν degrees of freedom, is defined as

$$t_{(\nu)} \sim \frac{N_{(0,1)}}{\sqrt{\chi^2_{(\nu)}/\nu}}$$

The t distribution is symmetrical, with a mean of 0 and a variance of $\nu/(\nu - 2)$. The kurtosis is positive, decreasing to zero as ν approaches infinity and the t distribution approaches the normal. Some of the distributions in Figure 1-2 are examples of the t distribution.

The t distribution is important generally in statistics because it is central to the commonly used t test. The t distribution and the t test, however, should not be confused — the one is a probability distribution, the other is a statistical test that happens to make use of that distribution.

The t test takes advantage of the unique statistical properties of the normal distribution, described above. In particular, if the population distribution is normal, then several things are true: First, the sample mean is normally distributed with a mean of μ and a variance of σ^2/n, so that

$$(\overline{X}_. - \mu)\,\sqrt{n} \sim N_{(0,\sigma^2)} \sim \sigma N_{(0,1)}.$$

Second, by Equation 2-3, the estimate of the population variance is proportional to chi-square, so that

$$\hat{\sigma} \sim \sigma\,\sqrt{\chi^2_{(n-1)}/(n-1)}$$

Third, the sample mean and variance are independent, allowing us to take their ratio and get

$$\frac{(\overline{X}_. - \mu)\,\sqrt{n}}{\hat{\sigma}} \sim \frac{\sigma N_{(0,1)}}{\sigma\,\sqrt{\chi^2_{(n-1)}/(n-1)}} = t_{(n-1)},$$

because the σs in the numerator and denominator cancel out. The t test for a single sample tests the null hypothesis H_0: $\mu = \mu^*$ by substituting μ^* for μ and comparing the result with the tabled values of the t distribution.

The t test for the difference between the means of two groups makes use of the fact that if the two groups have equal population means and equal population variances, then

$$\frac{\overline{X}_{1.} - \overline{X}_{2.}}{\sqrt{1/n_1 + 1/n_2}} \sim \sigma N_{(0,1)} \tag{2-4}$$

$$\sqrt{\frac{(n_1 - 1)\hat{\sigma}_1{}^2 + (n_2 - 1)\hat{\sigma}_2{}^2}{n_1 + n_2 - 2}} \sim \sigma\,\sqrt{\chi^2_{(n_1 + n_2 - 2)}/(n_1 + n_2 - 2)} \tag{2-5}$$

The t test for the difference between means takes the ratio of the above two quantities and compares it with the t distribution with $(n_1 + n_2 - 2)$ degrees of freedom.

F DISTRIBUTION

The F distribution is defined as the ratio of two independent chi-squares, each divided by its degrees of freedom:

$$F_{(\nu_1, \nu_2)} \sim \frac{\chi^2_{(\nu_1)}/\nu_1}{\chi^2_{(\nu_2)}/\nu_2}$$

The values ν_1 and ν_2 are commonly called the numerator and denominator degrees of freedom, respectively.

The mean of F is

$$E[F_{(\nu_1, \nu_2)}] = \nu_2/(\nu_2 - 2),$$

which is approximately one if ν_2 is large. The formula for the variance is complicated, but the variance tends to decrease as ν_1 and ν_2 increase.

One useful property of the F distribution is that the reciprocal is also distributed as F, with the degrees of freedom reversed:

$$1/F_{(\nu_1, \nu_2)} \sim F_{(\nu_2, \nu_1)}.$$

Another important property of the F distribution is that both the chi-square and the t distribution are special cases of it. The relationship of F to these other two distributions is

$$F_{(\nu_1, \infty)} \sim \chi^2_{(\nu_1)}/\nu_1; \tag{2-6}$$

$$F_{(1, \nu_2)} \sim [t_{(\nu_2)}]^2. \tag{2-7}$$

Consequently, any table of the F distribution also contains, implicitly, tables of the chi-square and t distributions. The F distribution can theoretically be used to test the null hypothesis that two different populations have the same variance. Under that null hypothesis, and the assumption that the population distributions are normal, the ratio of the sample variances has the F distribution. The assumption of normal population distributions is critical to the test, however; if that assumption is violated, the test is not valid. Since the populations can seldom be assumed to be precisely normal, the test is seldom valid. The real importance of the F distribution is the subject of the rest of this book.

NONCENTRAL F

Just as there is a noncentral version of the chi-square distribution, so is there a noncentral version of the F distribution. In fact, the two noncentral distributions are very closely related; the noncentral F is defined like central F except that the numerator is a noncentral chi-square divided by its degrees of freedom. The noncentral F will be important for some of the power calculations discussed in later chapters.

TABLES AND INTERPOLATION

The appendix contains tables of the normal, chi-square, t, and F distributions. Since the distributions are continuous, however, only selected values are included. Other values can be found by complex interpolation methods (simple linear interpolation usually does not work well).

One easy and fairly accurate way to interpolate using graph paper is illustrated in Figure 2-1. Here we find the significance level of an F ratio obtained in Chapter 3. The obtained F is 4.4 with $\nu_1 = 2$, $\nu_2 = 12$. From the F table we find and plot the tabled F ratios versus their significance levels for 2 and 12 degrees of freedom. We then draw a smooth curve through these points. Finally, we read from the graph that $F = 4.4$ has a significance level of about .04 or, more accurately, .036.

The graph in Figure 2-1 is on semi-log paper. Ordinary graph paper will do, but the curve drawn on semi-log paper is usually closer to a straight line (generally, the straighter the curve, the easier it is to draw). If obtain-

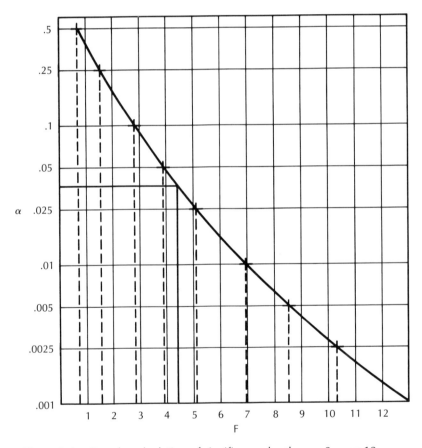

Figure 2-1. Sample calculation of significance level; $\nu_1 = 2$, $\nu_2 = 12$, and $F = 4.4$.

able, "normal-curve" paper (with values on the Y axis spaced so that the normal integral plots as a straight line) is usually even better. If both degrees of freedom are small (say, less than five), log-log paper may work best.

Accurate interpolation between different degrees of freedom can be accomplished by linear interpolation among the reciprocals of the degrees of freedom. To facilitate this, degrees of freedom listed in the tables in the appendix are mostly in multiples of either 60 or 120.

If, for example, we want to find the F needed for $\alpha = .05$, with 3 and 14 degrees of freedom, we solve for k in the equation

$$\frac{120}{14} = k \left(\frac{120}{12}\right) + (1 - k) \left(\frac{120}{15}\right),$$

or $8.57 = 10k + 8(1 - k)$, or $k \approx .3$. Then we have $F_{(3, 14)} = .3F_{(3, 12)} + .7F_{(3, 15)}$. For $\alpha = .05$, $F_{.05(3, 14)} = .3(3.49) + .7(3.29) = 3.35$.

3

Analysis of Variance, One-way, Fixed Effects

In the simplest version of the t test, the means of two independent groups of subjects are compared. The simplest form of the F test, or *analysis of variance*, is an extension of the t test to the comparison of more than two groups. Consider, for example, the data in Table 3-1. These are hypothetical test scores obtained by college students majoring in each of three different areas. Five randomly selected students from each area were given the same test — since the test is hypothetical and the data are imaginary, neither the nature of the test nor the areas represented need concern us. Just to add a touch of realism, let us assume that the scores are on a test of "intellectual maturity," and that the three courses are "History of Lichtenstein, 1901–1905" (A_1), "Modern Indian Basket Weaving" (A_2), and "Psychological Problems of Golf Widows" (A_3).

The experimenter is interested in the question: do students from different areas really tend to score differently on the test, or are the obtained differences due to the random selection of the subjects? That is, if the "true" mean of the scores in group A_i is μ_i, then the null hypothesis is

$$H_0: \mu_i = \mu \text{ for all groups } A_i. \tag{3-1}$$

ASSUMPTIONS

To apply the analysis of variance, the experimenter must first make some assumptions about the data. He must assume that the population of scores of all students within any area of study, A_i, is normally distributed with a mean, μ_i, and a variance, σ^2. He must further assume that the variance is

Table 3-1. Hypothetical data for independent samples from three groups of five subjects each.

	A_1 History	A_2 Art	A_3 Psychology	Sum	Mean
X_{ij}	5	1	5		
	3	2	4		
	5	2	7		
	6	0	5		
	1	3	2		
$\Sigma\, X_{ij}$	20	8	23	51	
$\overline{X}_{i.}$	4.0	1.6	4.6	10.2	3.4
$\overline{X}_{i.}^{2}$	16.0	2.56	21.16	39.72	
$\Sigma\, X_{ij}^{2}$	96	18	119	233	
$(n-1)\hat{\sigma}_i^2$	16.0	5.2	13.2	34.4	
$\hat{\sigma}_i^2$	4.0	1.3	3.3	8.6	2.867

	SS	MS	df
w	34.4	2.867	12
bet	25.2	12.600	2
t	59.6	4.257	14

$F_{(2,12)} = 12.6/2.867 = 4.395$

the same for all three areas, and that all scores are independent of each other. Put more precisely, he must assume that each score, X_{ij}, is sampled randomly and independently from a normally distributed population having a mean μ_i and a constant variance σ^2. In mathematical notation, this can be stated as

$$X_{ij} \sim N_{(\mu_i, \sigma^2)}.$$

This particular notation will be used throughout the book. It is intended only as a shorthand way of saying that X_{ij} is normally distributed with a particular mean and variance.

These same assumptions (applied to two groups) are required in the t test; we are simply generalizing them to where the experimenter is comparing I different treatment groups (for our example, $I = 3$). These assumptions may seem too restrictive to ever hold for real data—later we will consider the effects of violating them. For the moment, however, let us assume that they hold for the data in question.

The experimenter does not need to assume that either σ^2 or the μ_i are known. In fact, if the μ_i were known, his problem would be solved. However, the assumption that all three populations have the same variance, σ^2, is important. In fact, the F test is a comparison between two different estimates of σ^2 calculated in different ways. One of the two, based only on the variances *within* the individual groups, is always an unbiased estimate of σ^2. The other, based on the differences *between the means* of the groups, is unbiased if and only if the null hypothesis is true. If the null hypothesis is true, both of these estimates should be approximately equal, and their ratio should be approximately equal to one. If the null hypothesis is false, we would expect their ratio to be very different from one. When making an F test, therefore, we are in the unusual position of using the ratio of two variances to test for a difference between means.

ANALYSIS: EQUAL SAMPLE SIZES

To see how the t test is extended to an F test comparing two variances, we must first digress to an estimation problem. Suppose we have a sample of only two observations (X_1 and X_2) from a normal population, and we wish to estimate the variance of that population. The mean of a sample of two is

$$\overline{X} = (X_1 + X_2)/2,$$

so the variance is

$$s^2 = \{[X_1 - (X_1 + X_2)/2]^2 + [X_2 - (X_1 + X_2)/2]^2\}/2.$$

With a little algebra, this can be shown to be

$$s^2 = (X_1 - X_2)^2/4.$$

From Equation 1-1, the unbiased estimate of the population variance is then

$$\hat{\sigma}^2 = 2s^2 = (X_1 - X_2)^2/2$$

If we square the formula for the numerator of the t test (Eq. 2-4), assuming equal sample sizes, we get

$$(\overline{X}_{1.} - \overline{X}_{2.})^2(n/2),$$

which is just n times the unbiased estimate of variance that we would get if we regarded $\overline{X}_{1.}$ and $\overline{X}_{2.}$ as single observations from the same population instead of as the sample means from two different populations. (For simplicity, we will develop the F test first for equal sample sizes; later we will generalize to unequal sample sizes.)

Under what circumstances can we assume that $\overline{X}_{1.}$ and $\overline{X}_{2.}$ are two observations from the same population? Consider the assumptions necessary for the t test:

$$\overline{X}_{1.} \sim N_{(\mu_1, \sigma^2/n)}$$
$$\overline{X}_{2.} \sim N_{(\mu_2, \sigma^2/n)}.$$

If $\mu_1 = \mu_2 = \mu$, then $\overline{X}_{1.}$ and $\overline{X}_{2.}$ are identically distributed and, for all practical purposes, we can regard them as a sample of two observations from the same population. In other words, they can be regarded as being from the same population if and only if the null hypothesis is true. Furthermore, if the null hypothesis is true, then the square of the numerator of the t test is an estimate of n times the variance of $\overline{X}_{1 \bullet}$ and $\overline{X}_{2 \bullet}$. But the variance of $\overline{X}_{1 \bullet}$ and $\overline{X}_{2 \bullet}$ is σ^2/n, so the squared numerator of the t test is an unbiased estimate of σ^2.

The squared denominator of the t ratio is another estimate of σ^2, the average of the separate variance estimates from each sample (see Eq. 2-5). Thus, the t statistic squared is the ratio of two variance estimates; furthermore, this ratio has the F distribution with one degree of freedom in the numerator, since from Equation 2-7,

$$[t_{(\nu)}]^2 \sim F_{(1, \nu)}.$$

The generalization to three or more groups is straightforward. We find two estimates of σ^2—the first being the average of the estimates from the individual groups, the second being based on the variance of the group means. Under the null hypothesis that the population means of the groups are all equal, the ratio of the two estimates will have the F distribution. In the next two sections we will derive these two estimates of σ^2.

Mean Square Within

For the first estimate, we begin by considering each of the three samples separately. For each sample, we find the usual unbiased estimate of the population variance:

$$\hat{\sigma}_i^2 = \frac{\Sigma_j (X_{ij} - \overline{X}_{i \bullet})^2}{n - 1} \tag{3-2}$$

which, by algebraic manipulations, becomes

$$\frac{\Sigma_j X_{ij}^2 - (1/n)(\Sigma_j X_{ij})^2}{n - 1},$$

where $\overline{X}_{i \bullet}$ is the mean of the sample of observations from group A_i, and n is the number of observations in each sample. For group A_1, for example,

$$\hat{\sigma}_i^2 = \frac{96 - (1/5)(20^2)}{5 - 1} = 4.0$$

The other two $\hat{\sigma}_i^2$ are found in the same way.

It can be shown that the best overall unbiased estimate of σ^2 in this situation is the average of the separate estimates obtained from the three samples:

$$MS_w = (1/I) \Sigma_i \hat{\sigma}_i^2. \tag{3-3}$$

The symbol MS_w in Equation 3-3 refers to *mean square within*; it is so named because it is the mean of the squared errors within the samples. For the example of Table 3-1,

$$MS_w = (1/3)(4.0 + 1.3 + 3.3) = 2.867.$$

Let us now consider how MS_w is distributed. According to Equation 2-3, $(n-1)\hat{\sigma}_i^2/\sigma^2$ is distributed as chi-square with $(n-1)$ degrees of freedom. In addition, each $\hat{\sigma}_i^2$ is independent of each of the others because they are independent random samples. Therefore, if we define

$$SS_w = \Sigma_i \, (n-1)\hat{\sigma}_i^2$$
$$= (n-1) \, \Sigma_i \, \hat{\sigma}_i^2$$
$$= I(n-1)MS_w \text{ (from Eq. 3-3)} \qquad \text{(3-4)}$$

(SS_w refers to *sum of squares within*), then

$$SS_w/\sigma^2 = \frac{\Sigma_i \, (n-1)\hat{\sigma}_i^2}{\sigma^2} \sim \Sigma_i \, \chi^2_{(n-1)}$$

In other words, SS_w/σ^2 is the sum of I independent random variables, each distributed as chi-square. But the sum of I independent chi-square random variables is also distributed as chi-square, and its degrees of freedom are the sum of the degrees of freedom of the variables. That is, if we let N be the total number of observations in all of the groups combined, ($N = nI$), so that

$$\Sigma_i \, (n-1) = I(n-1) = N - I,$$

then

$$SS_w/\sigma^2 \sim \chi^2_{(N-I)}$$

and

$$SS_w \sim \sigma^2 \chi^2_{(N-I)}.$$

The symbol $\chi^2_{(N-I)}$ is similar to the notation used earlier for the normal distribution; it will be used frequently, and merely states that SS_w/σ^2 is distributed as chi-square with $(N - I)$ degrees of freedom. For the example of Table 3-1 this would mean that

$$SS_w = 12(2.867) = 34.40 \sim \sigma^2\chi^2_{(12)}.$$

Since MS_w is equal to SS_w divided by its degrees of freedom (see Eq. 3-4), it follows that MS_w is distributed as a chi-square variable divided by its degrees of freedom, times σ^2. That is,

$$MS_w \sim \frac{\sigma^2}{(N-I)} \chi^2_{(N-I)}. \qquad \text{(3-5)}$$

Mean Square Between

According to normal distribution theory, the mean of each sample is normally distributed with a mean equal to the population mean and a variance equal to the population variance divided by n:

$$\overline{X}_{i\cdot} \sim N_{(\mu_i, \sigma^2/n)}.$$

Thus, under the null hypothesis, for all groups, A_i,

$$\bar{X}_{i\bullet} \sim N_{(\mu, \sigma^2/n)}. \tag{3-6}$$

This means that all the $\bar{X}_{i\bullet}$ are normally distributed with the same mean and variance. We can treat them in effect as if they themselves were a random sample of I observations from the same population. We can estimate the variance of this population by calculating the sample variance in the usual way. We must remember, however, that since there are only I treatment groups, there are only I observations in the sample. The estimate of the variance is

$$\hat{\sigma}^2/n = \frac{\Sigma_i\ (\bar{X}_{i\bullet} - \bar{X}_{\bullet\bullet})^2}{I - 1}$$

which, by algebraic manipulations, becomes

$$\hat{\sigma}^2/n = \frac{\Sigma_i\ \bar{X}_{i\bullet}^2 - (1/I)\ (\Sigma_i\ \bar{X}_{i\bullet})^2}{I - 1},$$

where $\bar{X}_{\bullet\bullet} = (1/N)\ \Sigma_i\ \Sigma_j\ X_{ij}$ is the grand mean of all of the observations.

Multiplying the above estimate by n gives us an unbiased estimate of σ^2, MS_{bet}:

$$MS_{bet} = n(\hat{\sigma}^2/n) = \frac{n\ \Sigma_i\ (\bar{X}_{i\bullet} - \bar{X}_{\bullet\bullet})^2}{I - 1}$$

$$= \frac{n[\Sigma_i\ \bar{X}_{i\bullet}^2 - (1/I)\ (\Sigma_i\ \bar{X}_{i\bullet})^2]}{I - 1}. \tag{3-7}$$

Finally, just as we found it useful to define SS_w for calculational purposes, we shall define *sum of squares between* as

$$SS_{bet} = (I - 1)MS_{bet} = n\ \Sigma_i\ (\bar{X}_{i\bullet} - \bar{X}_{\bullet\bullet})^2 \tag{3-8}$$

Mean Square Total

A third estimate plays an important part in later calculational formulas. MS_{bet} is an estimate of σ^2 if and only if the null hypothesis is true. However, it is not the best estimate of σ^2 when the null hypothesis is true. The best estimate is found by treating the data as though they were all from the same population. We would then assume that the sample contained N observations, and the best estimate of σ^2 would be

$$MS_t = \frac{\Sigma_i\ \Sigma_j\ (X_{ij} - \bar{X}_{\bullet\bullet})^2}{N - 1}$$

which, by algebraic manipulations, becomes

$$MS_t = \frac{\Sigma_i\ \Sigma_j\ X_{ij}^2 - (1/N)\ (\Sigma_i\ \Sigma_j\ X_{ij})^2}{N - 1}. \tag{3-9}$$

The symbol MS_t refers to *mean square total*, and is based on the total of all observations. We can now define the *sum of squares total*:

$$SS_t = (N - 1)MS_t \sim \sigma^2\chi^2_{(N-1)}. \tag{3-10}$$

The last part of Equation 3-10 derives from the fact that under H_0, MS_t is regarded as an unbiased estimate of the variance based on a single sample of N observations from the same population. For the data of Table 3-1:

$$MS_t = \frac{233 - (1/15)(51)^2}{15 - 1} = 4.257,$$

$$SS_t = (15 - 1)(4.257) = 59.60.$$

The *F* Test

In general, MS_{bet} is the worst estimate of the three. It has two serious disadvantages not shared by the other two estimates. First, it is based only on the variability of the means of the samples, ignoring the variability of the observations within a sample. Thus, all of the information that can be gained from the variability of the observations within each sample (the variability that is used explicitly in calculating MS_w) is lost when MS_{bet} is used as an estimate of σ^2. Second, the estimate depends heavily on the assumption that all of the parent populations from which the samples were drawn have the same mean. As an estimate of σ^2, it is rather poor.

Nevertheless, these two disadvantages become advantages when we set out to test the null hypothesis stated earlier. Let us consider the ratio $F = MS_{bet}/MS_w$ under the assumptions we have made so far, including the null hypothesis (Eq. 3-1). Under these assumptions, MS_w and MS_{bet} are independent random variables. Their independence derives from the fact that MS_w depends only on the variances of the samples, whereas MS_{bet} depends only on their means. Since the mean and variance of a sample from a normal population are independent, MS_w and MS_{bet} are also independent. Their ratio, then, is

$$MS_{bet}/MS_w \sim \frac{\sigma^2\chi^2_{(I-1)}/(I - 1)}{\sigma^2\chi^2_{(N-I)}/(N - I)}$$

$$\sim \frac{\chi^2_{(I-1)}/(I - 1)}{\chi^2_{(N-I)}/(N - I)} \sim F_{(I-1, N-I)}.$$

That is, since the F distribution is defined as the distribution of the ratio of two independent χ^2 variables, each divided by its degrees of freedom, MS_{bet}/MS_w has the F distribution with $(I - 1)$ and $(N - I)$ degrees of freedom. Notice, however, that MS_{bet} is distributed as $\chi^2_{(I-1)}$ only if the null hypothesis is true. Hence, the ratio is distributed as F only if the null hypothesis is true. This suggests that we can test the null hypothesis by comparing the obtained ratio with that expected from the F distribution. The procedure is the same as that in a t test, except that the F distribution is used instead of the t distribution. If the obtained ratio varies too greatly from the expected ratio (which, for the F distribution, is approximately one, since MS_{bet} and MS_w are unbiased estimates of the same quantity and therefore should be about equal), we can conclude that the null hypothesis is false. For our example, $F_{(2,12)} = 12.6/2.867 = 4.395$. According to the F table, the probability of obtaining a ratio this large or larger, with two and twelve degrees of freedom, is smaller than .05 but not as small as .025. By using the graphic method described in Chapter 2 (pp. 18–19), we can determine that the probability is in fact about .04.

Before we can reject the null hypothesis with confidence, however, two important questions must be answered. The first is the question of whether a one-tailed or two-tailed test is appropriate. If a one-tailed test is appropriate, we can reject the null hypothesis at the 4 percent level. If it is not, we must consider the other tail of the distribution in calculating the significance level.

The other question involves the importance of the assumptions for the F test. In addition to the null hypothesis that the population means are equal, the F test also requires that all observations be sampled randomly and independently from normally distributed populations with the same variance σ^2. If the obtained value of F deviates significantly from its expected value, we can conclude with some confidence that one of the assumptions is false. Until we have examined the issues more carefully, however, we cannot confidently reject the null hypothesis. The large F ratio may be due to nonnormal populations, unequal population variances, or nonrandom sampling. The next two sections will take up these issues.

Expected Values of Mean Square Within and Between

The question of whether a one-tailed or two-tailed test is appropriate is most easily answered by considering the expected values of the two random variables (MS_w and MS_{bet}) that go into the calculation of F. Under the assumptions discussed above, including that of the null hypothesis, MS_w and MS_{bet} both have the same expected value, σ^2, since they are both unbiased estimates of the population variance. However, the expected value of MS_w is σ^2 even if the null hypothesis is false; based only on the variances of the samples, MS_w is independent of any assumptions we might make about the group means.

The expected value of MS_{bet}, on the other hand, depends in a very definite way on the true means of the populations from which the samples were drawn. To derive the expected value of MS_{bet}, we must first introduce the concept of a *linear model*, and we can best introduce the concept by deriving an appropriate linear model for the data under discussion.

Let us begin by letting μ_i represent the true mean of the ith treatment population. (Notice that μ_i represents the mean of the *population*, not the sample.) Then each observation can be represented as

$$X_{ij} = \mu_i + \epsilon_{ij} \tag{3-11}$$

where $\epsilon_{ij} \sim N_{(0,\sigma^2)}$. This is just another way of stating the assumptions made at the beginning of this chapter, but it shows more clearly that each observation can be thought of as a random deviation from its expected value, the population mean. Notice, now, that in Equation 3-11, X_{ij} is expressed as the sum of two variables. Whenever an observed value is expressed as a linear combination (in this case, a simple sum) of two or more other variables, we have a *linear model*. Linear models will become increasingly important as we consider more and more complicated experimental designs. They provide us with a means of separating out the various influences affecting a given observation. In this very simple case, for example, the linear model allows us to distinguish between the constant effect due to the group from which an observation is drawn (μ_i) and a random component due to the random selection of the observations from that population (ϵ_{ij}).

The linear model of Equation 3-11 is easier to use if it is expanded slightly. To expand it, we first define

$$\mu = \Sigma_i \, \mu_i / I$$

and

$$\alpha_i = \mu_i - \mu. \tag{3-12}$$

That is, the *grand mean*, μ, is the average of the population means, and α_i is the difference between each group mean and the grand mean. The linear model then becomes

$$X_{ij} = \mu + \alpha_i + \epsilon_{ij} \tag{3-13}$$

where, because of the definition of the α_i in Equation 3-12,

$$\Sigma_i \, \alpha_i = 0. \tag{3-14}$$

Since, $\mu_i = \mu + \alpha_i$ in terms of this model, the null hypothesis (Eq. 3-1) can be restated as

$$H_0: \alpha_i = 0, \text{ for all } i.$$

The quantities μ and α_i can be regarded as concrete though unobservable values. As concrete quantities, they can be estimated from the data. The best unbiased estimates of them are

$$\hat{\mu} = \overline{X}_{..}$$
$$\hat{\alpha}_i = \overline{X}_{i.} - \overline{X}_{..}.$$

It is interesting to note that Equation 3-7 can be rewritten as

$$MS_{\text{bet}} = \frac{n \, \Sigma_i \, \hat{\alpha}_i^2}{I - 1}.$$

Note that the size of MS_{bet} depends on the *squares* of the $\hat{\alpha}_i$.

The estimates $(\hat{\alpha}_i)$ can help us interpret the differences among the groups. For the data in Table 3-1, for example,

$$\hat{\alpha}_1 = 4.0 - 3.4 = 0.6$$
$$\hat{\alpha}_2 = 1.6 - 3.4 = -1.8$$
$$\hat{\alpha}_3 = 4.6 - 3.4 = 1.2.$$

From these we see that the most deviant group is A_2, whose mean is much smaller than that of either A_1 or A_3. This deviation is the main contributor to the large MS_{bet}, and therefore to the significant F. Although the deviation of A_2 is only half again as great as that of A_3, the *squared* deviation is more than twice as large. Consequently, group A_2 contributes more than twice as much to MS_{bet} and F.

We can now use Equation 3-13 to derive the expected value of MS_{bet}. We begin by deriving the expected value of SS_{bet}; the expected value of MS_{bet} can then be found by dividing by $I - 1$.

To find $E(SS_{bet})$, note first that, by Equation 3-13,

$$\overline{X}_{i.} = (1/n) \ \Sigma_j \ X_{ij} = (1/n) \ \Sigma_j \ (\mu + \alpha_i + \epsilon_{ij}),$$

which, when the summation and the division by n are carried out on each term, becomes

$$\overline{X}_{i.} = \mu + \alpha_i + (1/n) \ \Sigma_j \ \epsilon_{ij}. \tag{3-15}$$

By a similar derivation,

$$\overline{X}_{..} = \mu + (1/N) \ \Sigma_i \ \Sigma_j \ \epsilon_{ij}. \tag{3-16}$$

The α_i drop out of Equation 3-16 because $\overline{X}_{..}$ is found by summing over all groups and, consequently, by summing over all of the α_i; by Equation 3-14, the α_i sum to zero and drop out of the equation.

We will simplify the notation if we define

$$\bar{\epsilon}_{i.} = (1/n) \ \Sigma_j \ \epsilon_{ij} \sim N_{(0, \sigma^2/n)}, \tag{3-17}$$

$$\bar{\epsilon}_{..} = (1/N) \ \Sigma_i \ \Sigma_j \ \epsilon_{ij} \sim N_{(0, \sigma^2/N)}, \tag{3-18}$$

so that $\bar{\epsilon}_{i.}$ is the mean of the ϵ_{ij} in group A_i and $\bar{\epsilon}_{..}$ is the mean of the $\bar{\epsilon}_{i.}$. Equations 3-15 and 3-16 can then be rewritten:

$$\overline{X}_{i.} = \mu + \alpha_i + \bar{\epsilon}_{i.},$$
$$\overline{X}_{..} = \mu + \bar{\epsilon}_{..}.$$

We can now write SS_{bet} (see Eq. 3-8) as

$$\begin{aligned} SS_{bet} &= n \ \Sigma_i \ (\overline{X}_{i.} - \overline{X}_{..})^2 \\ &= n \ \Sigma_i \ [(\mu + \alpha_i + \bar{\epsilon}_{i.}) - (\mu + \bar{\epsilon}_{..})]^2 \\ &= n \ \Sigma_i \ [\alpha_i + (\bar{\epsilon}_{i.} - \bar{\epsilon}_{..})]^2 \end{aligned} \tag{3-19}$$

To get the expected value of Equation 3-19, we first expand the square (the term inside the brackets) to get

$$SS_{bet} = n \ \Sigma_i \ [\alpha_i^2 + (\bar{\epsilon}_{i.} - \bar{\epsilon}_{..})^2 + 2\alpha_i(\bar{\epsilon}_{i.} - \bar{\epsilon}_{..})].$$

We then distribute the summation sign over the individual terms:

$$SS_{bet} = n[\Sigma_i \alpha_i^2 + \Sigma_i (\bar{\epsilon}_{i.} - \bar{\epsilon}_{..})^2 + 2 \Sigma_i \alpha_i(\bar{\epsilon}_{i.} - \bar{\epsilon}_{..})].$$

Then, since the expected value of a sum is simply the sum of the expected values,

$$E(SS_{bet}) = n\{E(\Sigma_i \alpha_i^2) + E[\Sigma_i (\bar{\epsilon}_{i.} - \bar{\epsilon}_{..})^2] + 2E[\Sigma_i \alpha_i(\bar{\epsilon}_{i.} - \bar{\epsilon}_{..})]\}. \quad \text{(3-20)}$$

Finally, since the α_i are constants,

$$E(\Sigma_i \alpha_i^2) = \Sigma_i \alpha_i^2$$

and

$$E[\Sigma_i \alpha_i(\bar{\epsilon}_{i.} - \bar{\epsilon}_{..})] = \Sigma_i \alpha_i E(\bar{\epsilon}_{i.} - \bar{\epsilon}_{..}) = 0.$$

The last equation equals zero because, from Equations 3-17 and 3-18,

$$E(\bar{\epsilon}_{i.} - \bar{\epsilon}_{..}) = E(\bar{\epsilon}_{i.}) - E(\bar{\epsilon}_{..}) = 0 - 0 = 0.$$

Substituting these results into Equation 3-20, we get

$$E(SS_{bet}) = n\{\Sigma_i \alpha_i^2 + E[\Sigma_i (\bar{\epsilon}_{i.} - \bar{\epsilon}_{..})^2]\}.$$

Then, dividing $E(SS_{bet})$, term by term, by $(I - 1)$, we have

$$E(MS_{bet}) = n\{\Sigma_i \alpha_i^2/(I - 1) + E[\Sigma_i (\bar{\epsilon}_{i.} - \bar{\epsilon}_{..})^2/(I - 1)]\}. \quad \text{(3-21)}$$

Since $\bar{\epsilon}_{i.}$ is a normally distributed random variable and $\bar{\epsilon}_{..}$ is the mean of a sample of these random variables, $\Sigma_i (\bar{\epsilon}_{i.} - \bar{\epsilon}_{..})^2/(I - 1)$ behaves statistically like an unbiased estimate of the variance of the $\bar{\epsilon}_{i.}$. We cannot use this as an estimate, of course, because the $\bar{\epsilon}_{i.}$ are not directly observable. The fact that we cannot actually observe them, however, does not change their mathematical properties. Their variance has the same mathematical properties as any sample from a normal population. In particular,

$$E \Sigma_i (\epsilon_{i.} - \bar{\epsilon}_{..})^2/(I - 1) = \sigma^2/n, \quad \text{(3-22)}$$

as can be seen from Equation 3-17 and the fact that the function on the left of the equality in Equation 3-22 acts like an unbiased estimate of the variance of the $\bar{\epsilon}_{i.}$. Substituting Equation 3-22 into Equation 3-21, we get

$$\begin{aligned} E(MS_{bet}) &= n[\Sigma_i \alpha_i^2/(I - 1) + \sigma^2/n] \\ &= \sigma^2 + n \Sigma_i \alpha_i^2/(I - 1), \end{aligned} \quad \text{(3-23)}$$

which is what we set out to derive.

The first thing to note from Equation 3-23 is that if the null hypothesis ($\alpha_i = 0$ for all i) is true, then $E(MS_{bet}) = \sigma^2$, as we stated earlier. If the null hypothesis is false, a term must be added to σ^2 to find $E(MS_{bet})$. The added term, moreover, will always be positive because it is a positive constant times the sum of the squares of the α_i. Therefore, if the null hypothesis is false, the expected value of MS_{bet} will be larger than the expected value of MS_w, and the F ratio will tend to be larger than one. An obtained F smaller than one cannot be due to a false null hypothesis; it must be due to the violation of one or more of the other assumptions discussed above, to faulty arithmetical calculations, or to pure chance. It may be due to any of these causes, or to all of them, but an F smaller than one cannot be grounds for rejecting the null hypothesis. Consequently, in an analysis of variance, the F test is always one-tailed. In the case of our example, the actual significance level obtained is .04, and we arrive at that figure without considering the other tail of the F distribution. It remains, however, for us to assure ourselves that this abnormally large value of F has been caused by the violation of the null hypothesis rather than by the violation of one of the other assumptions we have made.

ROBUSTNESS OF F

None of the above assumptions necessary for an F test are ever fully satisfied by real data. The underlying populations from which the samples are drawn are never exactly normally distributed with precisely equal variances. If the violation of these assumptions were likely to result in a large F, then the F test would be a poor test of the null hypothesis. One could never know whether a significant F justified rejection of the null hypothesis or was due to the violation of one of the other assumptions. A statistical test which is not greatly affected by such extraneous assumptions is termed *robust*. We will examine the robustness of the F test with respect to each of the assumptions that must be made. The treatment here will necessarily be sketchy. A more thorough theoretical treatment is beyond the scope of this book (such a treatment can be found in Scheffé, 1959, Ch. 10).

Effects of Nonnormality

The most important departures from normality are likely to be either non-zero skewness or nonzero kurtosis (see pp. 5–6 for definitions of skewness and kurtosis.) These, of course, are not the only ways in which the distribution can be nonnormal, but other departures from normality are not likely to be very important. In general, only the kurtosis of the distribution is likely to have any appreciable effect on F. Although MS_w is an unbiased estimate of σ^2 no matter what the population distribution— i.e., MS_w, regarded as a random variable, has a *mean* of σ^2 no matter what shape the population distribution might have—the *variance* of MS_w depends on the kurtosis of the population. If the kurtosis is high, the resulting F will tend to be too small, and the data will appear to be less significant than they really are. The opposite result occurs if the population kurtosis is smaller than zero. In that case the F ratio will tend to be too large, and the null hypothesis may be rejected at a higher significance level than the data justify. Since the skewness of the distribution has little effect on MS_w, the effects of skewness on F can usually be ignored.*

---

*Neither the skewness nor the kurtosis have much effect on MS_{bet} if N is fairly large. The central limit theorem applies here, so that nonnormality of the population is unimportant.

Fortunately, in most cases the effects of nonzero kurtosis can be ignored as well. Box and Anderson (1955) have shown that when the kurtosis is not zero, an approximate F test can still be used. In the approximate test, F is calculated in the usual way but with different degrees of freedom. The degrees of freedom are found by multiplying both the numerator and the denominator degrees of freedom, calculated in the usual way, by $1 + Ku/N$, where Ku is the kurtosis of the population. To illustrate this, consider again the example in Table 3-1. These data were actually taken from a rectangular distribution with $Ku = -1.2$ instead of from a normal distribution. Consequently, instead of an F with 2 and 12 degrees of freedom, we should have multiplied each of these values by $1 + (-1.2)/15 = .92$. Our degrees of freedom would then have been

$$(2)(.92) = 1.84 \simeq 2 \text{ and } (12)(.92) = 11.04 \simeq 11.$$

With two and eleven degrees of freedom, our significance level would be about .039, which is very close to the actual value (to three decimals) of .037. Therefore, our previous value of $\alpha = .04$ will hold. As can be seen from this example, deviations from normality seldom have much effect on F, even if the deviations are large.

Of course, the correction factor just discussed will have little value in practical applications because the kurtosis of the population will not be known. However, it can be used as a rough guide in estimating how large an effect nonnormality is likely to have. In most applications the kurtosis will fall somewhere between plus or minus two, although in some extreme cases it may be as high as seven or eight (it is mathematically impossible for the kurtosis to be smaller than minus two). With such small values of kurtosis, in most applications even moderate sizes of N will make the correction factor so close to one that it can be ignored. Table 3-2 gives actual significance levels for a nominal significance level of .05, assuming

Table 3-2. Probabilities of Type I error with the F test for the equality of means of five groups of five each at the nominal five-percent significance level, approximated for the Pearson distribution by a correction on the degrees of freedom.

			Ku		
Sk^2	-1	-0.5	0	0.5	1
0	.053	.051	.050	.048	[a]
0.5	.052	.051	.050	.049	[a]
1	.052	.050	.049	.048	.048

Source: Box and Andersen, 1955, p. 14.
[a] The method of approximation used was unsatisfactory for these combinations of values.

five groups of five observations each taken from populations with varying amounts of skewness and kurtosis. For most practical applications, deviations as small as those in Table 3-2 can be ignored.

We conclude, then, that the F test is robust with respect to nonnormality.

Unequal Variances

The F test also tends to be robust with respect to the assumption of equality of variances as long as each group contains the same number of observations. The exact effects of unequal variances are difficult to calculate, but when the assumption of equal variances is violated, the obtained significance level is usually higher than it should be. Table 3-3 shows the *true* significance levels, i.e., α values, associated with an *obtained (apparent)* significance level of $\alpha = .05$ for some selected sample sizes and variance ratios. These data make it clear that some bias exists when the variances are unequal (usually the F ratio will not be quite as significant as it appears to be), but the bias is small if the variances in the different treatment groups do not vary too greatly.

Tests for Equality of Variances

A common practice in research using the analysis of variance has been to test the assumption of equality of variances before calculating F. If the result of the test for equality of variances is significant, the experimenter concludes that the F test cannot be made; if the result is not significant, the experimenter concludes that the F test is justified. Of course, this amounts to accepting the null hypothesis as true, and the appropriateness of accepting a null hypothesis, given nonsignificant statistical results, has

Table 3-3. Effects of inequality of variances on probability of Type I error with F test for equality of means at nominal five-percent level.

I	n	Group variances	True α
2	7	1:2	.051
		1:5	.058
		1:10	.063
3	5	1:2:3	.056
		1:1:3	.059
5	5	1:1:1:1:3	.074
7	3	1:1:1:1:1:1:7	.12

Source: Combined data from Hsu (1938) and Box (1954).

long been a subject of debate. Aside from this, however, there are other arguments against the commonly used methods of testing for equality of variances.

The three most commonly used tests are the Hartley, Cochran, and Bartlett tests. In the Hartley test the variance of each sample is calculated and the ratio of the largest to the smallest sample variance is compared with a set of tabled values. The Cochran test compares the largest sample variance with the average of all of the sample variances. The most sensitive test, the Bartlett test, compares the logarithm of the mean of the sample variances with the mean of their logarithms. All of these tests share a serious deficiency: they are all very sensitive to departures from normality in the population distribution. They all tend to mask existing differences in variance if the kurtosis is smaller than zero, or to exhibit nonexistent differences if the kurtosis is greater than zero. This deficiency is particularly serious because the F test itself is relatively insensitive to both non-normality and inequality of variance (when sample sizes are equal). If the population kurtosis is greater than zero, the experimenter may decide not to perform a perfectly valid F test. All three tests for inequality of variances are described in Winer (1971), with appropriate tables provided. I do not recommend their use, however. Scheffé (1959, pp. 83–87) has suggested an alternative test that is not sensitive to departures from normality. Unfortunately, it requires that the user make some preliminary guesses about the sizes of the variances. Scheffé recommends using it only when the equality of the variances is itself of theoretical interest; he does not recommend it as a preliminary test to the analysis of variance.

Transformations on the Data

In some cases the variances of the treatment groups have a known relationship to the means of the groups. Such is the case, for example, when each observation, X_{ij}, is a proportion. If the mean of a sample of proportions is close to one or zero, the variance will be small, whereas it tends to be larger if the mean is close to one-half. If the variances have a known relationship to the group means, it may be possible to transform the X_{ij} in such a way that the variances of the transformed X_{ij} are independent of their means. When the X_{ij} are proportions, for example, the *arcsin transformation*, $Y_{ij} = \arcsin \sqrt{X_{ij}}$, will make the variances approximately equal, independently of the means. (Tables of the arcsin transformation are in the appendix.)

Two other transformations are commonly used. If the variances can be assumed to be approximately proportional to the means of the distributions, then $Y_{ij} = \sqrt{X_{ij}}$ is an appropriate transformation. If the standard deviations are approximately proportional to the means of the distributions, i.e., if the variances are approximately proportional to the squares of the means, then $Y_{ij} = \log(X_{ij})$ is appropriate. Still other transformations can be found for various other relationships between the means and variances.

It should be remembered, however, that such transformations are not necessary in most cases for doing an analysis of variance. Furthermore, these transformations change the group means as well as the variances.*

Sometimes, however, a transformation is appropriate. For example, if each observation (X_{ij}) is itself an estimate of a variance, one might be undecided whether to perform the *F* test on the variance estimates themselves or on their square roots, which would be estimates of standard deviations. An appropriate transformation (for the considerations described above) is $Y_{ij} = \log(X_{ij})$. This is especially appropriate when the observations are variances because the difference between the logarithm of $\hat{\sigma}^2$ and the logarithm of $\hat{\sigma}$ (the estimated standard deviation) is only a multiplicative factor of two. Since multiplication by two is a linear transformation, and linear transformations do not affect *F*, the *F* test on the logarithms of the variance estimates will be the same as the test on the logarithms of the estimated standard deviations. The test is essentially a test of the equality of the geometric means of the variance estimates (or, equivalently, a test of the equality of the geometric means of the estimated standard deviations). In this case the transformation has some intuitive justification in addition to the purely mathematical basis.

A transformation may also be appropriate when the data are measured in time units, such as running times, reaction times, or numbers of trials required to learn a response. Such data are often highly skewed, with the variance increasing as a function of the mean. The reciprocals of the times can be interpreted as measures of speed, and they often (though not always) have more satisfactory statistical properties.

Independence of Observations

The *F* test clearly requires that each sample be randomly obtained, with each observation independent of all other observations. It might be assumed, however, that a test that is robust with respect to the other assumptions is robust with respect to this requirement as well. Such is not the case. If the observations are not independent random samples, the obtained value of *F* is likely to be strongly affected. The direction of the effect depends on the nature of the dependence between the observations. In most actual cases the dependencies tend to make MS_w smaller than it should be and *F* correspondingly too large (although the opposite can also occur). One is thus likely to reject the null hypothesis at a higher level of significance than is justified by the data. The only solution to this problem is to gather the data in such a way that one can be reasonably certain the assumption of independence is justified.

---

*Many times they change them in ways that are not intuitively meaningful. In the case of proportions, for example, the statement that the means of the proportions in different groups differ from each other is easily understood. The statement that the means of the arcsins of the square roots of the proportions in different groups differ from each other is more difficult to interpret.

POWER OF THE *F* TEST

We have found so far that the *F* test may be a useful test of the null hypothesis that the means of *I* independently sampled populations are all equal. In addition, we have found that the *F* test is relatively insensitive to departures from the normality and equal variance assumptions. The final problem is to determine just how sensitive it is to departures from the equality of means specified by the null hypothesis. If it is not sensitive to such departures, none of its other features can make it a useful test.

The sensitivity of *F* to the violation of the null hypothesis is called its *power*. The power of *F* is measured as the probability that the null hypothesis will be rejected when it is in fact false. In other words, the power is $1 - \beta$, where β is the probability of a Type II error. The power of the *F* test depends on the distribution of *F* when the null hypothesis is false. This distribution is *noncentral F*, and it depends importantly on four factors: the actual means of the treatment populations, the population variance, the significance level (α), and the number (n) of observations in each sample. To calculate the power exactly, all four of these quantities must be known. Since the first two cannot in practice be specified, it is impossible to calculate the power of the *F* test exactly. However, some things are known. First, for most sets of population means, a given σ^2, and a given α, and under the assumptions used in computing *F*, the *F* test is the most powerful test one can make (a discussion of this can be found in Scheffé, 1959, Chs. 1, 2). Also, it has been shown that, in most cases at least, when all groups have the same number of observations the power of the *F* test is not greatly affected by violation of the assumptions of normality and equal variance. Hence, it is the most powerful practical test of the null hypothesis.

Finally, even though the power of the *F* test can never be calculated exactly in actual practice, it can sometimes be estimated reasonably well. The estimate can then be used before the experiment to guess how large an *n* will be needed for a specified significance level and power. Suppose, for example, an experimenter is interested in comparing five different treatment populations ($I = 5$), and he wishes to detect the difference between the groups in the event that any of the treatment means differ from the others by more than six. If none of the treatment means differ from each other by more than six, he will consider the differences too small to be of any special concern; but if one of them differs from the others by more than six, he will want to be reasonably sure of detecting the difference. Let us say that he has set the significance level (α) at .05, and that he wants the probability of detecting a difference (the power of the test), if it is as large as six, to be at least .90. He estimates the population variance (σ^2) to be about 40. The number of observations (n) needed per group is then estimated.

The effects of σ^2 and the population means on the power of *F* can be summarized in the single parameter:

$$\phi' = \sqrt{\frac{\Sigma_i \ (\mu_i - \mu)^2}{I\sigma^2}} = \sqrt{\frac{\Sigma_i \ \alpha_i^2}{I\sigma^2}} \tag{3-24}$$

where α_i, in this case, refers to the α_i in the linear model of Equations 3-12 and 3-13 rather than the α level of significance. Unfortunately, it is common practice to use the symbol α to signify both. In this text when the symbol α has *no subscript* it will always refer to the α level of significance. The symbol α_i, *with subscript,* will refer to the values in the linear model of Equation 3-13. By comparing the expression of ϕ' with the first term to the right of the equality in Equation 3-23, one can see why Equation 3-24 is useful.

Feldt and Mahmoud (1958) have calculated charts giving the power of F for various values of I, n, α, ϕ', and P, where P is the power of the test. (These charts are reproduced here as Table A-10 in the appendix.) The charts give values of numerator degrees of freedom from one through four. To use them, we first find the value of ϕ' on the axis of the appropriate table for our degrees of freedom. We then find the point directly above it on the appropriate power curve and read the value of n on the Y axis. This is the minimum n needed to guarantee a test of power P with the given degrees of freedom, σ^2, and α. In our example we must first find the smallest value of ϕ' that can occur when two means differ from each other by six. Because ϕ' depends on the squares of the α_i, this smallest value will occur when one of the α_i is equal to -3, another is equal to $+3$, and all other α_i are equal to zero. We can then calculate ϕ' as

$$\phi' = \sqrt{\frac{[(3)^2 + (-3)^2]}{(5)(40)}} = 0.30.$$

In general, the minimum ϕ' for a given maximum difference between treatment means is found by setting one α_i value to $+\frac{1}{2}$ the specified difference, another α_i to $-\frac{1}{2}$ the difference, and all remaining α_i to zero. In the Feldt and Mahmoud table for $\nu = 4$ (see Table A-10, appendix) we first find $\phi' = .30$ on the x axis. We then find that the height of the curve $P = .90$, at the point $\phi' = .30$, is about 35. Therefore, the experimenter will need at least 35 observations in each group. Since this would require a total of 175 observations in all five treatment groups combined, the experimenter may have to consider relaxing his standards unless the data are relatively easy to obtain.

COMPUTATIONAL FORMULAS

Having established the F test as an appropriate and useful test of the null hypothesis that the means of several treatment groups are equal, and having examined some of the properties of this test, we are now ready to consider the best ways of calculating F for a given set of data.

It may be noticed in Table 3-1 that

$$SS_t = SS_w + SS_{bet}. \tag{3-25}$$

This will be true in general, as the following derivation shows:

$$SS_t = \Sigma_i \, \Sigma_j \, (X_{ij} - \overline{X}..)^2,$$

which by adding and subtracting $\overline{X}_i$.

$$= \Sigma_i \, \Sigma_j \, [(X_{ij} - \overline{X}_{i\bullet}) + (\overline{X}_{i\bullet} - \overline{X}_{\bullet\bullet})]^2,$$

and by expanding the square

$$= \Sigma_i \, \Sigma_j \, [(X_{ij} - \overline{X}_{i\bullet})^2 + (\overline{X}_{i\bullet} - \overline{X}_{\bullet\bullet})^2 + 2(\overline{X}_{i\bullet} - \overline{X}_{\bullet\bullet})(X_{ij} - \overline{X}_{i\bullet})],$$

and distributing the summations

$$
\begin{aligned}
&= \Sigma_i \, \Sigma_j \, (X_{ij} - \overline{X}_{i\bullet})^2 + n \, \Sigma_i \, (\overline{X}_{i\bullet} - \overline{X}_{\bullet\bullet})^2 \\
&\quad + 2 \, \Sigma_i \, (\overline{X}_{i\bullet} - \overline{X}_{\bullet\bullet}) \, \Sigma_j \, (X_{ij} - \overline{X}_i).
\end{aligned}
\tag{3-26}
$$

Now, since $\Sigma_j \, (X_{ij} - \overline{X}_{i\bullet}) = 0$, by reason of the fact that $\overline{X}_{i\bullet}$ is the mean of the X_{ij} in group A_i, the last term in Equation 3-26 can be dropped. This leaves

$$SS_t = \Sigma_i \, \Sigma_j \, (X_{ij} - \overline{X}_{i\bullet})^2 + n \, \Sigma_i \, (\overline{X}_{i\bullet} - \overline{X}_{\bullet\bullet})^2 \tag{3-27}$$

$$= (n-1) \, \Sigma_i \, \hat{\sigma}_i^2 + n \, \Sigma_i \, (\overline{X}_{i\bullet} - \overline{X}_{\bullet\bullet})^2. \tag{3-28}$$

The first term in this equation derives from the fact that

$$\hat{\sigma}_i^2 = \Sigma_j \, (X_{ij} - \overline{X}_{i\bullet})^2 / (n-1)$$

so that

$$\Sigma_j \, (X_{ij} - \overline{X}_{i\bullet})^2 = (n-1)\hat{\sigma}_i^2.$$

The assertion made in Equation 3-25 can now be verified by comparing the first term in Equation 3-28 with Equation 3-4 and the second term with Equation 3-8.

Equation 3-25 is a special case (for the one-way design) of an important general principle in analysis of variance. Every analysis of variance divides SS_t into two or more independent sums of squares whose total is SS_t. Since the sums of squares into which SS_t is divided are independent, the mean squares of some of them can be the denominators of F ratios used to test others, just as MS_w is the denominator for testing MS_{bet}.

In practice, one need only calculate two of the three sums of squares explicitly, since the third can be found from the relationship given in Equation 3-25. Generally, SS_t and SS_{bet} are the two that are most easily calculated; SS_w is then found by subtracting SS_{bet} from SS_t; SS_t can be calculated from Equations 3-9 and 3-10; and SS_{bet} can be calculated from

Table 3-4. Summary of calculations in one-way F test, equal ns; $t_{i\bullet} = \Sigma_j \, X_{ij}$; $T = \Sigma_i \, t_{i\bullet} = \Sigma_i \, \Sigma_j \, X_{ij}$.

	RS	SS	df	MS	F
m		T^2/N			
bet	$(1/n) \, \Sigma_i \, t_i^2$	$RS_{bet} - SS_m$	$I - 1$	$SS_{bet}/(I-1)$	MS_{bet}/MS_w
w		$RS_t - RS_{bet}$	$N - I$	$SS_w/(N-I)$	
t	$\Sigma_i \, \Sigma_j \, X_{ij}^2$	$RS_t - SS_m$	$N - 1$		

Table 3-5. Illustrations of calculations on data of Table 3-1; $t_1. = 20$; $t_2. = 8$; $t_3. = 23$; $T = 51$.

	RS	SS	df	MS	F	α
m		$(51)^2/15 = 173.4$				
bet	$(1/5)(20^2 + 8^2 + 23^2)$ $= 198.6$	$198.6 - 173.4$ $= 25.2$	2	$25.2/2$ $= 12.6$	$12.6/2.867$ $= 4.395$	.04
w		$233 - 198.6$ $= 34.4$	12	$34.4/12$ $= 2.867$		
t	233	$233 - 173.4$ $= 59.6$	14			

Equation 3-8. Equation 3-8 can be modified, however, so that SS_{bet} is expressed in terms of the sums of the variables instead of their means.

$$SS_{bet} = (1/n) \, \Sigma_i \, (\Sigma_j \, X_{ij})^2 - (1/N) \, (\Sigma_i \, \Sigma_j \, X_{ij})^2. \qquad \textbf{(3-29)}$$

The calculation of SS_t and SS_{bet} can be simplified by introducing the following terms:

$$t_i. = n\overline{X}_i. = \Sigma_j \, X_{ij},$$
$$T = \Sigma_i \, t_i. = N\overline{X}.. = \Sigma_i \, \Sigma_j \, X_{ij},$$
$$SS_m = T^2/N,$$
$$RS_{bet} = \Sigma_i \, t_i.^2/n, \text{ and}$$
$$RS_t = \Sigma_i \, \Sigma_j \, X_{ij}^2.$$

Then

$$SS_{bet} = RS_{bet} - SS_m,$$
$$SS_t = RS_t - SS_m, \text{ and}$$
$$SS_w = SS_t - SS_{bet} = RS_t - RS_{bet}.$$

Each $t_i.$ is the total of the observations in group A_i, and T is the grand total of all of the observations in the experiment. The term RS stands for *raw sum of squares*; it refers to a sum of squares based on the squared raw data values rather than on the squares of the deviations of the values from their means. The general procedure is to first calculate the $t_i.$, T, and RS_t. These values are then used to calculate RS_{bet} and SS_m. The calculation of the sums of squares follows. Table 3-4 summarizes this procedure. Table 3-5 illustrates the calculations for the data of Table 3-1.

ANALYSIS: UNEQUAL SAMPLE SIZES

So far, we have assumed that the same number of observations were sampled from each treatment group. In general, it is better if the number of observations in each group is the same (for reasons that will be evident later). However, this is not always possible. Circumstances beyond the control of the experimenter may dictate different sample sizes in different groups. In an experiment like that in Table 3-1, for example, the experimenter may choose five students from each field of study, but some of them may not show up to take the test. If one subject failed to show up in group A_1 and two subjects failed to show up in group A_3, the data might look like those in Table 3-6. In such a case, of course, one might question the validity of the samples, since those who failed to show up may be systematically different from those who did not. However, we cannot always meet the assumptions exactly. Frequently, we have to use our good judgment to decide whether the departures from the assumptions are large enough to make a difference. We will assume for the purposes of the example that the effects of the departures are small.

To derive the F test for this case, we have to change our notation slightly. As before, we will assume that I different groups are being compared, but the number of observations in each sample will be n_i, with the subscript denoting the fact that the sample sizes are different for the different groups. In this case, Equation 3-2 becomes

$$\hat{\sigma}_i^2 = \frac{\Sigma_j \ (X_{ij} - \overline{X}_{i.})^2}{n_i - 1}$$

Table 3-6. Hypothetical data for independent samples from three groups with unequal numbers of observations.

	A_1	A_2	A_3	Sum
	3	1	5	
	5	2	5	
	6	2	2	
	1	0		
		3		
$\Sigma \ X_{ij}$	15	8	12	35
$\overline{X}_{i.}$	3.75	1.6	4.0	2.92
$n_i \ \overline{X}_{i.}^2$	56.25	12.80	48.00	117.05
$\Sigma \ X_{ij}^2$	71	18	54	143
$(n_i - 1)\hat{\sigma}_i^2$	14.75	5.20	6.00	25.95
$\hat{\sigma}_i^2$	4.92	1.30	3.00	

	SS	df	MS	F	α
w	25.95	9	2.883		
bet	14.97	2	7.483	2.60	.13
t	40.92	11	3.720		

so that

$$(n_i - 1)\hat{\sigma}_i^2 \sim \sigma^2 \chi^2_{(n_i - 1)}$$

and

$$SS_w = \Sigma_i \ (n_i - 1)\hat{\sigma}_i^2 \sim \sigma^2 \chi^2_{(N-I)} \tag{3-30}$$

where, just as before, N is the total number of observations in all of the samples combined $(N = \Sigma_i \ n_i)$, so that $\Sigma_i \ (n_i - 1) = N - I$, and this is the reason for the number of degrees of freedom attributed to the χ^2 in Equation 3-30. A comparison of Equations 3-30 and 3-4 will show that the two are identical when the sample sizes are equal.

In the same way, Equation 3-8 can be generalized to

$$SS_{bet} = \Sigma_i \ n_i(\overline{X}_{i.} - \overline{X}_{..})^2.$$

In this case, however, it is not so obvious that SS_{bet}/σ^2 is distributed as χ^2 under the null hypothesis. The proof of this is rather difficult, but it can be outlined. We can begin by repeating the derivation found in Equations 3-26, 3-27, and 3-28, for the case of unequal ns. The individual steps are virtually identical to those of the previous derivation, but the end result is now

$$SS_t = \Sigma_i \ \Sigma_j \ (X_{ij} - \overline{X}_{..})^2 = \Sigma_i \ (n_i - 1)\hat{\sigma}_i^2 + \Sigma_i \ n_i(\overline{X}_{i.} - \overline{X}_{..})^2$$
$$= SS_w + SS_{bet} \tag{3-31}*$$

Under the null hypothesis, $MS_t = SS_t/(N-1)$ is still an unbiased estimate of σ^2, and $SS_t \sim \sigma^2\chi^2_{(N-1)}$, as before. Furthermore, the generalization to unequal sample sizes does not alter the fact that SS_w and SS_{bet} are independent, since the one is still based only on sample variances and the other is based only on sample means. We need a theorem that we will state but not prove: if the sum of two independent random variables is distributed as chi-square, and one of the random variables in the sum is also distributed as chi-square, then the other variable in the sum is distributed as chi-square. In particular, since SS_t/σ^2 and SS_w/σ^2 are both distributed as chi-square, then, from Equation 3-31 and the independence of SS_w and SS_{bet}, we can conclude that SS_{bet}/σ^2 is also distributed as chi-square. A corollary of this theorem is that the number of degrees of freedom of SS_{bet} is the difference between the degrees of freedom of SS_t and SS_w, so that

$$SS_{bet} \sim \sigma^2\chi^2_{(I-1)}.$$

The values of MS_{bet}, MS_w, and F are then found as before. Thus, the F test for samples with unequal ns is a generalization of the test with equal ns.

---

*The reader may find the derivation of Equation 3-31 to be a useful exercise.

Table 3-7. Summary of calculations in one-way F test, unequal ns; $t_{i.} = \Sigma_j X_{ij}$; $T = \Sigma_i t_{i.} = \Sigma_i \Sigma_j X_{ij}$.

	RS	SS	df	MS	F
m		T^2/N			
bet	$\Sigma_i \ (t_{i.}^2/n_i)$	$RS_{\text{bet}} - SS_{\text{m}}$	$I - 1$	$SS_{\text{bet}}/(I-1)$	$MS_{\text{bet}}/MS_{\text{w}}$
w		$RS_{\text{t}} - RS_{\text{bet}}$	$N - I$	$SS_{\text{w}}/(N-I)$	
t	$\Sigma_i \Sigma_j X_{ij}^2$	$RS_{\text{t}} - SS_{\text{m}}$	$N - 1$		

Computational Formulas

The computational procedure with unequal ns is similar to that with equal ns, and dependent on the identity

$$SS_{\text{bet}} = \Sigma_i \ (\Sigma_j X_{ij})^2/n_i - (\Sigma_i \Sigma_j X_{ij})^2/N$$

(which is the same as that in Eq. 3-29) as well as on the identities in Equations 3-9 and 3-31. The procedure is the same as that given for equal sample sizes, except that n_i is substituted for n in calculating RS_{bet}. The procedure is summarized in Table 3-7; the calculations (for the data in Table 3-6) are shown in Table 3-8. For the data from Table 3-6, the loss of subjects has made the differences somewhat less significant as would be expected (although this does not *necessarily* happen), since the smaller n results in less power.

Power of *F* with Unequal Sample Sizes

As with equal sample sizes, the distribution of the F ratio when the null hypothesis is false is noncentral F. In this case, however, we cannot evaluate the power in terms of a single value of n. Instead of ϕ', we must define the general *noncentrality parameter* ϕ:

$$\phi = \sqrt{\frac{\Sigma_i \ n_i \alpha_i^2}{I\sigma^2}}.$$

Table 3-8. Illustration of calculations for one-way F test (using data of Table 3-6); $t_{1.} = 15$; $t_{2.} = 8$; $t_{3.} = 12$; $T = 35$.

	RS	SS	df	MS	F	α
m		$35^2/12$ = 102.08				
bet	$15^2/4 + 8^2/5$ $+ 12^2/3 = 117.05$	$117.05 - 102.08$ = 14.97	2	14.97/2 = 7.483	7.483/2.883 = 2.60	.13
w		$143 - 117.05$ = 25.95	9	25.95/9 = 2.883		
t	143	$143 - 102.08$ = 40.92	11			

The parameter ϕ is not usually very useful in determining the n required for a given level of power because there is a different n_i for each sample and, usually, an experimenter does not plan to have unequal ns. However, it may occasionally be useful in evaluating the power of the test with a given, already obtained, sample. In such a case, charts giving the power of F as a function of ϕ and the degrees of freedom can be found in Scheffé (1959), along with directions for using them.

Robustness of *F* with Unequal Sample Sizes

Previously (pp. 31–34), we found that when sample sizes are equal, the F test is robust with respect to nonnormality and unequal variances. Although very little has been done to assess the effect of nonnormality with unequal sample sizes, the results appear to indicate that the effect of nonzero skewness is again negligible, and the effect of nonzero kurtosis, although slightly greater than with equal sample sizes, is still too small to be of concern in most applications.

The results are very different, however, with respect to inequality of variances. In this case, the effect may be quite large; the direction of the effect depends on the sizes of the variances in the larger samples as compared to those in the smaller samples. This is true because, as Equation 3-30 indicates, MS_w is a weighted average of the sample variances, with the weights being equal to $(n_i - 1)$. Since greater weight is placed on the variances of the larger samples, MS_w will tend to be larger if the samples are from populations with large variances rather than from populations with small variances. Since MS_w is in the denominator of F, a large MS_w will tend to result in a small F, and vice versa. Accordingly, if the larger samples are taken from the populations with the larger variances, MS_w will tend to be too large and F will tend to be too small. If, on the other hand, the variances of the populations from which the larger samples are taken tend to be small, F will tend to be too large. Table 3-9 shows the effects of inequality of variances on the significance level of the F test on two treatment groups when the samples are large. Table 3-10 and Table 3-11 show the same results for selected values of I and n_i and for selected variances. It is clear from these data that relatively small departures from the assumption of equal variances may have a sizable effect if the sample sizes are small, and that the effect is not greatly reduced by increasing the sizes of samples, so long as the ratios of the sample sizes remain constant.

A rough index of the effect of unequal variances on the distribution of F can be found by noting that in this case MS_w is a weighted average of the unbiased variance estimates $(\hat{\sigma}_i^2)$ for the individual groups (Eq. 3-30). If we let σ_i^2 be the "true" variance of group i, and

$$\bar{\sigma}_w^2 = \frac{\Sigma_i (n_i - 1)\sigma_i^2}{N - I},$$

so that $\bar{\sigma}_w^2$ is a weighted average of the true group variances, then MS_w is an unbiased estimate of $\bar{\sigma}_w^2$. That is, $E(MS_w) = \bar{\sigma}_w^2$.

Similarly, we can let

$$\bar{\sigma}_u^2 = \left(\frac{1}{I}\right) \Sigma_i \, \sigma_i^2,$$

so that $\bar{\sigma}_u^2$ is the unweighted average of the group variances. It can be shown that

$$E(MS_{bet}) = \bar{\sigma}_w^2 + \left(\frac{N-1}{N}\right)\left(\frac{I}{I-1}\right)(\bar{\sigma}_u^2 - \bar{\sigma}_w^2).$$

We can thus approximate the expected value of F as

$$E(F) \approx E(MS_{bet})/E(MS_w)$$

$$= \left[\bar{\sigma}_w^2 + \left(\frac{N-1}{N}\right)\left(\frac{I}{I-1}\right)(\bar{\sigma}_u^2 - \bar{\sigma}_w^2)\right]/\bar{\sigma}_w^2$$

$$= 1 + \left(\frac{N-1}{N}\right)\left(\frac{I}{I-1}\right)\left(\frac{\bar{\sigma}_u^2}{\bar{\sigma}_w^2} - 1\right) \qquad \text{(3-32)}$$

We can estimate $E(F)$ by estimating σ_w^2 and σ_u^2 and inserting these estimates into Equation 3-32. However, we already know that MS_w is an unbiased estimate of $\bar{\sigma}_w^2$, and it is easy to show that

$$MS_u = \left(\frac{1}{I}\right) \Sigma_i \, \hat{\sigma}_i^2$$

is an unbiased estimate of $\bar{\sigma}_u^2$. Inserting these values into Equation 3-32, we get

$$\lambda = 1 + \left(\frac{N-1}{N}\right)\left(\frac{I}{I-1}\right)\left(\frac{MS_u}{MS_w} - 1\right),$$

Table 3-9. Effect of unequal variances and unequal group sizes on true probability of Type I error for F test on two samples with large N.

n_1/n_2		0^a	0.2	0.5	σ_1^2/σ_2^2 1	2	5	∞^a
1	$E(F)$	1.00	1.00	1.00	1.00	1.00	1.00	1.00
	α	.050	.050	.050	.050	.050	.050	.050
2	$E(F)$	2.00	1.57	1.25	1.00	0.80	0.64	0.50
	α	.17	.12	.080	.050	.029	.14	.006
5	$E(F)$	5.00	2.60	1.57	1.00	0.64	0.38	0.20
	α	.38	.22	.12	.050	.014	.002	10^{-5}
∞	$E(F)$	∞	5.00	2.00	1.00	0.50	0.20	0.00
	α	1.00	.38	.17	.050	.006	10^{-5}	.00

Source: α values taken from Scheffé, 1959, p. 340.
[a]Unattainable limiting cases to show bounds.

where λ is an estimate of $E(F)$ under the null hypothesis. This estimate is independent of the validity of the null hypothesis itself, since it is based only on the variances of the individual groups and not on their means. In general, then, if λ is close to one, the F test is probably not greatly affected by the effects of possibly unequal variances. If it is larger than one, the obtained value of F is likely to be too large, so that the obtained level of significance is greater than it should be. If it is smaller than one, the obtained F is likely to be too small, and the obtained significance level will not be as high as it should be. The approximate values of $E(F)$ under the null hypothesis (Eq. 3-32) are included in Tables 3-9, 3-10, and 3-11 to illustrate the effects of departures from one on the validity of the F tests.

Table 3-10. Effect of unequal variances on probability of Type I error with F test at nominal five-percent significance level for two samples of selected sizes.

n_1	n_2		0^a	0.1	0.2	0.5	σ_1^2/σ_2^2 1.0	2.0	5.0	10.0	∞^a
15	5	$E(F)$	3.38	2.58	2.12	1.43	1.00	0.70	0.49	0.41	0.32
		α	.32	.23	.18	.098	.050	.025	.008	.005	.002
5	3	$E(F)$	1.88	1.66	1.50	1.22	1.00	0.82	0.68	0.62	0.56
		α	.22	.14	.10	.072	.050	.038	.031	.030	.031
7	7	$E(F)$	1.00	1.00	1.00	1.00	1.00	1.00	1.00	1.00	1.00
		α	.072	.070	.063	.058	.050	.051	.058	.063	.072

Source: α values from Hsu, 1938.
[a]Unattainable limiting cases to show bounds.

Table 3-11. Effect of inequality of variances on probability of Type I error with F test at nominal five-percent level.

I	σ_i^2	n_i	N	$E(F)$	α
3	1,2,3	5,5,5	15	1.00	.056
		3,9,3	15	1.00	.056
		7,5,3	15	1.20	.092
		3,5,7	15	0.80	.040
3	1,1,3	5,5,5	15	1.00	.059
		7,5,3	15	1.35	.11
		9,5,1	15	1.93	.17
		1,5,9	15	0.60	.013
5	1,1,1,1,3	5,5,5,5,5	25	1.00	.07
		9,5,5,5,1	25	1.28	.14
		1,5,5,5,9	25	0.73	.02
7	1,1,1,1,1,1,7	3,3,3,3,3,3,3	21	1.00	.12

Source: α values from Box, 1954.

OTHER POSSIBLE F TESTS

You may have noticed that degrees of freedom in the F test add up to one less than the number of observations—one degree of freedom has been lost somewhere. The usual explanation is that this degree of freedom is lost in estimating the grand mean. In reality, it is not lost—it is simply not used. It can be used, however, and in some cases it is appropriate to use it. Consider the quantity SS_m, defined earlier:

$$SS_m = T^2/N.$$

To begin with, T is the sum of N normally distributed random variables, so it is distributed as

$$T \sim N_{(N\mu, N\sigma^2)}, \tag{3-33}$$

where

$$\mu = (1/N) \, \Sigma_i \, n_i \mu_i.$$

That is, μ is the weighted average of the μ_i.

Now, consider the null hypothesis

$$H_0: \mu = 0. \tag{3-34}$$

According to Equation 3-33, this null hypothesis is equivalent to

$$H_0: E(T) = 0.$$

But this, combined with Equation 3-33, implies that

$$T^2/N\sigma^2 \sim \chi^2_{(1)},$$

since it is the square of a normally distributed random variable with a mean of zero and a variance of one. Consequently,

$$SS_m \sim \sigma^2 \chi^2_{(1)}.$$

Furthermore, since SS_m has only one degree of freedom, we can define

$$MS_m = SS_m/1 = SS_m,$$

which is distributed as σ^2 times χ^2 divided by its degrees of freedom. Finally, MS_m is independent of MS_w for the same reason that MS_{bet} is independent of MS_w, so that under the null hypothesis (Eq. 3-34),

$$MS_m/MS_w \sim F_{(1, N-1)}.$$

This F ratio uses the "lost" degree of freedom to test the null hypothesis that the population grand mean is zero (Eq. 3-34). It should be pointed out that MS_m and MS_{bet} are also independent of each other, because MS_m is based only on the average of the sample means and MS_{bet} is based on their variance. In this case we say that the two null hypotheses (Eqs. 3-1 and 3-34) are *independent null hypotheses*. Note, however, that since both F

tests use MS_w in the denominator, the tests of these hypotheses are not independent even though the hypotheses are. If, by chance, MS_w is an underestimate or an overestimate of σ^2, both F tests will be affected in the same way.

The test of the grand mean can be modified somewhat if one wishes to test a more general null hypothesis:

$$H_0: \mu = \mu^*,$$

where μ^* is any value specified by the experimenter. By a derivation like that above, we find that under the null hypothesis:

$$SS_m^* = (T - N\mu^*)^2/N \sim \sigma^2\chi_{(1)}^2.$$

Finally, the original F test on the differences between groups can be modified to test the null hypothesis that each group has a specified population mean, μ_i^*. In such a case,

$$SS_{bet}^* = \Sigma_i \, n_i(\overline{X}_{i.} - \mu_i^*)^2 \sim \sigma^2\chi_{(I)}^2.$$

Notice here that there are I degrees of freedom rather than $I - 1$. This null hypothesis is not independent of Equation 3-34. We define

$$MS_{bet}^* = SS_{bet}^*/I,$$

so that under the null hypothesis $H_0: \mu_i = \mu_i^*$,

$$MS_{bet}^*/MS_w \sim F_{(I, N - I)}.$$

These tests are seldom used in practice. Nevertheless, they illustrate how the degrees of freedom may be used in different ways. The next chapter will illustrate more imaginative ways of using the available degrees of freedom to make significance tests.

Exercises

1. An experimenter studying the effects of drugs on behavior of mental patients divided 30 patients randomly into six groups of five patients each. The control group (G_1) received no drugs, four experimental groups (G_2 through G_5) each received a different type of tranquilizer, and one group (G_6) received a new "psychic energizer." The data were before and after differences on a specially designed personality inventory (positive scores indicate improvement). The data are as follows:

A_1	A_2	A_3	A_4	A_5	A_6
1.3	2.5	1.0	0.5	2.7	2.2
2.5	−1.1	−2.9	−1.0	2.0	3.7
−1.0	−0.2	−1.1	3.3	3.3	2.9
0.4	1.7	−1.1	1.2	2.3	3.2
2.8	2.1	1.6	3.0	3.2	3.0

(a) Do an analysis of variance on the above data.

(b) Estimate the variance in each group and comment on the validity of the assumption of equal variances.

(c) Is this F test robust to the assumption of equal variance? Explain.

(d) Test $H_0: \mu = 0$.

(e) What is the power of the test in part d against $H_a: \mu = 1$?

2. An experimenter ran 5 groups of rats in a T maze. Group 1 had only olfactory and tactile cues, i.e., they had no visual or auditory cues; Group 2 had only visual and tactile cues; Group 3 had only visual and olfactory cues; Group 4 had visual, olfactory, and tactile cues. Group 5 also had visual, olfactory, and tactile cues, but were given shocks when they made an error. The data below are for the trial of last error, so low scores indicate good performance.

A_1	A_2	A_3	A_4	A_5
7	5	9	7	10
8	4	11	12	7
5	4	6	8	3
9	6	8	5	12
10	3	7	11	13

(a) Do an analysis of variance on these data.

(b) Estimate the variance for each group. Comment on the validity of the assumption of equal variances in this design and the effect of its violation on the validity of the analysis of variance.

(c) What transformation might be used to make the variances more nearly equal, but such that the transformed values would still be intuitively meaningful?

(d) Make the transformation you suggested in part c, and perform the analysis of variance on the transformed data. Compare your results with those in (a).

3. Refer to the data in Table 3-1. Suppose that a fourth group had also been run and the data were A_4: 3, 1, 2, 5, 4.

(a) Do an analysis of variance on the four groups.

(b) Suppose only the first three observations had been taken in Group 4, i.e., suppose the data are A_4: 3, 1, 2. Do the analysis of variance again on the four groups.

(c) If you analyzed the data correctly, the data in part a are not significant at the .05 level, but those in part b are. By adding data, we reduced the significance level (in comparison with the smaller amount of data in Table 3-1) and then, by removing some of the added data, we increased it again. Explain this paradox. What conclusions can you draw for application to practical problems of experimental design?

4. An experimenter wants to set his power at .9 for detecting a difference as large as 10 between *any* two means. He estimates $\sigma^2 = 50$, and $\alpha = .05$.

(a) How large, approximately, must n be if $I = 4$? How large must N be?

(b) How large, approximately, must n and N be if $I = 3$? Compare these values with those in part a.

5. Derive Equation 3-31.

For the remaining questions:

I. Do an analysis of variance on the data assuming that all of the assumptions for a one-way test were met.

II. Interpret the results, commenting on which differences are most likely to have contributed to the significant effect (if there was one).

III. Tell which, if any, of the assumptions of a one-way analysis of variance were most probably violated and comment on the effect each violation is likely to have had on the validity of the analysis.

IV. If there are both positive and negative scores in the data, test the null hypothesis that the grand mean is zero and comment on the meaning of such a test for the particular set of data being tested.

---

6. Four males and four females were each given two forms of a short achievement test. The scores were:

	Males		Females	
	A_1	A_2	A_3	A_4
m_1	3	2	3	1
m_2	3	3	1	2
m_3	1	2	2	0
m_4	3	1	1	1

Analyze the results, following instructions I-III above. Assume that this is a one-way analysis of variance with four groups and four subjects per group.

7. Mr. Gallop was curious as to whether locality was a significant factor in the political opinions of New York families. He polled three different areas within the city: the area surrounding New York University (A_1), a slum area (A_2), and a ritzy section (A_3). In each section he randomly selected three families; in each family he asked both the husband and wife how satisfied they were with current efforts at school integration. Answers were scored on a 7-point scale from +3 (deliriously happy) to −3 (thoroughly nauseated). A score of 0 indicated indifference. Thus, six scores (2 from each family) were obtained in each area. The scores were:

A_1	A_2	A_3
−3	−1	+2
−3	−2	+3
−1	+2	+1
−2	+2	0
+1	−3	+3
0	0	+3

8. An experimenter studied the reaction times of politicians at a cocktail party; subjects were sober (A_1), moderately drunk (A_2), and completely sloshed (A_3). Because his data were highly skewed (one subject was unable to react at all), the experimenter decided that it would be more legitimate to rank the reaction times and do an analysis of variance on the ranks instead of on the original reaction times. The ranks, for the only seven subjects he could test, are as follows:

A_1	A_2	A_3
1	4	7
	5	6
	2	
	3	

9. A sociologist conducting a survey asked several college-age youths whether they liked, disliked, or felt neutral toward their parents. The persons surveyed fell into the six groups shown below. Each respondent got a score of +1 if he liked his parents, −1 if he did not like them, and 0 if he felt neutral toward them. The scores were:

Large College		Small College		Noncollege	
Male	Female	Male	Female	Male	Female
−1	0	1	1	0	−1
0	−1	0	1	−1	0
−1	−1	−1	1	0	0
−1	0	−1	−1	0	1
−1	+1	0	1	0	0
0	−1	0	1	1	0
+1	+1	1	1	0	0
−1	−1	1	1	1	
0	−1	0		1	
+1		0		0	

4

Comparing Groups

The F test discussed in Chapter 3 really tells us very little about the differences among the groups. If it is significant, we can be reasonably confident that some differences exist between the population means of the I groups being compared. We cannot tell, however, what the differences are or whether they are large enough to be important. In this chapter we will discuss a number of additional techniques, related to the F test, that will help us answer these questions.

PROPORTION OF VARIANCE ACCOUNTED FOR

The question of whether the differences are large enough to be important can be partially answered by estimating ω^2, the *proportion of variance accounted for*. If ω^2 is small, the differences between the group means are small relative to the variability of the observations within each group. In such a case the differences, even though they may be significant, are not likely to be practically or scientifically very important. A large value of ω^2 implies that the differences between the groups are large relative to the variability within groups, and the overlap between the scores in the different groups is small.

To derive ω^2, we first consider the entire population of observations from which the I groups have been drawn. That is, we "combine" the I different populations into one large population and calculate the mean and variance of this composite population. Alternatively, imagine that we randomly choose an observation by a two-step process: first, we randomly select one of the I groups and then we randomly select an observation from that group. Our task is to find the mean and variance of the population from which the observation was selected.

To simplify the problem somewhat, we will assume that the initial probability of choosing an observation from a given group, A_i, is n_i/N, the proportion of the total number of observations that are in group A_i in the sample. In this case, it is easy to see that

$$E(X_{ij}) = (1/N) \Sigma_i \, n_i\mu_i = \mu,$$
$$E(X_{ij} - \mu)^2 = E[(X_{ij} - \mu_i) + (\mu_i - \mu)]^2$$

In the second equation, the formula on the right is obtained from the formula on the left by adding and subtracting μ_i. If we then distribute the square, we get

$$E(X_{ij} - \mu)^2 = E(X_{ij} - \mu_i)^2 + E(\mu_i - \mu)^2$$
$$+ 2E(X_{ij} - \mu_i)(\mu_i - \mu).$$

However, for the normal distribution the differences within a sample are independent of its mean, so that

$$E(X_{ij} - \mu_i)(\mu_i - \mu) = E(X_{ij} - \mu_i)\,E(\mu_i - \mu) = 0. \qquad \textbf{(4-1)}$$

Consequently,

$$E(X_{ij} - \mu)^2 = E(X_{ij} - \mu_i)^2 + E(\mu_i - \mu)^2.$$

The term $E(\mu_i - \mu)^2$ needs some explanation, since in ordinary applications of statistics the μ_i are not random variables. Remember that in selecting an observation we randomly selected a group and then randomly selected an observation from that group. Randomly selecting a group is the same as randomly selecting a value of μ_i, i.e., the population mean of the group selected. In this sense, μ_i and, consequently, $(\mu_i - \mu)$ can be regarded as random variables, and their means and variances calculated.

If we let σ_e^2 be the variance of the observations within a given group (the variance that in Chapter 3 we called σ^2), and

$$\sigma_{bet}^2 = E(\mu_i - \mu)^2 = (1/N) \Sigma_i \, n_i(\mu_i - \mu)^2$$
$$= (1/N) \Sigma_i \, n_i \, \alpha_i^2 \qquad \textbf{(4-2)*}$$

then, by Equation 4-1,

$$\sigma_t^2 = E(X_{ij} - \mu)^2 = \sigma_e^2 + \sigma_{bet}^2. \qquad \textbf{(4-3)}$$

The total variance of the observations is the sum of two variances: the variance of the *observations within each group* and the variance of the *group means about the grand mean*. If the latter is large with respect to the former, there will be relatively little overlap between the observations in the I groups. If it is small, the variability of the group means will be small

---

*See Equation 3-12.

compared to the random variability of the observations themselves. The ratio

$$\omega^2 = \sigma_{bet}^2/\sigma_t^2$$

is the proportion of the total variance that can be "accounted for" by differences between the means of the I groups.

The exact value of ω^2 depends on the true values of σ_{bet}^2 and σ_t^2. It can be estimated, however, by estimating σ_{bet}^2 and σ_t^2 and taking the ratio of these estimates.

From Equations 3-23 and 4-2 we can see that

$$E(MS_{bet}) = \sigma_e^2 + \left(\frac{N}{I-1}\right)\sigma_{bet}^2 \qquad (4\text{-}4)$$

for equal ns. A similar derivation from Equation 3-31 shows that Equation 4-4 is true for unequal sample sizes as well. We can rewrite Equation 4-4 as

$$E(SS_{bet}) = (I-1)\,\sigma_e^2 + N\,\sigma_{bet}^2.$$

Since MS_w is an unbiased estimate of σ_e^2, a little algebra reveals that

$$\hat{\sigma}_{bet}^2 = (1/N)[SS_{bet} - (I-1)MS_w] \qquad (4\text{-}5)$$

is an unbiased estimate of σ_{bet}^2. By Equation 4-3 and the fact that MS_w is an unbiased estimate of σ_e^2, it follows that

$$\begin{aligned}\hat{\sigma}_t^2 &= \hat{\sigma}_{bet}^2 + MS_w \\ &= (1/N)(SS_t + MS_w)\end{aligned} \qquad (4\text{-}6)$$

is an unbiased estimate of σ_t^2. The last equality in Equation 4-6 follows from some algebra plus the fact (noted in Chapter 3) that $SS_{bet} + SS_w = SS_t$.

The ratio of Equation 4-5 to Equation 4-6 provides a reasonably unbiased estimate of ω^2:

$$\begin{aligned}\hat{\omega}^2 &= \hat{\sigma}_{bet}^2/\hat{\sigma}_t^2 \\[4pt] &= \frac{SS_{bet} - (I-1)MS_w}{SS_t + MS_w}.\end{aligned} \qquad (4\text{-}7)$$

For the data in Table 3-1,

$$\hat{\omega}^2 = \frac{25.2 - (2)(2.867)}{59.6 + 2.867} = .312.$$

Note that ω^2 is a different kind of quantity than the F derived in Chapter 3. Any difference between the means, no matter how small, will be significant if the N is large enough. The size of N, however, has no effect on ω^2. The quantity F is a measure of the statistical significance of the differences; the quantity ω^2 is a measure of their practical or scientific importance. An important difference may be insignificant if the N is too small, and a very unimportant difference may be significant if the N is large. In general, F and $\hat{\omega}^2$ are related by the equation

$$\hat{\omega}^2 = \frac{F - 1}{F - 1 + N/(I - 1)}.$$

Note also that $\hat{\omega}^2$ is an approximately unbiased estimate of ω^2 (the bias is very small if N is reasonably large). Although ω^2 can never be smaller than zero, its *estimate*, $\hat{\omega}^2$, can be negative. Since $\hat{\omega}^2$ will vary around its true value, it may be negative if ω^2 is close to zero; the above equation shows that $\hat{\omega}^2$ is negative whenever F is less than one.

There are two ways to handle negative $\hat{\omega}^2$, and there is no general agreement among statisticians as to which is better. One is to report all negative values as zero, the idea being that negative $\hat{\omega}^2$ are impossible and so negative estimates should not be reported. The other way is to report all estimates exactly as they are obtained; the reasoning behind this approach is that the more negative $\hat{\omega}^2$ is, the stronger is the evidence that ω^2 is close to zero, and the reader of the research report should be given that evidence.

GRAPHS AND FIGURES

One of the most useful (and at the same time, most neglected) ways of comparing groups is by means of a graph. Statistical consultants are always asked, "What do these results mean?" If the answer is simple enough—and it usually is—a graph will make it clear. For example, Figure 4-1 is a graph of the means for the data in Table 3-1. It is clear from this figure that F is significant primarily because the scores of Group A_2 tend to be much lower than those of Groups A_1 and A_3, which are more nearly equal. Of course, these observed differences may be due to chance, just as the significant F may be due to chance. Nevertheless, if a real difference does exist, Figure 4-1 will tell us where that difference is likely to be.

Yet comparison of group means by a graph alone can be deceptive. In order to be confident of our interpretation, we need some way of testing whether the specific observed differences are due to chance. The remainder of this chapter is devoted to methods of comparing individual pairs of

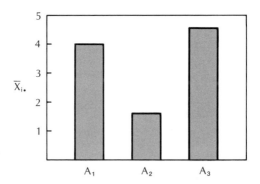

Figure 4-1. Group means from Table 3-1.

means and testing specific hypotheses about which differences are significant and which are not.

PLANNED COMPARISONS

Frequently, the experimenter has some ideas in advance about which group means are likely to be different and which are not. That is, in addition to the overall hypothesis that there will be some difference among the means, the experimenter has a notion about where those differences are likely to be. Consider, for example, the design in Table 4-1. Here four groups with eight subjects each are being compared. These are hypothetical data showing errors in a paired-associate learning task by subjects under the influence of two drugs. Group A_1 is a control group, not given any drug. Groups A_2, A_3, and A_4 are experimental groups; Group A_2 was given one drug, Group A_3 another, and Group A_4 was given both drugs. The results of an overall F test are given in the table; Figure 4-2 is a graph of results. The estimate of ω^2 for these data is .45.

The experimenter is really interested in answering three specific questions: On the average, do the drugs have any effect on learning at all? Do subjects make more errors if given both drugs than if given only one? Do the two drugs differ in the number of errors they produce?

Table 4-1. Hypothetical error data for four groups of eight subjects each.

	A_1 (No drug)	A_2 (Drug 1)	A_3 (Drug 2)	A_4 (Both drugs)	Total
X_{ij}	1	12	12	13	
	8	6	4	14	
	9	10	11	14	
	9	13	7	17	
	7	13	8	11	
	7	13	10	14	
	4	6	12	13	
	9	10	5	14	
ΣX	54	83	69	110	316
$\overline{X}$	6.750	10.375	8.625	13.750	9.875
ΣX^2	422	923	663	1532	3540
$\hat{\sigma}^2$	8.214	8.839	9.696	2.786	7.384

	RS	SS	df	MS	F	α
m		3120.50	1			
bet	3333.25	212.75	3	70.917	9.60	.01
w		206.75	28	7.384		
t	3540.00	419.50	31			

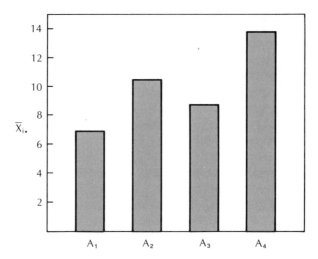

Figure 4-2. Group means from Table 4-1.

The first question essentially asks whether the mean for Group A_1 differs from the average of the means for groups A_2, A_3, and A_4. That is, the experimenter wishes to test the null hypothesis

$$H_0(1): \mu_1 = (\mu_2 + \mu_3 + \mu_4)/3.$$

An equivalent way to state this hypothesis is

$$H_0(1): \mu_1 - (1/3)\mu_2 - (1/3)\mu_3 - (1/3)\mu_4 = 0. \tag{4-8}$$

In other words, the hypothesis says that a particular linear combination of the true group means has a value of zero.

To develop a test of this hypothesis, we first find the distribution of the same linear combination of the obtained means:

$$C_1 = \overline{X}_{1.} - (1/3)\overline{X}_{2.} - (1/3)\overline{X}_{3.} - (1/3)\overline{X}_{4.}.$$

Since the obtained sample means are independently distributed as

$$\overline{X}_{i.} \sim N_{(\mu_i, \sigma_e^2/8)},$$

C_1 is normally distributed with mean and variance equal to

$$E(C_1) = \mu_1 - (1/3)\mu_2 - (1/3)\mu_3 - (1/3)\mu_4,$$
$$V(C_1) = (1 + 1/9 + 1/9 + 1/9)\sigma_e^2/8 = \sigma_e^2/6$$

(see pp. 6–7). Furthermore, if the null hypothesis $H_0(1)$ is true, then, according to Equation 4-8, $E(C_1) = 0$, and $H_0(1)$ can be restated as

$$H_0(1): C_1 \sim N_{(0, \sigma_e^2/6)} \sim \sqrt{(\sigma_e^2/6)}\, N_{(0,1)},$$

or, since the square of a normal deviate is distributed as chi-squared,

$$H_0(1): C_1^2 \sim (\sigma_e^2/6)\chi_{(1)}^2.$$

Then, if the null hypothesis is true,

$$6C_1^2 \sim \sigma_e^2 \chi_{(1)}^2, \tag{4-9}$$

and combining Equations 4-9 and 3-5, we have

$$6C_1^2/MS_w \sim \frac{\sigma_e^2 \chi_{(1)}^2}{\sigma_e^2 \chi_{(N-I)}^2/(N-I)} \sim F_{(1,N-I)}.$$

We can use an F test on the null hypothesis (Eq. 4-8). If the F is significantly larger than one, we reject the null hypothesis.

When the null hypothesis is false,

$$E(6C_1^2) = \sigma_e^2 + 6\psi_1^2,$$

where $\psi_1 = \mu_1 - (1/3)\mu_2 - (1/3)\mu_3 - (1/3)\mu_4$, so that a one-tailed test is appropriate.

For the data of Table 4-1, the values actually obtained are $C_1 = -4.167$ and $F = 14.11$; this value of F, with 1 and 28 degrees of freedom, is significant well beyond the .01 level.

Equivalently, since the square root of F with one degree of freedom in the numerator is distributed as t, we could also test $H_0(1)$ with

$$t_{(N-I)} = C_1 \sqrt{6/MS_w} = -3.76.$$

Under the null hypothesis this has the t distribution with $N - I$ degrees of freedom. The result is the same, of course, with either the t test (two-tailed) or the F test (one-tailed).

General Theory

The calculations discussed in the previous pages can be generalized to any linear comparison:

$$H_0(k): \Sigma_i c_{ik} \mu_i = 0.$$

The c_{ik} here are constants that define the desired comparison; for $H_0(1)$, $c_{11} = 1$, and $c_{21} = c_{31} = c_{41} = -1/3$.

The theory for the most general case, the possibility that there may be unequal numbers of observations in the different groups, will be given first. It will then be derived in a simpler form for equal sample sizes.

In general, the $\overline{X}_i$. are independently distributed as $N_{(\mu_i, \sigma_e^2/n_i)}$. A linear combination $C_k = \Sigma_i c_{ik} \overline{X}_i$. will then be normally distributed as

$$C_k \sim N_{(\Sigma_i c_{ik}\mu_i, \sigma_e^2 \Sigma_i c_{ik}^2/n_i)}, \tag{4-10}$$

so that, under the null hypothesis,

$$C_k' = C_k/\sqrt{\Sigma_i c_{ik}^2/n_i} \sim N_{(0,\sigma_e^2)} \sim \sigma_e N_{(0,1)}.$$

The zero mean is derived from the null hypothesis that $E(C_k)$ is zero; dividing C_k by a constant will not change this.

Since, in general,

$$MS_w \sim \sigma_e^2 \chi_{(N-I)}^2/(N-I)$$

the ratio of C_k' to the square root of MS_w is the ratio of a normal variable to the square root of a chi-square variable divided by its degrees of freedom:

$$C_k'/\sqrt{MS_w} \sim \frac{\sigma_e N_{(0,1)}}{\sigma_e \sqrt{\chi^2_{(N-I)}/(N-I)}} \sim t_{(N-I)}. \tag{4-11}$$

That is, the ratio of Equation 4-11 is distributed as t with $(N-I)$ degrees of freedom. The F distribution derived above follows from the general fact that

$$[t_{(N-I)}]^2 \sim F_{(1,N-I)}.$$

If the null hypothesis is not true, then the expected value of $(C_k')^2$ is

$$E(C_k')^2 = \sigma_e^2 + \psi_k^2/[\Sigma_i\ (c_{ik}^2/n_i)], \tag{4-12}$$

where

$$\psi_k = \Sigma_i\ c_{ik}\mu_i. \tag{4-13}$$

Power calculations for determining sample size, etc., can be made from the charts described in the previous chapter, using

$$\phi_k^2 = \frac{\psi_k^2}{\sigma_e^2\ \Sigma_i\ (c_{ik}^2/n_i)}.$$

If the same number (n) of observations is taken in all groups, then

$$\bar{X}_{i\bullet} \sim N_{(\mu_i,\sigma_e^2/n)},$$

and Equation 4-10 becomes

$$C_k \sim N_{(\Sigma_i\ c_{ik}\mu_i,\ (1/n)\sigma_e^2\ \Sigma_i\ c_{ik}^2)}$$

so that

$$C_k' = C_k\ \sqrt{n/\Sigma_i\ c_{ik}^2} \tag{4-14}$$

and

$$E(C_k')^2 = \sigma_e^2 + n\psi_k^2/(\Sigma_i\ c_{ik}^2).$$

Estimates of power can be found from the charts in Scheffé (1959) using

$$\phi_k^2 = \frac{n\psi_k^2}{\sigma_e^2\ \Sigma_i\ (c_{ik}^2)},$$

or from the appendix here, using

$$(\phi_k')^2 = \frac{\psi_k^2}{\sigma_e^2\ \Sigma_i\ (c_{ik}^2)}.$$

We can now apply these formulas to the other two experimental questions asked at the beginning of this section. The first, whether subjects make more errors if given both drugs rather than only one, can be answered by testing the null hypothesis

$$H_0(2): \mu_4 - (1/2)\mu_2 - (1/2)\mu_3 = 0. \tag{4-15}$$

The other can be answered by testing the null hypothesis

$$H_0(3): \mu_2 - \mu_3 = 0.$$

These two tests, along with the test of $H_0(1)$, are summarized in Table 4-2. Given these results, the experimenter can conclude with some confidence that administering the drugs does degrade performance and that both drugs together have a greater degrading effect than either drug separately. He cannot be confident that there is a difference between the effects of the two drugs.

Before moving on to other theoretical issues, here are some computational simplifications. Since the null hypotheses that we will discuss always specify that some linear function of the means is zero, hypotheses will not be changed if every c_{ik} is multiplied by the same constant. If, for example, in Equation 4-15 we multiply through by two, we will have

$$H_0(2): 2\mu_4 - \mu_2 - \mu_3 = 0, \tag{4-16}$$

which is obviously equivalent, mathematically, to Equation 4-15. Since the hypotheses are equivalent, the tests ought to be equivalent, and this is the case. The obtained value of C_k will differ, but C_k' will be the same. For example, for Equation 4-16 we have:

$$C_2 = 8.5$$
$$(C_2')^2 = (8.5)^2(8/6) = 96.33,$$

which is the same as the value in Table 4-2 (see Eq. 4-14). (You may find it interesting to prove that, in general, the value of $(C_k')^2$ is not changed when all of the c_{ik} are multiplied by the same constant.) The advantage of Equation 4-16 over Equation 4-15 is that the c_{ik} values are all integers, which are generally easier to work with than fractions. Most linear combinations that are of interest to the experimenter can be expressed in such a way

Table 4-2. Tests of three planned comparisons on the data in Table 4-1.

$H_0(1): \mu_1 - (1/3)\,\mu_2 - (1/3)\,\mu_3 - (1/3)\,\mu_4 = 0$

$$(C_1')^2 = \frac{8(-4.167)^2}{(4/3)} = 104.2$$

$F = 14.1, \; t = -3.76, \; \alpha < .01$

$H_0(2): \mu_4 - (1/2)\,\mu_2 - (1/2)\,\mu_3 = 0$

$$(C_2')^2 = \frac{8(4.25)^2}{(3/2)} = 96.33,$$

$F = 13.0, \; t = 3.61, \; \alpha < .01$

$H_0(3): \mu_2 - \mu_3 = 0$

$$(C_3')^2 = \frac{8(1.75)^2}{2} = 12.25$$

$F = 1.66, \; t = 1.29, \; \alpha = .20$

---

that the c_{ik} values are all integers; the preceding discussion shows that this can be done without changing the nature of the statistical test.

Another simplification may be achieved if, instead of finding the minimum significance level for each F, the experimenter decides to set a significance level and either accept or reject each null hypothesis accordingly. (Personally, I do not recommend this; see pp. 8–9.) It is easy to show from Equation 4-11 that the experimenter should reject the null hypothesis if and only if

$$|C_k'| \geq t_{\alpha(N-1)} \sqrt{MS_w},$$

where $t_{\alpha(N-1)}$ is the value of t needed to reach the α level of significance. Equivalently, he should reject the null hypothesis if and only if

$$(C_k')^2 > F_{\alpha(1,N-1)}MS_w.$$

The advantage of this simplification is that the term on the right can be calculated once for all significance tests.

Confidence Intervals for Planned Comparisons

In addition to testing the null hypothesis, it is possible to derive confidence intervals for the ψ_k. The above derivations make it clear that, with probability $1 - \alpha$,

$$C_k - S \leq \psi_k \leq C_k + S,$$

where

$$S = t_{\alpha(N-1)} \sqrt{MS_w \, \Sigma_i \, (c_{ik}^2/n_i)}.$$

For the case of equal ns this becomes

$$S = t_{\alpha(N-1)} \sqrt{MS_w(1/n) \, \Sigma_i \, c_{ik}^2},$$

For the three null hypotheses of Table 4-2, the 95-percent confidence intervals are

$$-6.44 \leq \psi_1 \leq -1.90$$
$$+1.84 \leq \psi_2 \leq +6.66$$
$$-1.03 \leq \psi_3 \leq +4.53$$

Necessary Assumptions and Their Importance

The same assumptions as in the ordinary F test are made in planned comparisons. The observations are assumed to be independently normally distributed with a constant variance. In addition, with one exception, the points made previously also apply to the robustness of planned comparisons. The violation of the assumption of normality generally has little effect on the result of a planned comparison unless the kurtosis of the population distribution is extremely small or extremely large and N is small. The violation of independence, on the other hand, has very large effects, just as it does with the overall F test. The effect of violating the assumption of equal variances requires some discussion.

The discussion will be simplified if at first we limit it to simple comparisons between two groups, such as $H_0(3)$ above. For this case the planned comparison is like an ordinary t test for the difference between two groups. In both tests the denominator is an estimate of the population standard deviation. In the t test, however, the estimate is obtained only from the observations in the two samples being compared. In a planned comparison the researcher takes advantage of the assumption of equality of variances to use the observations in all I groups. He thus obtains a more powerful test, as can be seen in the difference between the degrees of freedom in the two cases. In the t test the number of degrees of freedom is $2(n-1)$ for the case of equal numbers of observations in all groups; in the planned comparison the number is $(N-I) = I(n-1)$.

When the assumption of equal variances is violated, MS_w, as we have seen, is an estimate of the weighted mean of the population variances. However, the mean of all I population variances may not be an appropriate error term for comparing only two of them. In that case the error term should be an estimate of the average of only the variances of the two groups being compared. If, for example, the two specific groups being compared have population variances larger than the average for all I groups, MS_w will underestimate the average for those two groups. The obtained t (or F) will then be too large and may be significant more often than it should. The opposite will be true if the two groups in question have population variances smaller than the average over all I groups.

More complicated comparisons produce additional difficulties. Even when all of the means are involved in a comparison, they usually are weighted differently. In testing $H_0(1)$, for example, the mean of Group A_1 is weighted three times as heavily as the means of the other three groups. If the variance of the observations in Group A_1 is larger than the variances in the other three groups, the F ratio will be too large. The problem exists, moreover, even though all n_i are equal.

One solution is to find a transformation on the data that will make the variances in all I groups approximately the same. The advantages and disadvantages of such a transformation were discussed in Chapter 3.

Another solution is an approximate F test. When the variances are not all equal, the expected value of C_k^2 is

$$E(C_k^2) = \Sigma_i (c_{ik}^2 \sigma_i^2 / n_i) + \psi_k^2,$$

where σ_i^2 is the variance of population A_1; ψ_k^2 is defined in Equation 4-13. The null hypothesis specifies that ψ_k^2 is zero, so that when the null hypothesis is true,

$$E(C_k^2) = \Sigma_i c_{ik}^2 \sigma_i^2 / n_i.$$

We can form an appropriate F ratio if we can find a denominator that depends only on the variance within groups and has the same expected value as C_k^2 under the null hypothesis. The obvious answer is

$$D_k = \Sigma_i c_{ik}^2 \hat{\sigma}_i^2 / n_i.$$

The ratio C_k^2/D_k is distributed approximately as $F_{(1, d_k)}$, where d_k, the denominator degrees of freedom, is approximately

$$d_k = \frac{D_k^2}{\Sigma_i c_{ik}^4 \hat{\sigma}_i^4 / (n_i^3 - n_i^2)}.$$

Note, however, that d_k and the F ratio are both only approximate. The approximations hold well for large n_i; they probably hold well for moderate n_i but not too well if the n_i are very small. A good rule of thumb is that d_k should be at least 20.

When the n_i are equal, the above equations reduce to

$$D_k = (1/n) \, \Sigma_i \, c_{ik}^2 \hat{\sigma}_i^2$$

$$d_k = \frac{(n^3 - n^2) \, D_k^2}{\Sigma_i \, c_{ik}^4 \hat{\sigma}_i^4}.$$

The sample variance in Table 4-1 for Group A_4 is considerably smaller than those for the other three groups. Although the obtained differences could easily arise from chance variations, they may well lead us to suspect that the true population variances are different. The results of the approximate F tests on $H_0(1)$, $H_0(2)$, and $H_0(3)$ are shown in Table 4-3. A comparison with Table 4-2 shows that *for these particular data* the principle difference is a loss of degrees of freedom when the variances are not assumed to be equal.

Table 4-3. Tests of three planned comparisons on the data in Table 4-1 (equality of variances not assumed).

$H_0(1)$: $\mu_1 - (1/3) \, \mu_2 - (1/3) \, \mu_3 - (1/3) \, \mu_4 = 0$

$\qquad C_1^2 = (-4.167)^2 = 17.36$

$\qquad D_1 = (1/8) [8.214 + (1/9) (8.839 + 9.696 + 2.786)] = 1.323$

$\qquad d_1 = \dfrac{(8^3 \quad 8^2) (1.323)^2}{8.214^2 + (1/81) (8.839^2 + 9.696^2 + 2.786^2)} = 11.3$

$\qquad F = 13.1, \; \alpha < .01$

$H_0(2)$: $\mu_4 - (1/2) \, \mu_2 - (1/2) \, \mu_3 = 0$

$\qquad C_2^2 = (4.25)^2 = 18.06$

$\qquad D_2 = (1/8) [2.786 + (1/4) (8.839 + 9.696)] = .9275$

$\qquad d_2 = \dfrac{(8^3 - 8^2) (.9275)^2}{2.786^2 + (1/16) (8.839^2 + 9.696^2)} = 20.8$

$\qquad F = 19.5, \; \alpha < .01$

$H_0(3)$: $\mu_2 - \mu_3 = 0$

$\qquad C_3^2 = (1.75)^2 = 3.0625$

$\qquad D_3 = (1/8) (8.839 + 9.696) = 2.317$

$\qquad d_3 = \dfrac{(8^3 - 8^2) (2.317)^2}{8.839^2 + 9.696^2} = 14.0$

$\qquad F = 1.32, \; \alpha = .25$

When the test is a simple comparison between the means of two groups, e.g., A_1 and A_2, the above approximation is identical to the approximate t test proposed by Satterthwaite (1946). (The square root of the approximate F is of course distributed approximately as t.)

Independent Hypotheses

There are several advantages to formulating the null hypotheses for different planned comparisons in such a way that the validity of each null hypothesis is unrelated to the validity of any others. The three null hypotheses in our previous example were formulated in this way. If $H_0(1)$ is false, we know that drugs have some effect on learning, although $H_0(1)$ says nothing about whether the effect of taking both drugs simultaneously is different from the average of the effects of the drugs taken singly. That is, the validity of $H_0(1)$ is unrelated to the validity of $H_0(2)$, and vice versa; $H_0(1)$ involves the *average* of μ_1, μ_2, and μ_3, whereas $H_0(2)$ involves *differences* among them. The two hypotheses $H_0(1)$ and $H_0(2)$ are *independent*. In the same way, $H_0(3)$ is independent of both $H_0(1)$ and $H_0(2)$ because $H_0(1)$ and $H_0(2)$ both involve the average of μ_2 and μ_3, and $H_0(3)$ involves their difference. Planned comparisons that test independent hypotheses are said to be *orthogonal*.

Advantages of Independent Tests

One important advantage of orthogonal planned comparisons is that they allow a maximum amount of information to be extracted from each test. Each new test says something new about the data. Suppose that in the example in Tables 4-1 and 4-2 the experimenter had also tested the null hypothesis that learning is the same in Group A_4 (in which the subjects took both drugs) as in the control group, A_1; he might then be wasting his time on the obvious. Since the results of the test on $H_0(1)$ make it clear that the drugs do have an effect on learning, it is likely that both drugs taken together will have an effect. Instead, by testing the null hypothesis $H_0(2)$ that the mean for Group A_4 is the same as the average of the means of Groups A_2 and A_3, he is asking a question that is independent of that asked by $H_0(1)$.

More easily interpreted results are another advantage of orthogonal planned comparisons. Consider, for example, the case in which the experimenter tests both $H_0(1)$ and the null hypothesis

$$H_0(4): \mu_4 - \mu_1 = 0$$

proposed in the last paragraph. It is conceivable that he might obtain data for which he could reject $H_0(4)$ but not $H_0(1)$, although the obtained means for Groups A_2, A_3, and A_4 were all higher than those for Group A_1. Table 4-4 contains data for which $H_0(4)$ can in fact be rejected at the .05 level, whereas $H_0(1)$ cannot be rejected even at the .25 level. The experimenter is now in the peculiar position of having one test that tells him he can confidently conclude that the drugs affect learning and another which tells

Table 4-4. Hypothetical error data illustrating inconsistent results with nonindependent tests.

	A_1	A_2	A_3	A_4	Total
	5	0	10	4	
	4	8	3	5	
	4	3	5	9	
X_{ij}	5	9	8	10	
	8	7	4	10	
	7	5	2	7	
	9	7	5	13	
	2	7	9	14	
ΣX	44	46	46	72	208
$\overline{X}$	5.50	5.75	5.75	9.00	
ΣX^2	280	326	324	736	1666

	RS	SS	df	MS	F	α
m		1352.0	1			
bet	1419.0	67.0	3	22.33	2.53	.08
w		247.0	28	8.821		
t	1666.0	324.0	31			

$H_0(1)$: $3\mu_1 - \mu_2 - \mu_3 - \mu_4 = 0$

$$(C_1')^2 = \frac{8(-4.0)^2}{12} = 10.67, \ F = 1.21, \ \alpha = .30$$

$H_0(4)$: $\mu_4 - \mu_1 = 0$

$$(C_4')^2 = \frac{8(3.5)^2}{2} = 49.00, \ F = 5.55, \ \alpha = .03$$

him he cannot! How much weight should he assign to each test in reaching his conclusion? This entire problem can be avoided by initially formulating the null hypotheses so that they are independent. Each result can then be interpreted on its own merits, relatively independently of the other statistical tests. Note the word "relatively." Although the hypotheses may be independent, the tests themselves generally are not *completely* independent (for reasons to be made clear later).

Constructing Independent Tests

Formulating orthogonal comparisons is usually not difficult. A mathematical test for orthogonality is easily derived. Comparisons on two hypotheses, $H_0(k)$ and $H_0(k')$, are orthogonal if their associated linear combinations, C_k and $C_{k'}$, are statistically independent. Since C_k and $C_{k'}$ are both normally distributed (cf. Eq. 4-10), they will be independent if and only if

$$E(C_k C_{k'}) = E(C_k) E(C_{k'}). \tag{4-17}$$

But it can be shown that

$$
\begin{aligned}
E(C_k C_{k'}) &= E(\Sigma_i\, c_{ik} \overline{X}_{i.})(\Sigma_{i'}\, c_{i'k'} \overline{X}_{i'}) \\
&= \sigma_e^2\, \Sigma_i\, c_{ik} c_{ik'}/n_i + E(C_k) E(C_{k'}).
\end{aligned} \tag{4-18}
$$

From Equations 4-17 and 4-18 we can see that C_k and $C_{k'}$ will be statistically independent if and only if

$$\Sigma_i\, c_{ik} c_{ik'}/n_i = 0. \tag{4-19}$$

For the case where all groups contain the same number of observations this simplifies to

$$\Sigma_i\, c_{ik} c_{ik'} = 0. \tag{4-20}$$

The importance of this derivation is that when two or more comparisons are orthogonal, their C_k values are independent and vice versa. Thus, their C_k' values (the numerators of the calculated t values) are also independent. This has led some to assert that the statistical tests themselves are independent. The statistical tests cannot be independent unless the calculated t ratios are independent, and as long as all of the t ratios use the same denominator, MS_w, they cannot be independent. To put it more directly, it is possible that by pure chance MS_w might be smaller than expected. All of the ts would then tend to be too large. The opposite could also be true, making the ts too small. If N is relatively large, however, the danger of a spuriously small or large MS_w is not very great. We can say, then, that with a large N the t tests are nearly independent.

The three hypotheses tested in Table 4-2 can now be checked mathematically for independence. Table 4-5 gives the c_{ik} values for each of the three independent hypotheses as well as $H_0(4)$. The reader can readily verify that all the sums of the cross-products of all c_{ik} are zero, with the exception of the comparison of $H_0(4)$ with $H_0(1)$ and $H_0(2)$, for which the sums of the cross-products are $-4/3$ and $+1$ respectively.

Table 4-5. Tabular representation of planned comparisons to facilitate tests for independence (the values in the cells are the c_{ik}).

	A_1	A_2	A_3	A_4
$H_0(1)$	1	$-1/3$	$-1/3$	$-1/3$
$H_0(2)$	0	$-1/2$	$-1/2$	1
$H_0(3)$	0	1	-1	0
$H_0(4)$	-1	0	0	1

Cautions Concerning Independent Tests

Orthogonal planned comparisons have many desirable qualities. These qualities, along with some rather elegant mathematical relationships described later in this chapter (pp. 67–70 and pp. 72–73), have led many statisticians to stipulate that only orthogonal planned comparisons can be legitimately tested. These statisticians claim that somehow nonorthogonal planned comparisons violate basic statistical laws and therefore that their significance levels cannot be trusted. The logic of their arguments is not clear, however.

There are times when the questions the experimenter wishes to ask cannot be answered very readily by orthogonal planned comparisons. In other cases orthogonal planned comparisons provide only indirect answers, whereas nonorthogonal tests give more direct answers. For the data in Table 4-1, for example, the question asked by hypothesis $H_0(1)$ might be answered more directly by two planned comparisons:

$$H_0(1a): \mu_1 = \mu_2$$
$$H_0(1b): \mu_1 = \mu_3.$$

These are not orthogonal because if by chance $\overline{X}_1$, should be much smaller than μ_1, that statistical deviation would affect both planned comparisons. In a sense, however, they answer more directly the question that $H_0(1)$ is supposed to answer.

Moreover, statisticians seldom distinguish clearly between statistical independence (defined in Eq. 4-19) and logical independence (involving independent questions asked of the data). Statistical independence, as Equation 4-19 shows, depends on the n_i values; logical independence involves the nature of the questions asked. A test involving the difference between two means, like $H_0(3)$, is logically independent of a test involving only the average of those two means; $H_0(1)$ and $H_0(2)$, for example, are logically independent of $H_0(3)$. Logical independence can be defined mathematically by Equation 4-20; it is equivalent to statistical independence if the n_i are equal, but is usually not equivalent when the n_i are unequal.

When statisticians assert that orthogonality is essential, they usually have in mind statistical independence. All of the mathematical discussion of orthogonality in this text, for example, refers to statistical rather than logical independence. Statistically orthogonal tests have all of the elegant properties that we discuss, while logically orthogonal tests have them only if all groups have the same number of observations. Paradoxically, however, logical independence is usually more important to the experimenter because logically independent tests yield more interpretable results.

The question of whether tests should be made orthogonal, either statistically or logically, must be answered in terms of the specific data that the experimenter is gathering and the specific questions that he intends to ask of those data. The possibly more direct answers obtainable from nonorthogonal tests must be weighed against the advantages of greater efficiency and interpretability in orthogonal tests. A blanket prohibition of nonorthogonal tests tends to make statistics the master rather than the servant of the experimenter.

Relationship to Ordinary *F*

For most comparisons of interest to the experimenter the c_{ik} will sum to zero. This is true of all of the hypotheses in Table 4-5, as an inspection of the c_{ik} values in that table will show. When the c_{ik} sum to zero, the comparisons are independent of the value of the grand mean, as we will soon show (pp. 71–72).

A linear comparison in which the c_{ik} sum to zero is called a contrast, and there is an important relationship between orthogonal contrasts and the ordinary *F* test. The first clue to this relationship comes from the fact that we found three orthogonal contrasts for testing in Table 4-1 and the ordinary *F* test for the data in that table had three degrees of freedom in the numerator. The second clue comes from adding the three values of $(C'_k)^2$ associated with the three independent contrasts. This sum is $(C'_1)^2 + (C'_2)^2 + (C'_3)^2 = 104.167 + 96.333 + 12.25 = 212.75$, which, as Table 4-1 shows, is exactly equal to SS_{bet}. These two facts can be generalized to any one-way analysis of variance; the total number of possible orthogonal contrasts will always be equal to $(I - 1)$, the degrees of freedom, and the sum of the $(C'_k)^2$ values will always be equal to SS_{bet}.

This fact throws new light on the nature of both the *F* test and the planned comparisons. Since C'_k is normally distributed with a mean of zero and a variance of σ_e^2 when the null hypothesis is true,

$$(C'_k)^2 \sim \sigma_e^2 \chi^2_{(1)},$$

and the $(C'_k)^2$ are independent when the C_k are independent. Since the sum of $I - 1$ independent chi-square variables is itself distributed as chi-square, the sum of the $(C'_k)^2$ values for the $I - 1$ independent contrasts when the null hypotheses for all of the contrasts are true will be distributed as chi-square with $I - 1$ degrees of freedom. But, as we have already seen, this sum is exactly equal to SS_{bet}, which does indeed have $I - 1$ degrees of freedom. Since the sum of the $(C'_k)^2$ values is distributed as chi-square if and only if all $I - 1$ of the null hypotheses are valid, the overall *F* test can be thought of as a kind of combined test on all $I - 1$ null hypotheses simultaneously. However, since the $I - 1$ independent contrasts actually tested can be chosen in an infinite number of ways, the overall *F* test can also be thought of as a kind of general test of the null hypothesis that the true value of every possible contrast is zero.

It must be emphasized, however, that a test of a specific contrast may be significant even though the overall *F* test is not (the opposite may also be true). Such is the case, for example, with the data in Table 4-4, where the overall *F* value of 2.43 is not significant but the test of $H_0(4)$ is significant. In the past, some statisticians have argued that it is not permissible to make planned comparisons unless the results of an overall *F* test are first found to be significant. As can be seen here, both the overall F test and the planned comparisons are based on the same statistical theory and there is no reason why one should have priority over the other. In fact, the procedure advocated in the past may well prevent the researcher from finding interesting differences in his data because the overall F test has less power than a planned comparison in testing the specific differences in which he is interested.

This does not mean, on the other hand, that one can make comparisons willy-nilly. In particular, the choice of planned comparisons cannot be based on the data obtained. If one first inspects the data and then chooses comparisons by noting which contrasts are likely to be large, one is deliberately biasing the t statistics in favor of rejecting the null hypotheses; the t ratios then will not have the t distribution. The choice of which comparisons to make cannot be influenced in any way by the data actually obtained. In fact, it is preferable to choose planned comparisons before any data are obtained. If Groups A_2, A_3, and A_4 in Table 4-1 were simply groups of people given three different drugs, and if the experimenter had no a priori reason to believe that one drug would retard learning more than another, it would not be legitimate to test $H_0(3)$. There would be no better reason to test the hypothesis that Group A_4 is different from Groups A_2 and A_3 than there would be to test the hypothesis that A_2 is different from A_3 and A_4, or the hypothesis that A_3 is different from A_2 and A_4. The differences in the obtained data do not justify testing $H_0(3)$ over these other two hypotheses. Similarly, $H_0(2)$ would not be legitimately tested by a planned comparison either. Only $H_0(1)$ could be legitimately tested.

The overall F test thus remains useful when few or no natural planned comparisons can be made. Although only one legitimate planned comparison was possible in the above example, an overall F test could still determine whether there were any significant differences among the means.

Moreover, the relationships between contrasts and the overall F (discussed above) can be used to improve the F test even for the above example. Suppose the experimenter faces the situation in which Groups A_2, A_3, and A_4 are given three different drugs, no one of which can be expected a priori to have any specifically different effect than the other two. One possibility would be for the experimenter to make an F test on the four groups and, if the test were not significant, discontinue testing. In such a case, however, a significant difference between the control group and the other three groups might be masked by the overall test. The experimenter should thus make a planned comparison on $H_0(1)$ even if the overall F test is not significant. If the planned comparison is significant, the experimenter is back in the predicament described earlier; one test denies the existence of differences while the other test asserts their existence. The theory discussed above provides the experimenter with a way out of his dilemma. More specifically, instead of the overall F test, he can make two independent statistical tests: one on the null hypothesis that the scores for the control group are the same as those under drugs, i.e., $H_0(1)$, and the other on the general null hypothesis that there are no differences between the three groups receiving drugs.

Recall that SS_{bet} for the general F test for this case involves the sum of three $(C'_k)^2$ values, each independently distributed. Any three values corresponding to three orthogonal contrasts will do (note that they must correspond to contrasts, that is, the c_{ik} must sum to zero). First, we test the contrast implied by $H_0(1)$ and find it significant for the data of Table 4-1. Next, we note that $H_0(2)$ and $H_0(3)$ are both independent of $H_0(1)$ and hence would be legitimate contrasts to test if there were a priori reasons for testing them. Unfortunately, in the hypothetical situation we are now considering there are no a priori reasons.

Suppose, though, that we had originally done our experiment only on Groups A_2, A_3, and A_4. We would then have three groups, and our F test with two numerator degrees of freedom would use the sum of the chi-squares of two orthogonal contrasts. Since hypotheses $H_0(2)$ and $H_0(3)$ involve only differences between these latter three groups, they would be acceptable as the two orthogonal contrasts in question. What all of this complicated reasoning means is that if an ordinary F test were done only on Groups A_2, A_3, and A_4, the resulting SS_{bet} would be

$$SS_{A_2A_3A_4} = (C'_2)^2 + (C'_3)^2.$$

This can of course be verified:

$$SS_{A_2A_3A_4} = (1/8)(83^2 + 69^2 + 110^2) - (1/24)(83 + 69 + 110)^2$$
$$= 2968.750 - 2860.167 = 108.583;$$
$$(C'_2)^2 + (C'_3)^2 = 96.333 + 12.250 = 108.583.$$

Thus, the general F test on the three remaining groups corresponds to the simultaneous testing of two contrasts, both of which are independent of $H_0(1)$. In making the test, however, we should use the MS_w found for the F test with all four groups, rather than calculate a new MS_w based only on the three groups being tested. This is because MS_w is simply an estimate of σ_e^2, and we should in general, use the best estimate we have. (This assumes, of course, that all of the σ_i^2 are equal; if we have reasons to doubt that assumption, MS_w should be based only on the three groups being tested.) Certainly an estimate based on all four groups will be better than one based on only three. The F test is

$$MS_{A_2A_3A_4} = 108.583/2 = 54.292,$$
$$F_{(2,28)} = 54.292/7.384 = 7.35,$$
$$\alpha < .01.$$

The experimenter has thus made two independent tests and can reject both null hypotheses with a high degree of confidence. He can confidently conclude both that the drugs affect learning and that the drugs differ in the degree of their effects on learning.

Actually, it was not necessary in the analysis above to directly calculate either $SS_{A_2A_3A_4}$ or C'_2 and C'_3. From the fact that the values for all three comparisons sum to SS_{bet} and the values for the latter two sum to $SS_{A_2A_3A_4}$, we have

$$SS_{A_2A_3A_4} = SS_{bet} - (C_1')^2 = 212.750 - 104.167 = 108.583.$$

These results are merely a special case of a general principle. Suppose one has a set of observations from I groups and he wishes to test r independent contrasts. If r is smaller than $I - 1$, the experimenter can also make an independent test of the null hypothesis that there are no differences other than the r differences already tested. He does this by finding

$$SS_{rem} = SS_{bet} - \sum_1^r (C_k')^2.$$

Then SS_{rem} must correspond to the sum of $(I - 1 - r)$ additional potential tests of contrasts, and so SS_{rem} must be distributed (under the null hypothesis of no additional differences) as chi-square with $(I - 1 - r)$ degrees of freedom. Consequently, the test using

$$F_{(I-1-r, N-I)} = MS_{rem}/MS_w,$$

where

$$MS_{rem} = SS_{rem}/(I - 1 - r),$$

will be independent of the r contrasts originally tested.

Interpreting Multiple Significance Tests

Even though the tests are relatively independent, some caution is required in interpreting the results of a number of significance tests. The problem arises from the fact that significant results can arise by chance. When a number of significance tests are made, the probability that one or more will arise by chance becomes considerably higher. Specifically, if r independent significance tests are made and all r null hypotheses are true, the probability is $1 - (1 - \alpha)^r$ that one or more of the tests will be significant. If $\alpha = .05$ and $r = 3$, the probability of obtaining one or more significant results even though all three null hypotheses are true is $1 - (1 - .05)^3 = .143$.

One can protect oneself by requiring a higher significance level when a number of comparisons are made. If the significance level for each of r independent tests is set at $\alpha^* = 1 - (1 - \alpha)^{1/r} \approx \alpha/r$, the probability will be $1 - \alpha$ that all significance tests are nonsignificant.

Even when the r tests are not independent, a significance level of α/r on each test will guarantee, with a probability of at least $1 - \alpha$, that no true null hypothesis will be rejected. The use of a significance level of α/r was first proposed by R. A. Fisher (1935)* and is sometimes referred to as *Fisher's method* of multiple comparisons. By setting the significance level

---

*See also Miller (1966).

for each test at α/r, we set the *expected number* of significant results (assuming that all null hypotheses are true) equal to α. Alternatively, one may calculate the conventional significance levels and use judgment in interpreting the pattern of results. Less weight would then be given to an occasional significant result if it occurred along with a number of results that were nonsignificant. Either solution to the problem is reasonable. Too often, however, the problem is simply ignored.

Planned Comparisons and the Grand Mean

In the previous chapter we saw that SS_m could be used to test the null hypothesis H_0: $\mu = 0$. Now, consider the following planned comparison:

$$H_0(m): \Sigma_i\, n_i\mu_i = 0. \tag{4-21}$$

This is a perfectly legitimate planned comparison, and it can be tested in the usual way. (Notice, however, that while it is a comparison it is not a contrast, since $\Sigma_i\, n_i \neq 0$.) However, when we make the planned comparison, we find that C_m is simply N times the grand mean. Furthermore, for C'_m we have

$$C'_m = C_m / \sqrt{\Sigma_i\, n_i^2/n_i} = (\Sigma_i\, n_i\overline{X}_{i\cdot}) / \sqrt{\Sigma_i\, n_i}$$
$$= N\,\overline{X}_{\cdot\cdot} / \sqrt{N} = \sqrt{N}\,\overline{X}_{\cdot\cdot},$$

and

$$C'^2_m = N\,\overline{X}_{\cdot\cdot}{}^2 = SS_m.$$

In other words, the null hypothesis that the grand mean of all of the groups is zero, is a planned comparison.

Even though this test is seldom used, its relationship to other planned comparisons, and particularly contrasts, helps to clarify the relationship between the F test and planned comparisons.

It is not difficult to prove that every contrast is orthogonal to $H_0(m)$. From Equations 4-19 and 4-21 we can see that a particular planned comparison is orthogonal to $H_0(m)$ if and only if

$$\Sigma_i\, c_{ik}n_i/n_i = \Sigma_i\, c_{ik} = 0.$$

That is, a given planned comparison is orthogonal to $H_0(m)$, and hence independent of the value of the grand mean, if and only if it is a contrast.

Earlier, I said that the overall F test corresponds to a test of the null hypothesis that all *contrasts* are zero. I also said that the total number of orthogonal *contrasts* that could be tested is only $I - 1$. In fact, it is possible to test I different orthogonal planned comparisons. If, however, we wish our planned comparisons to be independent of the grand mean as well as each other (that is, to be orthogonal contrasts), we will only be able to find $I - 1$. The Ith planned comparison in that case would be the test of the grand mean.

A point made in some textbooks is that all planned comparisons must be contrasts. As we have seen, only orthogonal contrasts can legitimately be used as components of SS_{bet}. However, there is no mathematical reason why any planned comparison should be thought of as illegitimate whether it is a contrast or not. In practice, planned comparisons are almost always contrasts, but there is no mathematical reason that they must be.

ω^2 for Planned Comparisons

Just as for the overall test, every difference tested for by a contrast, i.e., every ψ_j, can be said to account for a proportion of the variance in the data. The proportion accounted for by a given ψ_j, i.e., ω_j^2, can be estimated in a manner similar to the estimation of ω^2 in the overall F test.

The theory for this rests on the fact (too complicated to derive here) that just as

$$SS_{bet} = \Sigma_k (C_k')^2, \tag{4-22}$$

so also

$$\sigma_{bet}^2 = \Sigma_k (\psi_k')^2,$$

where

$$(\psi_k')^2 = \psi_k^2 / [\Sigma_i (c_{ik}^2/n_i)]$$

and the ψ_j are the values of $I - 1$ independent contrasts (Eq. 4-13). Therefore, Equation 4-3 can be extended to

$$\sigma_t^2 = \sigma_e^2 + \Sigma_k (\psi_k')^2,$$

and

$$\omega_j^2 = (\psi_k')^2 / \sigma_t^2$$

is the proportion of the total variance accounted for by the linear contrast ψ_k.

Applying to Equation 4-12 a derivation similar to that on p. 53, we find that

$$(\hat{\psi}_k')^2 = (1/N) [(C_k')^2 - MS_w]$$

is an unbiased estimate of $(\psi_k')^2$. Consequently, from Equation 4-6,

$$\hat{\omega}_k^2 = \frac{(C_k')^2 - MS_w}{SS_t + MS_w} \tag{4-23}$$

can be used as an estimate of the proportion of the variance accounted for by ψ_k. For the contrasts tested in Table 4-1, the $\hat{\omega}_k^2$ are

$$\hat{\omega}_1^2 = .23$$
$$\hat{\omega}_2^2 = .21$$
$$\hat{\omega}_3^2 = .01.$$

Furthermore, this method of estimating proportion of variance accounted for can be extended to the case in which r different independent contrasts are combined into one general F test. In that case the estimate of the proportion of variance accounted for by all of the r contrasts making up the overall test is the sum of the $\hat{\omega}_k^2$ values for the r individual contrasts. It is relatively easy to show, from Equations 4-22 and 4-23, that the sum of the $\hat{\omega}_k^2$ values for all $I - 1$ independent contrasts is equal to the $\hat{\omega}^2$ calculated for the overall F test, as indeed it should be.

POST HOC COMPARISONS

Unfortunately, a complete data analysis usually requires more than a simple overall F test or a limited number of tests on independent planned comparisons. Some of the most important discoveries in science are "after the fact" discoveries — unanticipated relationships found in the data. Such relationships cannot be tested by the planned-comparison technique because the choice of a comparison on the basis of what appears to be a real difference among the data would introduce a strong bias in favor of rejecting the null hypothesis. As a result, the true significance level would not in actuality be as high as it appears to be. Good techniques for making *post hoc comparisons* (comparisons of relationships that become evident only after the data have been studied) would be valuable statistical tools. Such methods do exist, but they have serious shortcomings. They have very little power — in many cases they will not find a significant difference unless it is large enough to be obvious without a test. In some cases their validity is dubious.

Nevertheless, these methods are sometimes useful in post hoc data analyses. We will discuss four approaches to post hoc comparisons here. They are the method of adjusted significance level, Fisher's method, the Scheffé, or S, method (Scheffé, 1959, pp. 66–72), and the Tukey, or T, method (see Scheffé, 1959, pp. 73–75). We will discuss only the formulas and general rationale of the more complicated S and T tests; the derivations of the formulas are too complicated to present in detail.

Adjusted Significance Levels

The method of adjusted significance levels is a completely intuitive approach that, in an ideal world, would probably be the best. It consists simply of adjusting the significance level required for each post hoc comparison according to a reasonable judgment of the prior plausibility of the data. If, in the absence of the data, we would have expected a large difference, then we would choose a rather liberal α (perhaps as large as .05 or .10) for rejection of the null hypothesis. If, on the other hand, we would have expected the difference to be very small or nonexistent, we would reject the null hypothesis only if the obtained significance level were very small (perhaps even less than .001 for some null hypotheses). The difficulty with this approach is that it depends heavily on the experimenter's ability to judge his own data objectively and to make judgments that readers of his research are likely to accept. Since neither of these can be assumed in actual practice, this method is not generally recommended (or even mentioned in most texts). The following three methods are essentially modifications of this approach, adding the objectivity that is lacking in the method of adjusted significance levels.

Fisher's Method

Fisher's method was discussed previously in connection with multiple planned comparisons. It can be adapted, in a somewhat limited way, to post hoc comparisons. Usually, before the data are gathered the experimenter has some idea of which comparisons are likely to be interesting. With a list of r such possible comparisons, he can apply Fisher's method to his data, using a significance level of α/r (even though the possible tests are not all orthogonal). Frequently, for example, the experimenter expects to test differences only among pairs of means. Although particularly interested in testing the difference between the largest and smallest means in the sample, he might also wish to test other differences that appear to be large.

To guarantee that his list of potential tests will contain a test of the difference between the largest and smallest means, he must include all tests of simple differences between means in the list. The number of such differences is equal to the number of pairs of groups. For the data in Table 4-1, for example, the four groups can be paired in six different ways, so $r = 6$. The actual significance levels would then be found by doing standard planned comparisons on all of the differences in which we are interested, and multiplying the significance levels found in the F table by 6. The results of such tests on all six pairs of means are shown in Table 4-6, along with the results that would have been obtained by planned comparisons on each pair of means. The listed significance levels using Fisher's method are each exactly 6 times as large as those that would be obtained by planned comparisons. Keep in mind, however, that these are only upper limits on the significance levels; the true α values may be somewhat smaller.

So long as the list of potential tests is small, Fisher's method does not sacrifice too much power. If the list is large, however, Fisher's method has very little power. Suppose the experimenter decides to test differences among all pairs of means in the experiment. If there are only three groups, he will have to multiply each significance level by 3; if there are four groups, the multiplier will be 6, as we saw above; if there are five groups, the multiplier is 10; and if there are six groups, the multiplier is 15.

The Scheffé Method

The S and T methods both have the same rationale. In a sense, each is an attempt to extend Fisher's method to the case of an unlimited list of potential post hoc contrasts. For both methods the significance level is a kind of "protection level," such that if there are no differences among the true means, the probability of finding *any significant difference at all* will be no larger than α. In other words, if there are no differences among the population means, the probability is $1 - \alpha$ that every single contrast that might possibly be tested will fail to be significant at the α level. The fact that these tests set a high protection level against finding *any* significant result accounts for their general lack of power. If their use is exploratory, and any significant result found by these methods is likely to be tested again in another experiment, a more generous significance level than usual may be in order for these tests (but see pp. 83–84). The Scheffé method is presented first because it is most similar to the tests discussed above.

We found previously that in testing a contrast $(C_k')^2$ was proportional to chi-square with one degree of freedom. The S method simply consists of treating each $(C_k')^2$ as if it were proportional to chi-square with $I - 1$ degrees of freedom instead. What this means, basically, is that we can make a post hoc test of any contrast that may interest us by computing

$$F'_{(I-1, N-I)} = \frac{(C_k')^2}{(I-1)MS_w}$$

and looking up the resulting F' in an ordinary F table with $I - 1$ and $N - I$ degrees of freedom.

The method will be illustrated with a number of tests of this kind on the data of Table 4-1. We will first use the S method to test the differences between all possible pairs of means. To test the difference between Groups A_1 and A_2, for example, we find

$$(C'_{1,2})^2 = 8(10.375 - 6.75)^2/(1 + 1) = 52.5625,$$

$$F'_{(3, 28)} = \frac{52.5625}{(3)(7.384)} = 2.37,$$

$$\alpha = .10,$$

where α is the significance level of an F of 2.38 with 3 and 28 degrees of freedom. The relative lack of power of this method of post hoc comparisons can be seen from the fact that, if a planned comparison could have been made, the F would have been

$$F_{(1, 28)} = 52.5625/7.384 = 7.12,$$
$$\alpha = .02.$$

The results of applying the S test to all the pairs of means are shown in Table 4-6.

As another example of the S method, suppose that in inspecting the data of Table 4-1 the experimenter notes that the combined effect of the two drugs administered together appears to be greater than the sum of the effects of the two drugs administered singly. More specifically, it appears to him that

$$(\overline{X}_{4.} - \overline{X}_{1.}) > (\overline{X}_{2.} - \overline{X}_{1.}) + (\overline{X}_{3.} - \overline{X}_{1.}).$$

Table 4-6. Post hoc tests by three methods, compared with planned tests on differences between pairs of means in Table 4-1.

		Comparison					
		A_4 vs A_1	A_4 vs A_3	A_2 vs A_1	A_4 vs A_2	A_3 vs A_1	A_2 vs A_3
	C_k:	7.000	5.125	3.625	3.375	1.875	1.750
Planned	F	26.54	14.23	7.12	6.17	1.90	1.66
tests	α	<.001	<.001	.013	.020	.19	.22
Fisher's	F	26.54	14.23	7.12	6.17	1.90	1.66
method	α	<.006	<.006	.078	.12	–	–
S	F'	8.85	4.74	2.37	2.06	0.63	0.55
method	α	<.001	.009	.10	.13	–	–
T	t'	7.29	5.33	3.77	3.51	1.95	1.82
method	α	<.01	<.01	.06	.09	–	–

To test this, he sets up the null hypothesis

$$H_0: (\mu_4 - \mu_1) = (\mu_2 - \mu_1) + (\mu_3 - \mu_1),$$

which is algebraically equivalent to the hypothesis

$$H_0: \mu_4 - \mu_3 - \mu_2 + \mu_1 = 0.$$

This form of the null hypothesis is a contrast that is tested with

$$(C')^2 = 8(13.750 - 8.625 - 10.375 + 6.750)^2/(1 + 1 + 1 + 1)$$

$$= 4.50,$$

$$F'_{(3, 28)} = \frac{4.50}{(3)(7.384)} < 1.$$

This difference is nonsignificant, and therefore we cannot conclude that the effect of giving both drugs simultaneously is greater than the sum of the effects of the individual drugs.

If the user decides to set a significance level in advance, the calculations for the S method can be made somewhat simpler, just as with planned comparisons. The technique consists of setting the significance level (α), finding the value (F_α) required to reach that significance level, and then rejecting the null hypothesis if and only if

$$|C'_k| > \sqrt{(I - 1)MS_w F_{\alpha(I - 1, N - I)}}.$$

Similarly, $1 - \alpha$ confidence limits can be found by the formula

$$C_k - S' \le \psi_j \le C_k + S',$$

where

$$S' = (I - 1)MS_w F_{\alpha(I - 1, N - I)} \, \Sigma_i(c_{ik}^2/n_i).$$

It should be pointed out here that the above results are valid only if all of the post hoc comparisons are contrasts — that is, if the c_{ik} values sum to zero. If it should happen that the experimenter wishes to test some comparisons that are not contrasts, he can do so, but he must treat $(C'_k)^2$ as if it has I degrees of freedom instead of $I - 1$. All of the above equations will hold true in this case if $I - 1$ is replaced by I wherever it appears. Note, however, that it is necessary in this case to treat all post hoc comparisons the same way even though only one or a few are not contrasts. This is necessary to maintain the "protection level" effect of the significance level used. The power of the tests will of course be reduced accordingly.

It should be remembered that, just as with planned comparisons, the validity of post hoc comparisons depends somewhat on the validity of the assumption of equal variances. Because of the general conservatism of the tests, violations of the assumption probably have less effect than in planned comparisons. Nevertheless, if the variances are unequal, the results of post hoc comparisons may be suspect. Unfortunately, there is no way to modify the Scheffé method for unequal variances. An appropriate transformation on the data (were one to exist) might make the variances more equal.

Relationship Between *S* and *F*

The S method is related to the overall F test in an interesting and useful way. It can be proved that the significance level of F is the highest attainable significance level with any post hoc comparison by the S method. That is, if the overall F test is just significant at the α level, then there is exactly one post hoc contrast that is also significant at the α level. The c_{ik} for this contrast are $n_i(\overline{X}_{i.} - \overline{X}_{..})$. Furthermore, there does not exist a post hoc contrast with a significance level higher than α.

Although a post hoc test significant at the α level does exist, the experimenter will probably not be interested in testing that contrast. Nevertheless, this information can influence the decision to test any post hoc contrasts. If the experimenter is interested only in results that reach a particular level of significance, and the overall F test does not reach this level of significance, then no post hoc contrast will reach it. For example, for the data of Table 4-4, no post hoc contrast, tested by the S method, can reach a significance level higher than .08.

The *T* Method of Post Hoc Contrasts

The T method of making post hoc comparisons differs considerably from the previous methods. It is strictly limited to tests of contrasts, and is based on a distribution known as the *Studentized range distribution*. The Studentized range distribution is defined as follows: Let z_i be a normally distributed random variable with a mean of zero and a variance of 1, and suppose that k values of z_i are sampled randomly and independently. Let $z_{\max}$ be the largest value sampled and $z_{\min}$ be the smallest, so that $z_{\max} - z_{\min}$ is the range of the z_i values in the sample. Let y^2, independent of the z_i, have the chi-square distribution with ν degrees of freedom. Then

$$t'_{(k,\nu)} = \frac{z_{max} - z_{min}}{\sqrt{y^2/\nu}}$$

has the Studentized range distribution with parameters k and ν. The .10, .05, and .01 levels of significance of the Studentized range distribution are tabled in the appendix.

The T test cannot be used unless all of the groups contain the same number (n) of observations. The test is made by calculating the statistics

$$C''_k = \frac{\sqrt{n}\, C_k}{(1/2)\, \Sigma_i\, |c_{ik}|}$$

$$t_{(I,N-I)} = |C''_k| / \sqrt{MS_w},$$

and finding the significance of t' in the table of the Studentized range distribution. Notice that C''_k differs from C'_k in only one way: instead of dividing by the square root of the sum of the squares of the c_{ik}, C''_k is found by dividing by 1/2 the sum of the absolute values.

We will first illustrate the T method by testing the null hypothesis $H_0: 3\mu_4 - \mu_3 - \mu_2 - \mu_1 = 0$ on the data of Table 4-1. The calculations are

$$C'' = \sqrt{8}\ (15.5)/[(1/2)\,(3 + 1 + 1 + 1)] = 14.61,$$
$$t'_{(4,28)} = 14.61/\sqrt{7.384} = 5.37,$$
$$\alpha = .01.$$

The results of applying the T method to all of the differences between pairs of means are shown in Table 4-6.

Once again, if a specific significance level is chosen in advance, the calculations can be simplified by rejecting the null hypothesis if and only if

$$|C''_k| \geq t'_{\alpha(I,N-I)} \sqrt{MS_w},$$

and confidence intervals for ψ_j can be found by noting that with probability $1 - \alpha$

$$C_k - T' \leq \psi_j \leq C_k + T',$$

where

$$T' = \sqrt{MS_w/n}\ (1/2)\,(\Sigma_i\, |c_{ik}|)\ t'_{(I,N-I)}.$$

The robustness of the T method is not completely known. It appears, however, to be similar to planned comparisons; it is robust to the assumption of normality but not to the assumptions of random sampling and equal variances.

Relationship of the *T* Method
to an Overall Test of Significance

Just as the S method is closely related to the overall F test, the T method is closely related to a different overall test for differences. It can be shown that if the T method is used to test the difference between the largest and the smallest of the I obtained means, the result is an overall test of the null hypothesis that there are no differences between the means of the I populations.

For the data of Table 4-1, this would be the difference between Groups A_4 and A_1. The t' value of 7.28 is significant beyond the .01 level (see Table 4-6).

The theory behind this kind of overall test has been as explicitly derived as for the F test, and mathematically it is a perfectly legitimate test to use. Generally, however, it is less powerful than the F test, so it is usually not recommended.

The T method of testing for overall differences is related to individual tests by the T method in the same way that the overall F test is related to tests by the S method. The level of significance of the difference between the largest and smallest obtained means is the highest significance level that can be attained in any test of a contrast by the T method. For the data of Table 4-4, for example, the overall T test is made on the difference between Groups A_4 and A_1:

$$C'' = \sqrt{8}\ (9.0 - 5.5)/[(1/2)(1 + 1)] = 9.898$$
$$t'_{(4.28)} = 9.899/\sqrt{8.821} = 3.33$$
$$\alpha > .10.$$

Since this difference is not significant at the .10 level, we are certain that no contrast, tested by the T method, would attain a level of significance as high as .10.

Increasing the Power of Post Hoc Comparisons

Since $1 - \alpha$ is the probability that all of the entire set of possible post hoc comparisons will be nonsignificant at the α level, the power attained by post hoc comparisons depends greatly on how large is the set of possible comparisons. By limiting the set of potential post hoc comparisons, it is possible to increase the power of the post hoc comparisons.

To see how this can work with the T method, consider the situation discussed on pp. 69–70, where A_2, A_3, and A_4 represent groups given three different drugs, with no a priori reasons for assuming that any specific drugs would have a greater effect than the others. We first made a planned comparison on the difference between the control group, A_1, and the mean of the three experimental groups, A_2, A_3, and A_4. Then we made an F test on the means of the three experimental groups. Presumably, the experimenter, having already tested for a difference between the control and experimental groups, would focus his attention on differences among the three experimental groups in making post hoc tests. In testing differences by the T method, he would calculate t' just as before; but because only the three groups are being compared, the result can now be regarded as $t'_{(3.28)}$, instead of $t'_{(4.28)}$ as in Table 4-6. The increase in power can be seen in the fact that if $t'_{(3.28)}$ is used, the difference between A_4 and A_2 (with a t' value of 3.51) is significant at the .05 level, rather than at the .10 level as in Table 4-6.

In general, if all of the potential contrasts to be tested by the T method are limited to some subset of r groups, then the value of t' can be regarded as $t'_{(r, N-1)}$ rather than $t'_{(I, N-1)}$, usually with a considerable increase in power.

A similar principle holds true for the S method. If all of the potential comparisons to be made by the S method are limited to a subset of r groups, then the $(C'_k)^2$ can be regarded as having $(r-1)$ instead of $(I-1)$ degrees of freedom. The test is then

$$F'_{(r-1, N-1)} = \frac{(C'_k)^2}{(r-1)MS_w}.$$

If, as postulated above, all post hoc tests were limited to differences between the means of the three experimental groups, the difference between Groups A_4 and A_2 would be tested by

$$(C'_k)^2 = (8)(3.375)^2/2 = 45.56,$$

$$F'_{(2, 28)} = \frac{45.56}{(2)(7.384)} = 3.09,$$

$$\alpha = .06.$$

By way of comparison, the same test, reported in Table 4-6 with $(I-1)$ degrees of freedom, was not significant at the .10 level. (To take advantage of this increase in power, however, one must choose the r groups among which the comparisons will be made before seeing the data.)

For the S method, this approach can be generalized further. If all of the potential post hoc contrasts are independent of a given set of r' orthogonal contrasts, then the $(C'_k)^2$ for the post hoc contrasts can be thought of as having $(I - r' - 1)$ instead of $(I - 1)$ degrees of freedom. In the example above, by limiting the post hoc comparisons to the three experimental groups, we made them all independent of the planned comparison on $H_0(1)$: $\mu_1 - (1/3)\mu_2 - (1/3)\mu_3 - (1/3)\mu_4 = 0$. Since our post hoc comparisons were all independent of one planned comparison, we could act as though the $(C'_k)^2$ had two degrees of freedom instead of three. (Once again, the r' orthogonal planned comparisons must be chosen before seeing the actual data.)

We can illustrate the extension of this principle to more complicated analyses by applying it to data in Table 4-7. These are also hypothetical data, obtained as follows. To investigate students' attitudes toward school, an educator gave a questionnaire to a number of students in the first, second, and third grades of his school. A high score on the questionnaire indicated a favorable attitude toward school. According to this school's policy, first grade pupils were divided into three classes on the basis of a reading-readiness test: "slow learners" (Group A_1), "moderate learners" (Group A_2), and "fast learners" (Group A_3). Children in the second and third grades were divided into two classes, but the division was random. Because the questionnaire required considerable time to administer, only four pupils from each class were tested. Table 4-7 contains the scores of the pupils from each of the seven classes. No overall F test was performed on

Table 4-7. Hypothetical data with seven groups of four subjects each.

	G_1			G_2		G_3		
A_1	A_2	A_3	A_4	A_5	A_6	A_7	Sum
35	41	41	31	19	31	32	
22	38	42	20	9	24	37	
37	73	43	31	40	26	45	
31	50	65	25	33	41	47	
ΣX	125	202	191	107	101	122	161	1009
$\overline{X}$	31.25	50.50	47.75	26.75	25.25	30.50	40.25	36.04
ΣX^2	4039	10954	9519	2947	3131	3894	6627	

$$SS_w = 2269.75 \qquad df = 21 \qquad MS_w = 108.08$$

Post hoc comparisons $[F'_{(4, 21)}]$

	G_1 vs G_2	G_1 vs G_3	G_3 vs G_2	A_2 vs A_1	A_3 vs A_1	A_2 vs A_3
C_k	17.167	7.792	9.375	19.25	16.50	2.75
F	3.27	0.67	0.81	1.71	1.26	0.03
α	.04	–	–	–	–	–

these data because the educator had a number of specific questions to ask by means of planned and post hoc comparisons. The mean square within for the data was 108.08. The first question the educator wished to ask was whether the average scores in the three grades were equal. His null hypothesis was

$$H_0(1): (\mu_1 + \mu_2 + \mu_3)/3 = (\mu_4 + \mu_5)/2 = (\mu_6 + \mu_7)/2.$$

This hypothesis is true if and only if the following null hypotheses are true (the reader should verify this statement):

$$H_0(1a): 4\mu_1 + 4\mu_2 + 4\mu_3 - 3\mu_4 - 3\mu_5 - 3\mu_6 - 3\mu_7 = 0,$$
$$H_0(2a): \mu_4 + \mu_5 - \mu_6 - \mu_7 = 0.$$

The $(C'_k)^2$ for these two contrasts are

$$(C'_{1a})^2 = 1067.86,$$
$$(C'_{2a})^2 = 351.56.$$

If $H_0(1)$ is true, the sum of these two values is distributed as chi square with two degrees of freedom, so

$$F_{(2, 21)} = \frac{(1067.86 + 351.56)/2}{(108.08)} = 6.57,$$

$$\alpha < .01.$$

Thus, the educator can confidently conclude that there are differences in attitude among the three grades.

The second question is whether there are attitude differences among students in different classes within the first grade:

$$H_0(2): \mu_1 = \mu_2 = \mu_3.$$

The numerator sum of squares for this hypothesis can be found by regarding the three first-grade classes as the only three levels in the experiment and calculating SS_{bet} on only A_1, A_2, and A_3. The result is

$$F_{(2,21)} = \frac{867.17/2}{108.08} = 4.01$$

$\alpha = .04.$

Thus, the educator can conclude with some confidence that there are differences in attitude among the pupils in the different classes of the first grade.

The educator does not expect to find differences between classes in the second and third grades, but he has 2 degrees of freedom left out of the original 6 and has decided to use them to test for these differences:

$$H_0(3): \mu_4 - \mu_5 = 0;$$
$$F_{(1,21)} = 4.50/108.08 < 1.0$$

$$H_0(4): \mu_6 - \mu_7 = 0;$$
$$F_{(1,21)} = 190.12/108.08 = 1.76$$

$\alpha = .20.$

There appear to be few or no differences between classes in the second and third grades. The educator would now like to make post hoc tests for specific differences between grades and between classes within the first grade. Since none of his tests will be between classes in the second and third grades, the comparisons actually tested will all be independent of $H_0(3)$ and $H_0(4)$. Consequently, he can regard the $(C_k')^2$ as having 4 degrees of freedom, i.e., $6 - 2$, instead of 6. In testing the difference between grades one and two by the S method, he gets

$$H_0(1,2): 2\mu_1 + 2\mu_2 + 2\mu_3 - 3\mu_4 - 3\mu_5 = 0$$

$$(C_{1,2}')^2 = 1414.5$$

$$F_{(4,21)} = \frac{1414.5}{(4)(108.08)} = 3.27$$

$\alpha = .04.$

If, instead, the educator had used 6 degrees of freedom, the result would have been:

$$F_{(6,21)} = \frac{1414.5}{(6)(108.08)} = 2.18$$

$\alpha = .10.$

The difference in power is obvious. The results of the remaining tests, both on differences between grades and on differences between classes, are shown in Table 4-7. The difference between grades one and two is the only significant difference.

Comparison of the Methods

The greatest power is achieved when a relatively small number of tests (preferably orthogonal) is selected in advance, and tests are made by either ordinary planned comparisons or by Fisher's method. However, if the number of potential planned comparisons is large, or if the experimenter does not want to limit himself to a specific list, the S and T methods can be used.

For a given set of contrasts, the choice between the S and T methods depends on a number of factors. First, it should be remembered that the T method is more limited as to the kinds of data and comparisons for which it can be used. For the T method to be used, all of the groups must contain the same number of observations and the comparisons must all be contrasts. If either of these requirements is not met, the T method cannot be used.

If both requirements are met, the choice of a method depends primarily on the kinds of comparisons being made. The T method was designed initially for studying simple differences between means, and it was later extended to more complex tests of contrasts. The S method was designed initially to test more general kinds of comparisons. Consequently, the T method is somewhat more sensitive to simple differences between means than is the S method, but the S method is more sensitive in testing more complex comparisons. If most or all of the contrasts in which the experimenter is interested are simple differences between means, the T method is probably the best choice because it is more powerful in testing such differences; this can be seen by comparing the results in Table 4-6. If, however, the experimenter is also interested in testing comparisons that are not simple differences, the S method has more power.

Critique of the S and T Methods

Unfortunately, there is a problem with the application of either the S or the T method to multiple comparisons. The problem lies in the fact that both methods limit the probability of obtaining *one or more* significant results to α. However, they say nothing about *how many* significant results one might obtain, given that we have already erroneously obtained one.

Consider, for example, an experiment with three groups and a very large n. Unknown to us, of course, all three population means are equal. We wish to test all pairs of means by the T method, using $\alpha = .05$. Suppose that this experiment is one of those 5 percent in which we obtain a significant difference. What then will be the probability of erroneously obtaining a second significant result? By some rather complicated calculations we can show that the probability of finding a second significant difference, *conditional on having found the first,* is .14. (For $\alpha = .01$, the probability would be .10.) Having found one significant result, there is an unexpectedly high probability of finding a second. Moreover, if the tests are not limited to simple differences between means, this conditional probability may be much higher. (It *can* be higher than .9, though that is highly unlikely.)

Although Type I errors are relatively rare with the S and T methods, they tend to occur in bunches when they do occur. One is therefore wise to be cautious about experiments in which a large number of unexpected T or S tests are significant.

Other Post Hoc Comparison Methods

A number of other methods of making post hoc comparisons have been devised — some of these are described in Scheffé (1959). Two methods that are widely used should be mentioned: one is the Duncan test, the other is the Newman-Keuls. Both are limited to tests among simple differences between means, but both are more powerful than either the S method or the T method. Consequently, they tend to be used more widely than either the S or the T methods.

Unfortunately, however, both tests suffer severely from the problem described in the previous section. Applying the Newman-Keuls method to the example in the previous section, the probability of a second significant result, conditional on erroneously obtaining the first, is .40 when $\alpha = .05$. The same conditional probability for the Duncan method is even higher. A complete description of both the Newman-Keuls and the Duncan tests can be found in Winer (1971, pp. 196–200).

Confidence Intervals for Individual Means

Scheffé (1959, p. 79) has presented a method of finding confidence intervals for the means of the I groups in which the probability is $1 - \alpha$ that all of the confidence intervals will simultaneously cover the true means of the I groups. The method is based on still another distribution, known as the *Studentized maximum modulus*. This distribution has been tabled, but the tables are not widely available (the interested reader is referred to Scheffé, 1959).

Exercises

1. For the data in Table 3-1:

(a) Test the following null hypotheses, by planned comparisons, finding both α and $\hat{\omega}^2$:

$$H_0(1): \mu_1 = \mu_2;$$
$$H_0(2): \mu_3 = (\mu_1 + \mu_2)/2.$$

(b) Add the following data, from a fourth group, A_4: 3, 1, 2, (these are the same as in Problem 3b, Chapter 3) and test the following two null hypotheses:

$$H_0(1): \mu_1 = \mu_2;$$
$$H_0(2): \mu_2 = (\mu_1 + 2\mu_4)/3.$$

(c) Using the analysis already performed in Chapter 3, how could you have computed the value of F for $H_0(2)$ in part a without first calculating C^2 or $(C')^2$?

2. The following data are the mean scores from an experiment on three groups, with nine observations in each group:

	A_1	A_2	A_3
m_i	12	-6	0

(a) Assuming that $MS_w = 100$, perform a planned comparison on each of the following hypotheses finding α and $\hat{\omega}^2$:

$$H_0(1): \mu = -3;$$
$$H_0(2): 3\mu_2 = \mu_3 + 2\mu_1;$$
$$H_0(3): \mu_1 = 2\mu_3.$$

(b) Which pairs of tests in part a are orthogonal? Which are not orthogonal?

(c) A fourth comparison, orthogonal to $H_0(2)$ and $H_0(3)$ in part a, can be made. What would its significance level be? (Hint: it is not necessary to find the comparison to determine its significance level.)

(d) Test H_0: $\mu_1 = 0$. Since the mean of the third group is zero, this would appear to be the same test as $H_0(3)$ of part a; is it? Why, or why not?

3. For the data in Problem 4, Chapter 3:

(a) Assuming that the assumptions of a one-way analysis of variance are met, test the following:

$$H_0(1): \mu_1 + \mu_2 = \mu_3 + \mu_4;$$
$$H_0(2): \mu_1 + \mu_3 = \mu_2 + \mu_4;$$

$H_0(3)$: There are no other differences, orthogonal to the above, among the means of the four groups.

Test each as a planned test, finding α and $\hat{\omega}^2$.

(b) Can you find a way to test each of the above hypotheses without assuming that the two scores obtained from a single individual are independent? (Hint: with a little ingenuity, each of the first two tests can be reduced to a t test.)

4. For the data in Problem 6, Chapter 3:

(a) Test $H_0(1): \mu_1 = \mu_2$, assuming that all of the assumptions are met.

(b) Find the contrast that is statistically orthogonal to $H_0(1)$.

5. For the data in Problem 7, Chapter 3:

(a) Do a planned comparison on the null hypothesis that, overall, there is no difference between the average scores of males and females, finding both α and $\hat{\omega}^2$.

(b) Find another test that is logically orthogonal to the test in part a, and tell whether it is also statistically orthogonal.

6. For the data in Problem 1, Chapter 3:

(a) Test the following two null hypotheses, using planned comparisons and finding both α and $\hat{\omega}^2$:

$$H_0(1): \mu_1 = (\mu_2 + \mu_3 + \mu_4 + \mu_5)/4 = \mu_6$$

(no overall difference due to type of drug given).

$$H_0(2): \mu_2 = \mu_3 = \mu_4 = \mu_5$$

(no differences in the effects of the four different tranquilizers).

(b) Perform post hoc tests (.10 level) by the T method on each of the differences tested by $H_0(2)$; again, assume that these are the only tests the experimenter had intended to make.

(c) Perform a post hoc test on the hypothesis

$$H_0(3): \mu_5 = (\mu_2 + \mu_3 + \mu_4)/3,$$

assuming that the only post hoc tests the experimenter had intended to make were among the four tranquilizer groups. Do the test twice, first with the S method and then with the T method.

(d) Would it be possible for any contrast among the four tranquilizer groups to be significantly different beyond the .01 level by the T method? Explain.

7. For the data in Problem 2, Chapter 3:

(a) Test the following null hypotheses (use planned comparisons, finding both α and $\hat{\omega}^2$):

$$H_0(1): \mu_4 = \mu_5$$

(shock has no effect).

$$H_0(2): \mu_1 = \mu_2 = \mu_3$$

(the three types of cues are equally important in learning the discrimination).

(b) What is the highest significance level that could possibly be obtained from the above data using the S method for post hoc comparisons? Using the T method? (It may not be necessary to determine *which* planned comparison is most significant in order to answer the question.)

(c) After looking at the data, the experimenter noted that the mean for Group 2 was lower than the mean for Groups 1 or 3. Using the S method, do a post hoc test to see if this difference is significant.

(d) Do the test in part c again, assuming that prior to running the experiment the experimenter had decided to limit all post hoc tests to differences between Groups 1, 2, and 3.

(e) Repeat parts c and d, using the T method.

8. The T method of post hoc comparisons is used on all pairs of means in an experiment having I groups. It happens that all necessary assumptions for these tests are met, and the overall null hypothesis ($\mu_i = \mu$ for all A_i) is true. If all tests are made at the .05 level, is it true that:

(a) The probability that at least two tests are significant is likely to be greater than, less than, or equal to .0025? (Explain.)

(b) The expected number of significant results is likely to be greater than, less than, or equal to .05($I - 1$)? (Explain.)

9. Prove that multiplying each c_{ik} in a planned comparison by a constant, $b \neq 0$, does not change the value of $(C'_k)^2$.

10. Find the $\hat{\omega}^2$ for each set of data in Problems 1, 2, 3, and 7, Chapter 3.

11. Refer to Problem 8, Chapter 3. If the experimenter chooses his n so that his power is exactly .9, under the conditions given, what is the smallest ω^2 for which his power would be .9 if

(a) $I = 4$

(b) $I = 3$

(Hint: both power and ω^2 depend on $\Sigma_i \, \alpha_i^2$.)

5

Two-way Analysis of Variance

In many cases a simple one-way analysis of variance, with or without planned comparisons, is the best way to analyze data. Sometimes, however, simple one-way models are not appropriate; even when they are appropriate, more complicated methods are usually better-suited to the needs of the experimenter.

Consider, for example, the data in Table 5-1. Here scores were obtained from six groups of three subjects each. A simple one-way analysis of variance indicates that the differences among the groups are significant at the .02 level. (Of course, a set of planned comparisons could also have been devised and tested if the experimenter had so wished.)

MAIN EFFECTS

The data in Table 5-1 are hypothetical improvement scores made by patients in two psychiatric categories (Group A_1 are schizophrenics, Group A_2 are depressives) under each of three different tranquilizer drugs, B_1, B_2, and B_3. This design suggests the arrangement of the data in Table 5-2, where the rows are the two psychiatric categories and the columns are the three types of drugs. Each observation is the difference between that patient's scores on an emotional adjustment scale before and after being given the drug. The experimenter wishes to ask two questions. In general, is there a difference in the improvement of the schizophrenics and that of the depressives when given tranquilizing drugs? Is there a difference in general effectiveness among the three drugs?

Table 5-1. Hypothetical data with six cells and three observations per cell.

	G_1	G_2	G_3	G_4	G_5	G_6	Sum
X_{ij}	8 4 0	8 10 6	4 6 8	10 6 14	0 4 2	15 9 12	
ΣX	12	24	18	30	6	36	126
$\overline{X}$	4	8	6	10	2	12	7
ΣX^2	80	200	116	332	20	450	1198

	RS	SS	df	MS	F	α
m		882				
bet	1092	210	5	42.00	4.75	.02
w		106	12	8.83		
t	1198	316	17			

The A Main Effect

The first question can be answered as a planned comparison on the data in Table 5-1. The planned comparison would be

$$H_0(A): \mu_1 + \mu_2 + \mu_3 - \mu_4 - \mu_5 - \mu_6 = 0. \qquad (5\text{-}1)$$

Following the procedure outlined in Chapter 4, we obtain for this comparison the values $F = 2.04$ and $\alpha = .18$. Thus, there is not a very significant overall difference in improvement between schizophrenics and depressives under these treatments.

The B Main Effect

The second question can also be answered in terms of planned comparisons; here, however, two planned comparisons must be combined into a single test. Furthermore, it is not immediately obvious what those planned comparisons should be. The problem can be simplified by referring to Table 5-2 and relabeling the cells of that table. We define the two rows in Table 5-2 to be the two levels of Factor A and the three columns to be the three levels of Factor B; the levels of Factor A are the two categories of patients and the levels of Factor B are three types of drugs. We then relabel each group, or *cell*, in Table 5-2, to conform with the levels under which it is found. Thus, instead of being labeled G_1 as in Table 5-1, the first cell is labeled AB_{11} because it falls under the first level (row and column) of both Factors A and B. The second cell is labeled AB_{12} because it falls in the first level (row) of Factor A and the second level (column) of Factor B. The six cells will then be labeled as follows:

	B_1	B_2	B_3
A_1	AB_{11}	AB_{12}	AB_{13}
A_2	AB_{21}	AB_{22}	AB_{23}

Similarly, the true mean of the population sampled in cell AB_{ij} is designated μ_{ij}, and the obtained sample mean for that cell $\overline{X}_{ij}$. Finally, we need some notation to represent the averages of these values, taken over either or both of the subscripts. Accordingly, we define:

I = number of levels of Factor A ($I = 2$ in Table 5-2)

J = number of levels of Factor B ($J = 3$ in Table 5-2)

$\mu_{i\cdot} = (1/J) \, \Sigma_j \, \mu_{ij}$ (= the average of the μ_{ij}, taken over the subscript j)

$\mu_{\cdot j} = (1/I) \, \Sigma_i \, \mu_{ij}$ (= the average of the μ_{ij}, taken over the subscript i)

$\mu = (I/J) \, \Sigma_i \, \mu_{\cdot j} = (1/I) \, \Sigma_i \, \mu_{i\cdot} = (1/IJ) \, \Sigma_i \, \Sigma_j \, \mu_{ij}$ (= the overall mean of all of the μ_{ij})

$\overline{X}_{i\cdot\cdot} = (1/J) \, \Sigma_j \, \overline{X}_{ij\cdot}$ (= the average of the $\overline{X}_{ij\cdot}$, taken over the subscript j)

$\overline{X}_{\cdot j\cdot} = (1/I) \, \Sigma_i \, \overline{X}_{ij\cdot}$ (= the average of the $\overline{X}_{ij\cdot}$, taken over the subscript i)

$\overline{X}_{\cdot\cdot\cdot} = (1/J) \, \Sigma_i \, \overline{X}_{\cdot j\cdot} = (1/I) \, \Sigma_i \, \overline{X}_{i\cdot\cdot} = (1/IJ) \, \Sigma_i \, \Sigma_j \, \overline{X}_{ij\cdot}$ (= the overall mean of all the $\overline{X}_{ij\cdot}$)

With this notation, the second question can be phrased in terms of the null hypothesis

$$H_0(B): \mu_{\cdot j} = \mu, \text{ for all } j. \tag{5-2}$$

Table 5-2. Data from Table 5-1, rearranged.

	B_1	B_2	B_3	Row summary
A_1	8 4 0 $\Sigma = 12$ $\overline{X} = 4$	8 10 6 $\Sigma = 24$ $\overline{X} = 8$	4 6 8 $\Sigma = 18$ $\overline{X} = 6$	$\Sigma = 54$ $\overline{X} = 6$
A_2	10 6 14 $\Sigma = 30$ $\overline{X} = 10$	0 4 2 $\Sigma = 6$ $\overline{X} = 2$	15 9 12 $\Sigma = 36$ $\overline{X} = 12$	$\Sigma = 72$ $\overline{X} = 8$
Column summary	$\Sigma = 42$ $\overline{X} = 7$	$\Sigma = 30$ $\overline{X} = 5$	$\Sigma = 54$ $\overline{X} = 9$	$\Sigma = 126$ $\overline{X} = 7$

A test of an hypothesis such as Equation 5-2 is called a test of the *main effect* of Factor B, or simply the "B main effect" (similarly, the test of $H_0(A)$ (Eq. 5-1) is referred to as a test of the "A main effect").

To see how Equation 5-2 can be tested, consider the three obtained means, $\overline{X}_{\cdot 1 \cdot}$, $\overline{X}_{\cdot 2 \cdot}$, $\overline{X}_{\cdot 3 \cdot}$. Each $\overline{X}_{\cdot j \cdot}$ is the average of the I cell means, $\overline{X}_{ij \cdot}$. Since each $\overline{X}_{ij \cdot}$ is normally distributed as

$$\overline{X}_{ij \cdot} \sim N_{(\mu_{ij}, \sigma_e^2/n)},$$

it follows that each $X_{\cdot j \cdot}$ is distributed as

$$\overline{X}_{\cdot j \cdot} \sim N_{(\mu_{\cdot j}, \sigma_e^2/nI)}.$$

A comparison of this with Equation 5-2 shows that, under $H_0(B)$, the $\overline{X}_{\cdot j \cdot}$ are identically distributed independent random variables. In other words, Equation 5-2 can be restated as

$$H_0(B): \overline{X}_{\cdot j \cdot} \sim N_{(\mu, \sigma_e^2/nI)}, \text{ for all } j. \tag{5-3}$$

The formal identity between this null hypothesis and the one tested in Chapter 3 should be readily apparent from comparison of Equation 5-4 with Equation 3-1, and Equation 5-3 with Equation 3-6. From here on the derivation follows exactly the same steps as the derivation in Chapter 3 (pp. 22–27).

We first obtain an estimate of σ_e^2/nI (the variance of the $\overline{X}_{\cdot j \cdot}$), noting that under $H_0(B)$ the $\overline{X}_{\cdot j \cdot}$ can be regarded as a random sample from a single normally distributed population. The estimate is

$$S_b^2 = \frac{\Sigma_j (X_{\cdot j \cdot} - X_{\cdots})^2}{J - 1}.$$

We then note that

$$MS_b = nIS_b^2 \sim \sigma_e^2 \chi_{(J-1)}^2/(J - 1) \tag{5-4}$$

is an unbiased estimate (under the null hypothesis) of σ_e^2. (If you have trouble following this argument, reread pp. 22–27.) Under the null hypothesis, therefore,

$$MS_b/MS_w \sim F_{(J-1, N-IJ)}. \tag{5-5}$$

The denominator degrees of freedom for this F ratio are found from the fact that, in general, the degrees of freedom for MS_w are equal to the total number of observations (N) minus the total number of groups, which in this case is I times J. Also, just as previously, we define SS_b as

$$SS_b = (J - 1)MS_b = nI \Sigma_j (\overline{X}_{\cdot j \cdot} - \overline{X}_{\cdots})^2. \tag{5-6}$$

For the data in Table 5-2 these values are:

$$SS_b = (3)(2)[(7 - 7)^2 + (5 - 7)^2 + (9 - 7)^2] = 48$$
$$MS_b = 48/2 = 24$$
$$F_{(2, 12)} = 24/8.83 = 2.72$$
$$\alpha = .11$$

Relationships to Planned Comparisons

We stated earlier that the null hypothesis $H_0(B)$ could have been tested as two simultaneous contrasts. This follows from the fact, discussed in Chapter 4, that any test of differences between means with ν_1 degrees of freedom in the numerator can be factored into ν_1 different independent contrasts. It may be instructive to illustrate how this can be done with the data in Table 5-2. $H_0(B)$ can be factored in an infinite number of ways; the following two hypotheses illustrate one of these ways:

$$H_0(B_1): \mu_{11} + \mu_{21} - \mu_{12} - \mu_{22} = 0,$$
$$H_0(B_2): \mu_{11} + \mu_{21} + \mu_{12} + \mu_{22} - 2\mu_{13} - 2\mu_{23} = 0.$$

You might test these two contrasts as an exercise. When you do, you will find that

$$(C'_{B_1})^2 + (C'_{B_2})^2 = 12 + 36 = 48 = SS_b.$$

Similarly, instead of testing $H_0(A)$ as a contrast, we could have tested it using the theory developed for testing $H_0(B)$. This follows from the fact that Equation 5-1 can be stated as

$$H_0(A): \mu_{i.} = \mu \text{ for all } i. \tag{5-7}$$

In this form it is identical to Equation 5-2. Finally, since the choice of rows and columns in Table 5-2 was arbitrary, it follows that all of the theory for testing Equation 5-2 will be valid for testing Equation 5-7 if we simply reverse the roles of I and J and substitute $\overline{X}_{i..}$ for $\overline{X}_{.j.}$ throughout. When we do this, Equations 5-6, 5-4, and 5-5 become, respectively,

$$SS_a = nJ \Sigma_i (\overline{X}_{i..} - \overline{X}_{...})^2 \tag{5-8}$$

$$MS_a = SS_a/(I - 1) \tag{5-9}$$

$$MS_a/MS_w = F_{(I-1, N-IJ)}. \tag{5-10}$$

For the data of Table 5-2, we have the following values:

$$SS_a = (3)(3)[(6-7)^2 + (8-7)^2] = 18,$$
$$MS_a = 18/1 = 18,$$
$$F_{(1, 12)} = 18/8.83 = 2.04,$$

and this F ratio is identical to the one obtained in testing the contrast, Equation 5-1.

INTERACTION

We now have a new problem. When we tested for the main effect of Factor A, the results were not very significant; when we tested for the main effect of Factor B, the results were not very significant either. We are led to conclude from these tests that the data indicate no reliable differences either in the overall reactions of the two types of patients to the drugs or among the types of drugs in their overall effect on the patients. Yet, the general

F test that we initially made on all six groups combined was significant, indicating that reliable differences do exist among the means of the six groups. Somehow, we have missed those differences in the tests in the previous section.

An initial clue to the location of the differences we are looking for can be found by noting that the initial F test on the six groups had five degrees of freedom, and this corresponded to a simultaneous test of five orthogonal contrasts. In testing $H_0(A)$, we tested one contrast; the test of $H_0(B)$ was a simultaneous test of two orthogonal contrasts. Our tests so far have covered only three of the five possible orthogonal contrasts. (It is trivial to show that Equations 5-1 and 5-2 are independent null hypotheses so that all three of the contrasts already tested are orthogonal.)

Two degrees of freedom remain. We can therefore follow the theory of Chapter 4 and test for any remaining differences among the means of the six groups by calculating $SS_{rem} = SS_{bet} - SS_a - SS_b = 210 - 18 - 48 = 144$, which in this case will have two degrees of freedom. Accordingly,

$$MS_{rem} = 144/2 = 72$$
$$F_{(2, 12)} = 72/8.83 = 8.15$$
$$\alpha < .01.$$

There are significant differences among the means, independent of the differences tested for in the previous section.

But what are these differences? Are they meaningful? A more complete answer comes from Fig. 5-1. This figure is a plot of the obtained means in Table 5-2. (Here is another example of using a figure to show important relationships in the data.) It shows clearly that the relative effectiveness of a drug depends on the type of patient to which it is given. Or, alternatively, the two kinds of patients react differently to the three drugs. The statistician combines both of these sentences into one by saying there is an *interaction* between drugs and patients.

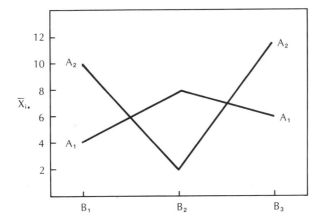

Figure 5-1. Cell means from Table 5-2.

The Model Equation

The nature of the interaction may become more clear if we resort to the strategy, used previously, of constructing a linear model. If we let

$$\alpha_i = \mu_{i.} - \mu, \tag{5-11}$$

$$\beta_j = \mu_{.j} - \mu, \tag{5-12}$$

and X_{ijk} = the kth observation in cell AB_{ij}, then

$$
\begin{aligned}
X_{ijk} &= \mu_{ij} + \epsilon_{ijk} \tag{5-13} \\
&= \mu + (\mu_{i.} - \mu) + (\mu_{.j} - \mu) + (\mu_{ij} - \mu_{i.} - \mu_{.j} + \mu) + \epsilon_{ijk} \\
&= \mu + \alpha_i + \beta_j + (\mu_{ij} - \mu_{i.} - \mu_{.j} + \mu) + \epsilon_{ijk}.
\end{aligned}
$$

It is easy to verify that $H_0(A)$ (the null hypothesis of no A main effect) is equivalent to the null hypothesis that all α_i are zero, and that $H_0(B)$, on the B main effect, is equivalent to the null hypothesis that all β_j are zero. Equation 5-13 shows, however, that even though α_i and β_j are both zero, the expected value of X_{ijk} may still differ from μ, the grand mean. The amount by which it can differ is defined to be

$$
\begin{aligned}
\alpha\beta_{ij} &= \mu_{ij} - \mu_{i.} - \mu_{.j} + \mu \tag{5-14} \\
&= (\mu_{ij} - \mu) - (\mu_{i.} - \mu) - (\mu_{.j} - \mu) \\
&= (\mu_{ij} - \mu) - \alpha_i - \beta_j.
\end{aligned}
$$

Thus, the interaction $\alpha\beta_{ij}$ is measured by the amount by which each population cell mean differs from the value we would expect from a knowledge of the row and column means only. Inserting Equation 5-14 into Equation 5-13, we obtain the basic model for a two-way analysis of variance:

$$X_{ijk} = \mu + \alpha_i + \beta_j + \alpha\beta_{ij} + \epsilon_{ijk}. \tag{5-15}$$

Interaction Mean Square

The sum of squares for interaction was obtained above as a residual sum of squares. However, a direct method for calculating it will now be shown. We will begin by reviewing the tests for main effects in light of Equation 5-15. We will then show the analogy between these tests and the test for interaction.

However, before we can review the tests for main effects, we must obtain estimates of the terms in Equation 5-15. On the basis of the definitions in Equations 5-11, 5-12, and 5-14, it is easy to show that the following sums are all identically zero:

$$\Sigma_i \alpha_i = \Sigma_j \beta_j = \Sigma_i \alpha\beta_{ij} = \Sigma_j \alpha\beta_{ij} = 0. \tag{5-16}$$

Then, averaging over the appropriate subscripts and eliminating sums that (according to Eq. 5-16) are zero, we can obtain the following equations:

$$\overline{X}_{ij.} = \mu + \alpha_i + \beta_j + \alpha\beta_{ij} + \overline{\epsilon}_{ij.}$$
$$\overline{X}_{i..} = \mu + \alpha_i + \overline{\epsilon}_{i..}$$
$$\overline{X}_{.j.} = \mu + \beta_j + \overline{\epsilon}_{.j.}$$
$$\overline{X}_{...} = \mu + \overline{\epsilon}_{...}$$

The expected value of each of the above terms can be found by eliminating the terms involving ϵ_{ijk}:

$$E(\overline{X}_{ij.}) = \mu + \alpha_i + \beta_j + \alpha\beta_{ij}$$
$$E(\overline{X}_{i..}) = \mu + \alpha_i$$
$$E(\overline{X}_{.j.}) = \mu + \beta_j$$
$$E(\overline{X}_{...}) = \mu.$$

Finally, from these equations we obtain the following unbiased estimates:

$$\hat{\mu} = \overline{X}_{...} = \mu + \overline{\epsilon}_{...} \tag{5-17}$$
$$\hat{\alpha}_i = \overline{X}_{i..} - \overline{X}_{...} = \alpha_i + \overline{\epsilon}_{i..} - \overline{\epsilon}_{...}$$
$$\hat{\beta}_j = \overline{X}_{.j.} - \overline{X}_{...} = \beta_j + \overline{\epsilon}_{.j.} - \overline{\epsilon}_{...} \tag{5-18}$$
$$\alpha\beta_{ij} = \overline{X}_{ij.} - \overline{X}_{i..} - \overline{X}_{.j.} + \overline{X}_{...} = \alpha\beta_{ij} + \overline{\epsilon}_{ij.} - \overline{\epsilon}_{i..} - \overline{\epsilon}_{.j.} + \overline{\epsilon}_{...}$$

Table 5-3 contains the estimates of the quantities for the data in Table 5-2.

Comparing these results with Equations 5-7, and 5-8, we can see that Equation 5-7 is equivalent to the null hypothesis $H_0(A)$: $\alpha_i = 0$ for all i. As can be seen from Equations 5-8, and 5-17, this hypothesis is tested by finding the sum of the squared estimates of the α_i and multiplying that sum by the number of observations averaged over to obtain the $\overline{X}_{i..}$ values:

$$SS_a = nJ \Sigma_i \hat{\alpha}_i^2.$$

Similarly, Equation 5-2 can be rewritten as

$$H_0(B): \beta_j = 0 \text{ for all } j,$$

and, according to Equations 5-4 and 5-18, this hypothesis is tested by summing the squared estimates of the β_j and multiplying the sum by the number of observations averaged over to obtain the $\overline{X}_{.j.}$ values:

$$SS_b = nI \Sigma_j \hat{\beta}_j^2.$$

These results suggest that the hypothesis

$$H_0(AB): \alpha\beta_{ij} = 0 \text{ for all } i, j,$$

i.e., that there is no interaction, can be tested by summing the squared estimates of the $\alpha\beta_{ij}$ and multiplying the sum by n, the number of observations averaged over to obtain the $\overline{X}_{ij.}$ values:

$$SS_{ab} = n \Sigma_i \Sigma_j \alpha\beta_{ij}^2 = n \Sigma_i \Sigma_j (\overline{X}_{ij.} - \overline{X}_{i..} - \overline{X}_{.j.} + \overline{X}_{...})^2. \tag{5-19}$$

We can see from Table 5-3 that for the data in Table 5-2

$$SS_{ab} = 3[(-2)^2 + 4^2 + (-2)^2 + 2^2 + (-4)^2 + 2^2] = 144.$$

Table 5-3. Estimates of μ, α_i, β_j, and $\alpha\beta_{ij}$. The values in the cells are estimates of $\alpha\beta_{ij}$. The marginal values are estimates of α_i (row margins), β_j (column margins), and μ (lower-right corner).

	B_1	B_2	B_3	$\hat{\alpha}_i$
A_1	-2	$+4$	-2	-1
A_2	$+2$	-4	$+2$	$+1$
β_j	0	-2	$+2$	7

Under the null hypothesis $H_0(AB)$ this is distributed as σ_e^2 times chi-square with $(I-1)(J-1)$ degrees of freedom. The proof of this is somewhat complicated, but it is possible to explain partially why there are $(I-1)(J-1)$ degrees of freedom. In general, in an I by J analysis of variance there are $(IJ-1)$ degrees of freedom available for testing contrasts. We already saw that $(I-1)$ of these are used in testing for the A main effect and that $(J-1)$ are used in testing for the B main effect. Consequently, the degrees of freedom which remain for testing the interaction are

$$(IJ-1) - (I-1) - (J-1) = (I-1)(J-1).$$

Another partial explanation of the degrees of freedom for the test of $H_0(AB)$ can be given in terms of the number of estimates that are "free to vary" (in fact, this explanation caused the parameters of F to be called "degrees of freedom"). If we consider the estimates, $\hat{\alpha}_i$, we will see that they must always sum to zero, just as the α_i sum to zero. Consequently, although the data are "free" to determine the values of the first $(I-1)$ estimates arbitrarily, the Ith estimate will always be determined from the other $(I-1)$ plus the restriction that the estimates sum to zero. We say therefore that only $(I-1)$ of the $\hat{\alpha}_i$ are "free" to vary, and the test of $H_0(A)$ has $(I-1)$ degrees of freedom. By the same reasoning, only $(J-1)$ of the $\hat{\beta}_j$ are free to vary, so the test of $H_0(B)$ has $(J-1)$ degrees of freedom. The estimates of the $\alpha\beta_{ij}$, however, must sum to zero over both the rows and columns. Consequently, only $(I-1)$ of the values in any given column are free to vary, and only $(J-1)$ are free to vary in any given row. Thus, only $(I-1)(J-1)$ of the $\alpha\beta_{ij}$ are free to vary. This reasoning may become clearer if you refer to Table 5-3. It should be clear in Table 5-3 that once $\alpha\beta_{11}$ and $\alpha\beta_{12}$ have been set, the rest of the $\alpha\beta_{ij}$ are determined by the restriction that both the rows and columns of the matrix must sum to zero.

The test for the interaction is made by calculating

$$MS_{ab} = SS_{ab}/(I - 1)(J - 1), \text{ and}$$
$$F_{[(I-1)(J-1), N-IJ]} = MS_{ab}/MS_{w}.$$

For the data in Table 5-2 we have the following:

$$MS_{ab} = 144/2 = 72$$
$$F_{(2, 12)} = 72/8.83 = 8.15$$
$$\alpha < .01.$$

We have thus divided the degrees of freedom into tests of three independent null hypotheses on two main effects and an interaction. Of course, these particular tests are not necessarily the ones that should be made — the questions that an experimenter wishes to ask may be better-answered by a different set of tests, e.g., planned comparisons. Nevertheless, these tests are appropriate for the questions an experimenter often does ask in a two-way analysis of variance.

Interpreting Interaction

The interpretation of a significant interaction is a problem commonly encountered in two-way analyses of variance. Unfortunately, no general answer can be given to the problem, since the scientific rather than the statistical importance of the interaction is usually at issue. Statistically, the meaning of a significant interaction can be answered easily in terms of a graph such as Figure 5-1. A general procedure for drawing such a graph is to represent the levels of Factor B as values on the x axis and plot the cell means at those levels as values on the y axis, making a separate curve for each level of Factor A. (Alternatively, of course, the levels of A could be represented on the x axis and a curve could be drawn for each level of B.) If there were no interaction, these curves would be parallel, like those in Figure 5-2; to the extent that they are not parallel, there is an interaction. However, the importance, or "meaning," of the fact that the lines are not parallel depends on the particular study. For example, Figure 5-1 shows that drug B_2 has a greater effect on schizophrenics than drugs B_1 and B_3, whereas the latter are more effective with depressives; that is the source of the interaction in this case.

INTERPRETING MAIN EFFECTS

Interpreting main effects in the presence of an interaction requires care. In the *absence* of an interaction, the tests on the main effects can be interpreted independently. Figure 5-2 illustrates data from the same kind of experiment as in Figure 5-1; in the data in Figure 5-2 both main effects are significant but there is no interaction. We can conclude that no matter which drug we use schizophrenics will respond better than depressives, and that no matter which patient we use it on drug B_3 will be most effective.

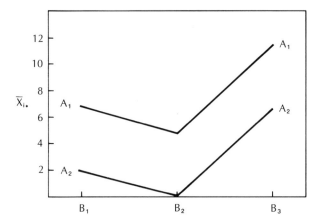

Figure 5-2. Sample data with both main effects but no interactions.

As a practical matter, the hospital could stock only drug B_3, and, if the drug were in short supply, give schizophrenics priority.

The situation is very different, however, if there is an interaction. Again, consider Figure 5-1. Even if we rejected the null hypothesis of no B main effect, we could not conclude that drug B_3 was most effective for all patients. The hospital might be wise to stock drug B_2 for schizophrenics and drug B_3 for depressives.

Neither can we conclude from the *absence* of an A main effect that there are no differences between types of patients. Figure 5-1 shows that there are clear and definite differences in the responses to particular drugs. There is no A main effect only because these differences average out when we average responses over drugs.

SIMPLE EFFECTS

When there is an interaction, tests of *simple effects* are sometimes more useful than tests of main effects and interactions. *Simple-effect* tests are one-way analyses of variance across levels of one factor, performed separately at each level of the other factor. We might, for example, test for the B main effect separately at each level of A. The test at each level can be made as though the other level of A were not even in the experiment. The results of the simple-effects test at each level of A are shown in Table 5-4. They show that there is a significant difference in the effects of drugs on depressives but there is no significant difference in the effects on schizophrenics. A practical result of these tests might be a decision to stock only drug B_3, since it works best on depressives, and all three drugs work about equally well on schizophrenics.

We can increase the power of these tests if we assume that the variances for the two types of patients are equal. For this we can use the MS_w from

Table 5-4. Tests of simple effects on Factor B.

	RS	SS	df	Level A_1 MS	F	α
m		324	1	324.0	40.5	.001
bet	348	24	2	12.00	1.50	.30
w		48	6	8.00		
t	396	72	8			

	RS	SS	df	Level A_2 MS	F	α
m		576	1	576	59.6	.001
bet	744	168	2	84	8.70	.02
w		58	6	9.67		
t	802	226	8			

the overall analysis as the denominator, increasing the degrees of freedom from 6 to 12. The test on A_1 is still not significant, but the test on A_2 is now

$$F_{(2, 12)} = 84/8.83 = 9.51$$
$$\alpha = .004.$$

The similarity of the separate MS_w values (8.00 and 9.67) to the overall MS_w of 8.83 lends some support to the assumption of equal variances, but we should also have good a priori reasons to expect equal variances.

Simple tests can also be made across levels of A for each level of B. For this particular experiment, however, they would probably have little practical value.

Note that tests of simple effects are not orthogonal to tests of main effects and interactions. The simple effects tested above, for example, are *not* orthogonal to either the B main effect or the interaction. In fact, they use exactly the same four degrees of freedom (and thus the same variation) as the B main effect and AB interaction. In standard analyses the total sum of squares for these two effects is $SS_b + SS_{AB} = 48 + 144 = 192$. In the simple-effects analysis, these are divided into $SS_{A_1} + SS_{A_2} = 24 + 168 = 192$.

There is no reason why either test should take precedence over the other. The choice of a standard or a simple-effects analysis depends on the purposes of the experimenter. However, if both are done, the cautions (pp. 66–67) about nonorthogonal tests apply.

Simple Tests on Means

The simple tests described above *are* orthogonal to both the A main effect and the grand mean. The total of two degrees of freedom in these tests can be divided into two tests on single means, in a manner analogous to the tests on simple effects. The two tests are on

$$H_0 (A_1): \mu_{1.} = 0$$
$$H_0 (A_2): \mu_{2.} = 0$$

In practical terms, they tell us whether on the average each type of patient benefits at all from the drugs.

These tests, in fact, have already been performed; they are the two tests on SS_m in Table 5-4. For these tests both null hypotheses are rejected, as the tests on the A main effect and grand mean would lead us to expect (for these particular data).

This discussion has not exhausted the possibilities for specific tests. In fact, the possibilities are limited only by the ingenuity of the experimenter. Remember that the experiment can be regarded as a one-way design, and all of the methods in Chapter 4 can be applied.

COMPUTATIONAL FORMULAS

Just as with the overall F test, the computation involved in finding mean squares can be simplified considerably by algebraic manipulations on Equations 5-6, 5-8, and 5-19. These manipulations are relatively straightforward, but too tedious to present here in their entirety. The results are summarized in Table 5-5. Table 5-6 shows the computations for the data in Table 5-2.

UNEQUAL SAMPLE SIZES

The above discussion assumed that every cell contained the same number of observations (n). If this is not the case, serious problems arise. To begin with, the formulas for SS_a, SS_b, and SS_{ab} become much more complicated.

Table 5-5. Computational formulas for two-way analysis of variance with equal ns.

$$t_{ij} = \Sigma_k X_{ijk}, \quad t_{i.} = \Sigma_j t_{ij}$$
$$t_{.j} = \Sigma_i t_{ij}, \quad T = \Sigma_i t_{i.} = \Sigma_j t_{.j}$$

	RS	SS	df
m		T^2/N	1
a	$(1/nJ) \, \Sigma_i \, t_{i.}^2$	$RS_a - SS_m$	$I - 1$
b	$(1/nI) \, \Sigma_j \, t_{.j}^2$	$RS_b - SS_m$	$J - 1$
ab	$(1/n \, \Sigma_i \Sigma_j \, t_{ij}^2$	$RS_{ab} - SS_a - SS_b - SS_m$	$(I-1)(J-1)$
w		$RS_t - RS_{ab}$ $= RS_t - SS_{ab} - SS_a - SS_b - SS_m$	$N - IJ$
t	$\Sigma_i \Sigma_j \Sigma_k \, X_{ijk}^2$	$RS_t - SS_m$	$N - 1$

Table 5-6. Summary of analyses on data in Table 5-2.

	RS	SS	df	MS	F	α
m		882	—	—	—	—
a	900	18	1	18.00	2.04	.19
b	930	48	2	24.00	2.72	.11
ab	1092	144	2	72.00	8.15	.01
w		106	12	8.83		
t	1198	316	17			

More importantly, however, these three sums of squares are not independent, since, according to Equation 4-19, independence of two tests depends partly on the relative numbers of observations in the cells. If the experimenter wishes to analyze his data as a two-way design, he should obtain the same number of observations in each cell.

Occasionally, however, the experimenter finds that he cannot quite reach the goal of having the same number of observations in every cell. In psychological studies, for example, human subjects may quit before the experiment is completed or animals may become sick or die. The results may be that some cells contain fewer observations than others. In such a case the experimenter can still make approximate tests of the main effects and the interactions if the numbers of observations in the cells are approximately equal.

To make the approximate tests, we first calculate MS_w, just as for a one-way design with unequal numbers of observations in the cells. To calculate SS_a, SS_b, and SS_{ab}, however, we must first find the mean of each cell and then calculate the sums of squares from these means, as though every cell had the same number of observations ($\bar{n}$). The value used for $\bar{n}$ in this case is the *harmonic mean* of the numbers of observations in the individual cells. If n_{ij} is the number of observations in cell AB_{ij}, then

$$\bar{n} = \frac{IJ}{\Sigma_i \Sigma_j (1/n_{ij})}$$
$$\overline{X}_{i..} = (1/J) \Sigma_j \overline{X}_{ij.}$$
$$\overline{X}_{.j.} = (1/I) \Sigma_i \overline{X}_{ij.}$$
$$\overline{X}_{...} = (1/I) \Sigma_i \overline{X}_{i.} = (1/J) \Sigma_j \overline{X}_{.j.} = (1/IJ) \Sigma_i \Sigma_j \overline{X}_{ij.}$$
$$SS_m = \bar{n} IJ \overline{X}^2$$
$$SS_a = \bar{n}J \Sigma_i (\overline{X}_{i..} - \overline{X}_{...})^2 = \bar{n}J \Sigma_i \overline{X}_{i..}^2 - SS_m$$
$$SS_b = \bar{n}I \Sigma_j (\overline{X}_{.j.} - \overline{X}_{...})^2 = \bar{n}I \Sigma_j \overline{X}_{.j.}^2 - SS_m$$
$$SS_{ab} = \bar{n} \Sigma_i \Sigma_j (\overline{X}_{ij.} - \overline{X}_{i..} - \overline{X}_{.j.} + \overline{X}_{...})^2$$
$$= \bar{n} \Sigma_i \Sigma_j \overline{X}_{ij.}^2 - SS_a - SS_b - SS_m.$$

The numbers of degrees of freedom for these quantities are the same as those in Table 5-5. Table 5-7 shows how these calculations might have been carried out if one observation each in cells AB_{13} and AB_{21} had been

Table 5-7. Data from Table 5-2 with one observation each missing in cells AB_{13} and AB_{21}.

	B_1	B_2	B_3	$\overline{X}_{i\cdot}$
A_1	8 4 0 $\Sigma = 12$ $\overline{X} = 4$	8 10 6 $\Sigma = 24$ $\overline{X} = 8$	4 8 $\Sigma = 12$ $\overline{X} = 6$	6
A_2	14 10 $\Sigma = 24$ $\overline{X} = 12$	0 4 2 $\Sigma = 6$ $\overline{X} = 2$	15 9 12 $\Sigma = 36$ $\overline{X} = 12$	8.67
$\overline{X}_{\cdot j}$	8	5	9	7.33

$$\bar{n} = \frac{(2)(3)}{4(1/3) + 2(1/2)} = \frac{18}{7} = 2.57$$

One-way analysis (MS_w)

	RS	SS	df	MS
bet	1044			
w		82	10	8.20
t	1126			

Two-way analysis

$SS_m = (2.57)(2)(3)(7.33)^2 = 828.50$

$SS_a = (2.57)(3)(6^2 + 8.67^2) - 828.50 = 28.61$
$\quad df_a = 1, MS_a = 28.61, F = 3.49, \alpha = .09$

$SS_b = (2.57)(2)(8^2 + 5^2 + 9^2) - 828.50 = 45.30$
$\quad df_b = 2, MS_b = 22.65, F = 2.76, \alpha = .14$

$SS_{ab} = (2.57)(4^2 + 8^2 + 6^2 + 12^2 + 2^2 + 12^2) - 45.30 - 28.61 - 828.50$
$\quad = 146.15$
$\quad df_{ab} = 2, MS_{ab} = 73.08, F = 8.91, \alpha < .01$

missing. (An exact, but much more complicated, analysis can be found in Scheffé, 1959, pp. 112–119, or Winer, 1972, pp. 404–422.)

EXPECTED MEAN SQUARES

The expected value of MS_w is, of course, σ_e^2, as in the one-way analysis of variance; ms_w is the same whether the design is treated as one-way or two-way, since it is simply the average of the cell variances.

Note that if we substitute $\overline{X}_{i..}$ for $\overline{X}_{i.}$ and (nJ) for n in Equation 3-19 (defining SS_{bet}), the result is Equation 5-8 (defining SS_a). It follows that derivation of the expected value of MS_a is essentially the same as that of MS_{bet} on pp. 27–31. The same holds true for MS_b since the labels A and B were arbitrarily assigned to the factors at the beginning. The results of the derivation are

$$E(MS_a) = \sigma_e^2 + nJ \, \Sigma_i \, \alpha_i^2/(I - 1) = \sigma_e^2 + nJ\left(\frac{I}{I-1}\right)\sigma_a^2 \quad \text{(5-20)}$$

$$E(MS_b) = \sigma_e^2 + nI \, \Sigma_j \, \beta_j^2/(J - 1) = \sigma_e^2 + nI\left(\frac{J}{J-1}\right)\sigma_b^2 \quad \text{(5-21)}$$

where

$$\sigma_a^2 = (1/I) \, \Sigma_i \, \alpha_i^2,$$
$$\sigma_b^2 = (1/J) \, \Sigma_j \, \beta_j^2.$$

The similarity of these equations to Equation 3-23 should be obvious. If we let

$$\tau_a^2 = \left(\frac{I}{I-1}\right)\sigma_a^2, \text{ and}$$

$$\tau_b^2 = \left(\frac{J}{J-1}\right)\sigma_b^2,$$

then Equations 5-20 and 5-21 become

$$E(MS_a) = \sigma_e^2 + nJ\tau_a^2, \text{ and}$$
$$E(MS_b) = \sigma_e^2 + nI\tau_b^2.$$

By defining τ_a^2 and τ_b^2 this way, we simplify the formulas for expected mean squares. The value of this simplification will be more apparent in later chapters.

Finding the expected value of MS_{ab} is more tedious, but the procedure is basically the same as that on pp. 27–31. The end result of the derivation is

$$E(MS_{ab}) = \sigma_e^2 + \frac{n}{(I-1)(J-1)} \, \Sigma_i \, \Sigma_j \, \alpha\beta_{ij}^2$$

$$= \sigma_e^2 + \frac{nIJ}{(I-1)(J-1)} \, \sigma_{ab}^2,$$

where

$$\sigma_{ab}^2 = (1/IJ) \, \Sigma_i \, \Sigma_j \, \alpha\beta_{ij}^2,$$

the variance of the $\alpha\beta_{ij}$. We can now define τ_{ab}^2 in a manner analogous to τ_a^2 and τ_b^2 above as

$$\tau_{ab}^2 = \frac{IJ}{(I-1)(J-1)} \, \sigma_{ab}^2$$

to obtain

$$E(MS_{ab}) = \sigma_e^2 + n\tau_{ab}^2.$$

POWER OF THE TESTS

For the overall F test (discussed in Chapter 3) we found that when the null hypothesis was false, the F ratio was distributed as noncentral F. The same is true for the F ratios used in testing $H_0(A)$, $H_0(B)$, and $H_0(AB)$. Furthermore, a simple general relationship holds between the power of a test and the expected value of the mean square in the numerator of the corresponding F ratio. In general, for any null hypothesis $H_0(G)$ (in the fixed-effects model) tested by the ratio MS_g/MS_w, where MS_g has ν_1 degrees of freedom,

$$\phi_g^2 = \left(\frac{\nu_1}{\nu_1 + 1}\right) [E(MS_g)/\sigma_e^2 - 1]. \tag{5-22}$$

For the case of equal n, we have the general function

$$(\phi_g')^2 = (1/n)\phi_g^2. \tag{5-23}$$

Power can be calculated by looking up ϕ_g in the graphs in Scheffé (1959) or, in the case of equal ns, by looking up ϕ_g' in Table A-10 in the appendix.

Applying Equations 5-22 and 5-23,

$$\phi_a^2 = nJ\sigma_a^2/\sigma_e^2, \quad \phi_a'^2 = J\sigma_a^2/\sigma_e^2$$
$$\phi_b^2 = nI\sigma_b^2/\sigma_e^2, \quad \phi_b'^2 = I\sigma_b^2/\sigma_e^2$$
$$\phi_{ab}^2 = n\sigma_{ab}^2/\sigma_e^2, \quad \phi_{ab}'^2 = \sigma_{ab}^2/\sigma_e^2.$$

ROBUSTNESS OF THE TESTS

The previous discussions on robustness of overall F tests and planned comparisons apply to all of the tests discussed in this chapter. In general, we can say that the tests are robust with respect to violations of normality but not with respect to violations of independence. In addition, the tests are not robust with respect to violations of the assumption of equal variances unless each cell contains the same (or approximately the same) number of observations.

In the previous chapters we found that, even when the ns were equal, F tests on planned comparisons generally were not robust with respect to the assumption of equal variances. Since the tests on main effects and interactions described in this chapter are composed of planned comparisons, we might suspect that they would not be robust either—fortunately, however, they are. Planned comparisons are not generally robust because in a planned comparison the means are weighted differently. A planned comparison in which all of the means were weighted equally (e.g., the test on the grand mean, or Eq. 5-1, where all of the weights have an absolute value of one) would be robust. Few planned comparisons meet this requirement, but planned comparisons can be combined in such a way that all of the means are weighted equally in the combined test. Tests on main effects and interactions meet this requirement. Hence, as long as the ns in the individual cells are equal or nearly equal, the F tests on main effects and interactions will be robust with respect to the assumption of equal variances.

TRANSFORMATIONS ON THE DATA

In Chapter 3 we considered the possibility of equalizing the variances in the cells by making appropriate transformations on the data values. We saw, for example, that when the standard deviations of the observations were proportional to their means, the transformation $Y = \log (X)$ would make them approximately equal. We noted, however, that the F ratio then tested the means of the transformed data rather than the means of the original observations.

Transformations may also be made on the data in the two-way analysis of variance. In this case, however, there is an additional complication: the transformation may have a large effect on the nature and size of the interaction. Consider the example in Tables 5-8 and 5-9. In Table 5-8 both

Table 5-8. Theortical data illustrating effects of transformations on interactions (untransformed data).

	B_1	B_2	Row summary
A_1	84 78 $\Sigma = 162$ $\overline{X} = 81$	23 27 $\Sigma = 50$ $\overline{X} = 25$	$\Sigma = 212$ $\overline{X} = 53$
A_2	26 24 $\Sigma = 50$ $\overline{X} = 25$	4 0 $\Sigma = 4$ $\overline{X} = 2$	$\Sigma = 54$ $\overline{X} = 13.5$
Column summary	$\Sigma = 212$ $\overline{X} = 53$	$\Sigma = 54$ $\overline{X} = 13.5$	$\Sigma = 266$ $\overline{X} = 33.25$

	RS	SS	df	MS	F	α
m		8844.5	1			
a	11965	3120.5	1	3120.5	347	$<.001$
b	11965	3120.5	1	3120.5	347	$<.001$
ab	15630	544.5	1	544.5	60.5	$<.002$
w		36	4	9.0		
t	15666	6793.5	7			

Table 5-9. Theoretical data illustrating effects of transformations on interactions (transformed data).

	B_1	B_2	Row summary
A_1	9.2 8.8 $\Sigma = 18.0$ $\overline{X} = 9.0$	5.2 4.8 $\Sigma = 10.0$ $\overline{X} = 5.0$	$\Sigma = 28.0$ $\overline{X} = 7.0$
A_2	5.1 4.9 $\Sigma = 10.0$ $\overline{X} = 5.0$	2.0 0.0 $\Sigma = 2.0$ $\overline{X} = 1.0$	$\Sigma = 12.0$ $\overline{X} = 6.0$
Column summary	$\Sigma = 28.0$ $\overline{X} = 7.0$	$\Sigma = 12.0$ $\overline{X} = 3.0$	$\Sigma = 40.0$ $\overline{X} = 5.0$

	RS	SS	df	MS	F	α
m		200.00	1			
a	232.00	32.00	1	32.00	58.7	.002
b	232.00	32.00	1	32.00	58.7	.002
ab	264.00	0	1	0	0	—
w		2.18	4	.545		
t	266.18	66.18	7			

main effects and the interaction are large and highly significant ($\alpha < .001$). Table 5-9 contains the same data as Table 5-8, but the values have been transformed by the function $Y = X^{1/2}$. The main effects are still large and significant, but the interaction has disappeared.

The effect that a transformation has on the interaction is not necessarily bad; in fact, it may be an advantage. In the example in Tables 5-8 and 5-9 the elimination of an interaction also eliminates the problems with interpretation of main effects (discussed on pp. 97–98). In any case, however, one should be aware that transformations on the data have an effect on the interaction.

POOLING SUMS OF SQUARES

If there is good reason to believe that for a particular set of data the interaction is zero or nearly zero in comparison to the error variance, it is possible to increase the power of tests against main effects by "pooling" SS_{ab} and SS_w. Under the null hypothesis $H_0(AB)$, that there is no interaction, $SS_{ab} \sim \sigma_e^2 \chi^2_{(IJ-I-J+1)}$ and SS_{ab} and SS_w are statistically independent, so that $SS_{\text{pooled}} = SS_{ab} + SS_w$ is distributed as σ_e^2 times the sum of two independent chi-square variables

$$SS_{\text{pooled}} \sim \sigma_e^2 \chi^2_{(IJ-I-J+1)} + \chi^2_{(N-IJ)}$$
$$\sim \sigma_e^2 \chi^2_{(N-I-J+1)},$$

and

$$MS_{\text{pooled}} = SS_{\text{pooled}}/(N - I - J + 1)$$

can be used as the denominator of the F ratio, sometimes with a sizeable increase in the degrees of freedom.

The same kind of derivation can be used to prove the general principle that the sum of squares for any effect (main effect, interaction, or even the grand mean) if the true value is zero, can be pooled with other SSs to increase the power of the tests of the remaining effects. Usually, however, the experimenter designs his study so that significant main effects are expected. Consequently, as a general rule, only the interaction term is pooled in this way.

In practice, one should be cautious about pooling any term with SS_w. Among some experimenters (and, unfortunately, some statisticians as well), the practice has arisen of routinely testing the interaction before testing the main effects. If the interaction is not significant at some previously specified significance level, it is automatically pooled in SS_w for the tests of main effects.

This practice is dangerous for two reasons. First, it unfairly biases the tests of main effects. In effect, one is waiting until the data have been obtained, then guessing which of two mean squares will produce a more significant result and using that mean square. The bias introduced by such a procedure should be obvious. Second, the significance level specified is usually fairly high; the .05 level, for example, is commonly specified. There is some danger that in such a case the test on the interaction will not be significant when in fact there is an interaction. As a result, MS_{pooled} may be larger than MS_w, and the difference may be great enough to more than offset the increase in degrees of freedom. The net result is that the experimenter loses power instead of gaining it by using MS_{pooled}.

Nevertheless, there may be times when there are good a priori reasons for assuming that there is no interaction, and in such a case MS_{pooled} may be used. The important point is that the decision to use MS_{pooled} should be based on considerations that are independent of the obtained data. Preferably, the decision should be made before the data are obtained. Furthermore, the experimenter should in theory be willing to abide by his decision to use MS_{pooled} even though the data strongly contradict the assumption of no interaction. In practice, however, this requirement is probably a little unrealistic, even if theoretically correct. In general, it is not reasonable for a person to abide by a decision after data have clearly shown that the decision was wrong. The theory is not seriously violated by the practice of using MS_w when the interaction is significant, even though the initial decision was to use MS_{pooled}. A good rule of thumb is to use MS_{pooled} only when there are strong a priori reasons to assume that there is no interaction and when the obtained data give you no good reason to doubt that assumption. The basic difference between this procedure and that criticized in the paragraph above is that the recommended procedure correctly uses a significant interaction as reason to reject the null hypothesis of no interaction. The procedure criticized in the above paragraph involves the rather dubious practice of accepting the null hypothesis merely because the test was not significant (with no a priori reasons for expecting no interaction).

ONE OBSERVATION PER CELL

There is one important case in which the assumption of no interaction must be made. In a two-way analysis of variance the total number of cells may be rather large. In a 4×6 analysis, for example, there are 24 cells. If the experimenter is limited in time and financing, he may be unable to obtain more than one observation in each cell. There are then no degrees of freedom for MS_w. However, if the experimenter can assume that there is no interaction, he can use MS_{ab} as the denominator in his F ratio. The tests for main effects would then be

$$MS_a/MS_{ab} = F_{(I-1, IJ-I-J+1)}$$
$$MS_b/MS_{ab} = F_{(J-1, IJ-I-J+1)}$$

This procedure, like that of pooling sums of squares, can be abused by being used routinely even when there is good reason to believe that there is an interaction. In this case, however, the abuse is somewhat excusable, since no valid test can be made if the assumption cannot be made. Furthermore, even though an interaction exists, the test may be better than none at all. The effect of a nonzero interaction on a test using MS_{ab} in the denominator of F is to reduce the probability that the test will be significant whether or not the effect really exists. Consequently, one can always be certain (even though there is an interaction) that the true α level is at least as small as that obtained by using MS_{ab}. If one obtains a significant result by this procedure, one can be confident that the results are in fact significant. If the obtained F is not significant, there is no way of telling whether it might have been significant by a valid test.

Exercises

1. The following data were obtained in an animal experiment designed to study the effects of two variables on measures of performance of rats in a maze test. Three strains of rats were used: bright, mixed, and dull.

Four rats from each strain were reared under free and restricted environ-
mental conditions. The data are number of errors made by each rat.

	Bright		Mixed		Dull	
Free	26	41	41	26	36	39
	14	16	82	86	87	99
Restricted	51	96	39	104	42	92
	35	36	114	92	133	124

(a) Do the analysis of variance on the above data.

(b) Graph the cell means and give a verbal interpretation of the nature of
the interaction.

(c) Test the following null hypotheses as planned comparisons:

$$H_0(a): \mu_{11} = \mu_{23} \text{ and } \mu_{21} = \mu_{13}$$
$$H_0(b): \mu_{12} = \mu_{22}$$
$$H_0(c): \mu_{11} = \mu_{21}$$

Use either the regular or the approximate method, whichever you consider
most appropriate.

2. In the following tables the subjects in A_1 are males and the subjects
in A_2 are females. Level B_1 is a control group, and B_2 through B_5 are different experimental treatments.

	B_1	B_2	B_3	B_4	B_5
A_1	4	10	5	9	15
	4	14	4	15	8
	9	14	9	11	15
	7	18	5	13	10
A_2	12	9	11	14	15
	2	12	14	8	2
	0	8	15	17	7
	10	3	16	17	8

(a) Do a two-way analysis of variance on the above data.

(b) Graph the cell means to show the nature of the interaction and comment
on what the graph shows about the interaction.

(c) Test the following two null hypotheses, using the results from part a
to reduce the number of additional computations to as few as possible:

$H_0(1)$: there is no difference between the overall mean for the
control group and the average of the four experimental
group means.

$H_0(2)$: there are no overall differences among the four experimental group means.

(d) Suppose the experimenter had told you that he had actually intended to make some orthogonal planned comparisons among the four experimental group means. All of his comparisons are contrasts. He will reject the null hypothesis only if his results are significant at the .01 level. Without actually making any tests, tell whether it would be possible for any of his tests to be significant.

(e) Repeat the tests in part c without the assumption of equal variances.

(f) From the data it appears that males have higher scores in B_2 and B_5 and females have higher scores in B_3 and B_4. Formulate a null hypothesis that would be appropriate to test for this difference with a single contrast, and test the contrast by both the Tukey method and the Scheffé method, assuming that the experimenter made no prior decision to limit the set of post hoc tests in the experiment.

(g) Repeat part f, assuming that the experimenter decided in advance to limit his post hoc tests to contrasts within the interaction.

(h) Repeat part f, assuming that the experimenter decided in advance to limit his post hoc tests to contrasts that were within the interaction and involved only the experimental groups B_2 through B_5.

3. For a master's thesis, Phil Psych ran a paired-associate learning study on subjects taking three different tranquilizers. He also used a fourth, control, group. Phil had four subjects in each group, and he analyzed the data as a simple fixed-effects one-way analysis of variance with four levels. However, his thesis advisor noted that half the subjects in each group were college freshmen and half were seniors. He suggested that Phil do a two-way analysis, with one factor being class ranking. When Phil reanalyzed the data as a two-way design, he obtained the following table:

	SS	df	MS	F	
Drugs	137.00	3	45.667	5.45	.01
Class	30.25	1	30.25	3.61	.10
Drug x Class	22.75	3	7.583	.91	—
Error	67.00	8	8.375		
Total	257.00	15			

(a) Using only this information, find the value of F and the significance level that Phil obtained in his original one-way analysis.

(b) When Phil discussed his problem with a statistician, the statistician agreed that class-ranking should be studied as a factor, but suggested that Phil should have been interested in testing the following null hypotheses about the main effect due to drugs:

$H_0(1)$: The average performance under the three drug conditions is the same as the performance of the control group.

$H_0(2)$: There is no difference between the three drugs in their effects on performance.

Given that the following scores were obtained for the *sums* (not the means) of the scores of the subjects under each condition, perform these two tests (remember to perform the tests on the data represented as a two-way, not a one-way, analysis):

	Drug		Control
A_1	A_2	A_3	A_4
52	60	42	74

(c) Test for differences in performances under the three drugs, using the Tukey method.

4. Reanalyze the data in Problem 1, assuming that there are only three scores (26, 41, and 14) in cell AB_{11}, and that there are only two scores (92 and 124) in cell AB_{23} (the lower right cell). Use the approximate method of analysis.

5. Twenty-four grossly overweight people participated in four differ-ent weight-reduction programs. There were four people in each of six age groups. Each of the four people in an age group participated in a different program, making a 4×6 design with only one observation per cell. The data are shown below (negative values indicate weight losses, positive values indicate weight gains). Analyze the data and interpret any significant effects.

				Age		
Program	21–25	26–30	31–35	36–40	41–45	46–50
1	−59	22	−29	10	−23	72
2	−7	25	7	−8	40	64
3	−21	−54	−1	−44	−22	79
4	−27	−2	−77	−46	−11	93

6

Random Effects

In the previous chapters we assumed that observations within a group were randomly chosen. The groups from which observations were to be taken, however, were tacitly assumed to have been the deliberate choice of the experimenter. For the data in Table 4-1, for example, the two drugs were assumed to have been chosen because the experimenter was particularly interested in those two drugs. They were not assumed to have been chosen randomly from a large population of potential drugs for the experiment. In some cases, however, the groups or "treatments" themselves may have been chosen randomly from a large number of potential treatments.

ONE-WAY MODEL

Consider the following modification of the problem in Chapter 3. A large university is considering a comprehensive survey of the "intellectual maturity" of students in all courses in the university. The survey, however, will be very complicated and expensive, requiring that every student in every course be tested. The university does not wish to undertake the study unless there is a good chance that large differences among the courses will be found.

The university decides to determine this with a pretest on ten courses. To avoid biasing the pretest, the ten courses are selected randomly from among all the courses in the university. Five students from each course are given the test of "intellectual maturity." The basic data are shown in Table 6-1.

Table 6-1. Hypothetical data on ten randomly selected university courses ($n = 5$).

	A_1	A_2	A_3	A_4	A_5
ΣX	340	368	251	267	292
$\overline{X}_{i\cdot}$	68.0	73.6	50.2	53.4	58.4
ΣX^2	23451	27943	13200	14609	17182

	A_6	A_7	A_8	A_9	A_{10}	Sum
ΣX	255	310	271	300	333	2987
$\overline{X}_{i\cdot}$	51.0	62.0	54.2	60.0	66.6	
ΣX^2	13411	19825	14941	18404	22882	185848

	RS	SS	df	MS	F	α
m		178443.38				
bet	181206.6	2763.22	9	307.02	2.65	.02
w		4641.40	40	116.04		
t	185848	7404.62	49			

$\hat{\sigma}_{bet}^2 = (307.02 - 116.04)/5 = 38.20$
$\hat{\sigma}_t^2 = 38.20 + 116.04 = 154.24$
$\hat{\omega}^2 = 38.20/154.24 = .25$

Model Equation and Assumptions

The model for this experiment, if X_{ij} is the score of the jth student in the ith class, is

$$X_{ij} = \mu + \alpha_i + \epsilon_{ij}, \tag{6-1}$$

where μ is the mean score of all students, α_i is the deviation of the ith course mean from the mean for all students (i.e., $\mu + \alpha_i$ is the mean score in the ith course), and ϵ_{ij} represents the deviation of the score of the jth student from the mean for his course.

Equation 6-1 is identical in form to Equation 3-13; however, there is an important difference. In Equation 3-13 the α_i were the effects of I specifically selected treatments; in Equation 6-1, since the treatments, i.e., the classes, are themselves chosen randomly, the α_i, like the ϵ_{ij}, are values of a random variable. Like the fixed-effects model, the random-effects model requires assumptions about the observations before any test can be made. In this case assumptions must be made about the distributions of both the ϵ_{ij} and the α_i. The assumptions are

$$\epsilon_{ij} \sim N_{(0, \sigma_e^2)} \tag{6-2}$$

$$\alpha_i \sim N_{(0, \sigma_{bet}^2)} \tag{6-3}$$

The assumptions that the ϵ_{ij} and the α_i have a mean of zero are inconsequential. That the α_i have a mean of zero follows from the fact that they are deviations from their own population mean (the mean of the population of potential courses), and the ϵ_{ij} have a mean of zero for a similar reason. The important assumption about the α_i is that they are normally distributed. The important assumptions about the ϵ_{ij} are that they are also normally distributed and that their variance does not depend on the particular course being considered. The first assumption about ϵ_{ij} is identical to the assumption made in Chapter 3 that the X_{ij} were normally distributed, and the second is identical to the assumption that the variance of X_{ij} was a constant across the I groups. (The assumption about the α_i is a unique feature of the random-effects model; it has no counterpart in Chapter 3.) Finally, it is necessary to assume that all α_i and ϵ_{ij} are independent. This is equivalent to the assumption that both the groups and the observations within the groups were sampled randomly.

These considerations, as well as the example given, may appear to be somewhat contrived. In fact, it is difficult to conceive of a useful experiment employing a one-way analysis of variance with a random-effects model. Occasionally, however, such situations can arise; in addition, random effects are importantly involved in some two-way and higher-way models. The primary purpose of this section is to develop a general understanding of random-effects models in a simpler, though admittedly somewhat artificial, context.

Characteristics of Random Effects

The random-effects model differs from the fixed-effects model principally in three ways. First, the fixed-effects model applies when the particular groups being compared have been chosen because they are of interest to the experimenter. In the random-effects model the particular groups being compared have been chosen randomly from a very large population of potential groups. In the experiment outlined above, for example, the "groups" are the courses, which have themselves been chosen randomly.

Second, since the groups being compared in a random-effects experiment have been chosen randomly, the experimenter is not interested in the means of the groups actually observed. He is interested, rather, in the information that these group means can give him about the population of potential groups from which they were chosen. In the above example, the university is no more interested in the mean of the ith course than it is in any other course in the entire university. Planned comparisons between the means of the groups tested are therefore uninteresting because the experimenter has no particular interest in specific comparisons between randomly-chosen courses. Instead, the experimenter's interest lies in such population variables as μ, σ_{bet}^2, and σ_e^2. The advantage of the random-effects experiment is that it enables the experimenter to generalize statistically beyond the I observations actually taken. In the experiment discussed in Chapter

3 the experimenter cannot generalize statistically beyond the three courses he actually tested. (Extrastatistical considerations, such as similarities between the courses tested and other courses, may allow him to make some tentative generalizations, but there is no statistical basis for such generalizations.) In the random design he can generalize statistically to all courses in the university.

Finally, the above considerations imply a difference between the fixed-effects and random-effects models in any "replication" of the experiment. To replicate a fixed-effects experiment, one would obtain more observations from the I groups used in the original experiment. To replicate a random-effects experiment, one would choose a new random sample of I groups from which to take observations. If the university wished to replicate its experiment, it would not retest the same ten courses. Instead, it would test a different random sample of ten courses.

Mean Squares

The values of μ, σ_{bet}^2, and σ_e^2 can be estimated from the data by first calculating $\overline{X}_{..}$, SS_m, MS_{bet}, and MS_w, using the same formulas developed for the fixed-effects model. However, SS_m and MS_{bet} are distributed differently in the random-effects and fixed-effects models. Since MS_w is based only on deviations within groups, it is distributed exactly the same in both models. That is,

$$MS_w \sim \frac{\sigma_e^2}{(N-I)}\chi_{(N-I)}^2 \tag{6-4}$$

$$E(MS_w) = \sigma_e^2,$$

where I is the total number of groups and N is the total number of observations in all groups, as before.

To derive the expected values of SS_m and MS_{bet} for the general case, it is essential that all I groups contain the same number of observations (n). Then,

$$\overline{X}_{i.} = (1/n)\ \Sigma_j\ X_{ij} = (1/n)\ \Sigma_j\ (\mu + \alpha_i + \epsilon_{ij})$$
$$= \mu + \alpha_i + \bar{\epsilon}_i.$$

Since the X_{ij} are normally distributed, the $\overline{X}_{i.}$ are also normally distributed with mean and variance

$$E(\overline{X}_{i.}) = \mu + E(\alpha_i) + E(\bar{\epsilon}_{i.}) = \mu$$
$$V(\overline{X}_{i.}) = V(\alpha_i) + V(\bar{\epsilon}_{i.}) = \sigma_{bet}^2 + \sigma_e^2/n.$$

Both of these results follow directly from Equations 6-2 and 6-3. Consequently, all of the $\overline{X}_{i.}$ are normally distributed with the same mean and variance:

$$\overline{X}_{i.} \sim N_{(\mu,\ \sigma_{bet}^2 + \sigma_e^2/n)}.$$

To find an unbiased estimate of the variance of the $\overline{X}_{i.}$, we calculate

$$S^2 = \frac{\Sigma_i (\overline{X}_{i.} - \overline{X}_{..})^2}{(I - 1)}.$$

Since S^2 is an estimate of a variance, based on a sample from a normally distributed population, it is distributed as

$$S^2 \sim \frac{(\sigma_{bet}^2 + \sigma_e^2/n)}{(I - 1)} \chi^2_{(I - 1)}.$$

From pp. 27–31, we can see that $MS_{bet} = nS^2$, so that

$$MS_{bet} \sim \frac{(\sigma_e^2 + n\sigma_{bet}^2)}{(I - 1)} \chi^2_{(I - 1)} \tag{6-5}$$

$$E(MS_{bet}) = (\sigma_e^2 + n\sigma_w^2). \tag{6-6}$$

From these values we can obtain the unbiased estimates of σ_e^2 and σ_{bet}^2:

$$\hat{\sigma}_e^2 = MS_w \tag{6-7}$$

$$\hat{\sigma}_{bet}^2 = (MS_{bet} - MS_w)/n.$$

Just as for the fixed-effects model, the total variance in the entire population of potential observations is

$$\sigma_t^2 = \sigma_e^2 + \sigma_{bet}^2,$$

so that we can define the proportion of variance accounted for as

$$\omega^2 = \sigma_{bet}^2/\sigma_t^2,$$

and an approximately unbiased estimate of ω^2 is

$$\hat{\omega}^2 = \hat{\sigma}_{bet}^2/(\hat{\sigma}_e^2 + \hat{\sigma}_{bet}^2) = \frac{SS_{bet} - (I - 1)MS_w}{SS_t - (n - 1)MS_w} \tag{6-8}$$

$$= \frac{F - 1}{F - 1 + n}$$

Table 6-1 shows these estimates for the university data.

Comparison with the Fixed-Effects Model

Comparisons between Equations 6-7 and 4-5 and between Equations 6-8 and 4-7 show differences that, though not usually large in practice, are nevertheless important theoretically. The differences lie in the difference in meaning of σ_{bet}^2 in the fixed-effects and random-effects models. In the fixed-effects model the "population" of α_i (whose variance is σ_{bet}^2) is a more or less artificial population consisting only of the means of the groups actually used in the experiment. In the random-effects model the population of α_i is much larger than the number of groups actually tested. In the fixed-effects model the α_i of the groups actually observed must sum to zero; in the random-effects model they need not do so. These differences result in a basic difference in the expected value of MS_{bet} in the two models, as a comparison of Equations 4-4 and 6-5 show.

By defining the population parameter, τ_{bet}^2, Equations 4-4 and 6-6 can be modified so that they are identical in form. If we let N_A be the total of all α_i in the population from which the groups were drawn, we can define

$$\tau_{bet}^2 = \left(\frac{N_A}{N_A - 1}\right) \sigma_{bet}^2.$$

Then, for the random-effects model, for which N_A can be assumed to be infinite,

$$\tau_{bet}^2 = \sigma_{bet}^2 \text{ (random-effects model).} \tag{6-9}$$

For the fixed-effects model, however, the total population consists only of the I groups actually studied. In this case $N_A = I$ and

$$\tau_{bet}^2 = \left(\frac{I}{I-1}\right) \sigma_{bet}^2 \text{ (fixed-effects model).} \tag{6-10}$$

Using Equation 6-10 to modify Equation 4-4 and Equation 6-9 to modify Equation 6-6, we find that for both models

$$E(MS_{bet}) = \sigma_e^2 + n\tau_{bet}^2. \tag{6-11}$$

The importance of the redefinition in Equation 6-11 (along with many other concepts discussed in this chapter) cannot be fully appreciated until more complex analyses of variance are discussed in later chapters.

It should be pointed out, however, that in many textbooks on the analysis of variance σ_{bet}^2 is defined in the same way as τ_{bet}^2 here. There are two difficulties with this practice. First, such a definition runs counter to that found in most statistics texts, and the reader could confuse the two. Second, it will no longer be true that $\sigma_t^2 = \sigma_e^2 + \sigma_{bet}^2$, and the meaning of ω^2 will be ambiguous. (In addition, in many texts $\sigma_a^2 + \sigma_b^2$ in the two-way design is defined in the same way as $\tau_a^2 + \tau_b^2$ is defined here, and similar problems result.)

The F Test

An F ratio can be computed for the random-effects model just as for the fixed-effects model, and again it is a test of the null hypothesis that there are no differences between groups. The null hypothesis that there are no differences between groups is equivalent to the hypothesis

$$H_0: \sigma_{bet}^2 = 0. \tag{6-12}$$

Under this null hypothesis, the same theory can be applied as in pp. 20–22 and 27–31 to show that

$$MS_w \sim \frac{\sigma_e^2}{(N-I)} \chi_{(N-I)}^2$$

$$MS_{bet} \sim \frac{\sigma_e^2}{(I-1)} \chi_{(I-1)}^2$$

and

$$MS_{bet}/MS_w \sim F_{(I-1, N-I)}.$$

Thus, the F test for the random-effects model is the same as for the fixed-effects model. (Table 6-1 shows the calculations for the university data.) However, the two tests differ in the distribution of the F ratio when the null hypothesis is false. For the fixed-effects model, the distribution is noncentral F (see pp. 42–43), from Equations 6-4 and 6-5 we can see that for the random-effects model

$$MS_{bet}/MS_w \sim \frac{\dfrac{(\sigma_e^2 + n\sigma_{bet}^2)}{(I-1)} \chi^2_{(I-1)}}{\dfrac{\sigma_e^2}{(N-I)} \chi^2_{(N-I)}}$$

$$\sim (1 + n\sigma_{bet}^2/\sigma_e^2) F_{(I-1, N-I)}$$

$$\sim [1 + n\omega^2/(1-\omega^2)] F_{(I-1, N-I)}.$$

When the null hypothesis is false, the F ratio is distributed as a constant times a central F.

The above equation shows that the power of the F test, for given degrees of freedom, depends entirely on the value of ω^2. Furthermore, the power can be calculated using an ordinary F table instead of the more complicated charts of noncentral F required in the fixed-effects model. For given values of n, ω^2, and degrees of freedom,

$$\frac{MS_{bet}/MS_w}{1 + n[\omega^2/(1-\omega^2)]} \sim F_{(I-1, N-I)}.$$

Therefore, for a given significance level (α), the power is found by finding $F_{\alpha(I-1, N-I)}$, the value of F (one-tailed) needed to reach the α significance level, and calculating

$$\frac{F_{\alpha(I-1, N-I)}}{1 + n[\omega^2/(1-\omega^2)]}.$$

The significance level of this ratio, with $(I-1)$ and $(N-I)$ degrees of freedom, is $1 - \beta$. (Normally, this ratio will be smaller than one. To find its significance level, we must find the significance level of $1/F$ with $(N-I)$ and $(I-1)$ degrees of freedom. The significance level obtained in this way is β, and the power is $1 - \beta$.)

Similar considerations lead to a more general null hypothesis test and a confidence interval for ω^2. Equation 6-12 is equivalent to the null hypothesis

$$H_0: \omega^2 = 0.$$

Instead of this, a more general null hypothesis,

$$H_0: \omega^2 = \omega^{*2}, \tag{6-13}$$

for any arbitrary value of ω^{*2}, can be tested. To do so, it is only necessary to note that under the null hypothesis

$$\frac{MS_{bet}/MS_w}{1 + n[\omega^{*2}/(1 - \omega^{*2})]} \sim F_{(I-1, N-I)}.$$

If ω^{*2} is larger than zero, either a one-tailed or a two-tailed test can be made. If a two-tailed test is made, the upper and lower limits of the $\alpha/2$ tails of the F distribution are used for a test at the α level. In the same way one can find a $1 - \alpha$ confidence interval for ω^2 by solving for ω^2 in the inequalities

$$F_{(1-\alpha/2)(I-1, N-I)} \leq \frac{MS_{bet}/MS_w}{1 + n\omega^2/(1 - \omega^2)} \leq F_{(\alpha/2)(I-1, N-I)}.$$

Solving for ω^2, we obtain

$$\frac{MS_{bet}/MS_w - G}{MS_{bet}/MS_w + (n-1)G} \leq \omega^2 \leq \frac{MS_{bet}/MS_w - H}{MS_{bet}/MS_w + (n-1)H},$$

where

$$G = F_{(\alpha/2)(I-1, N-I)}$$
$$H = F_{(1-\alpha/2)(I-1, N-I)}.$$

For the data in Table 6-1, the .95 confidence interval is

$$.285 < \frac{2.65}{1 + 5\omega^2/(1 - \omega^2)} < 2.45,$$

$$.016 < \omega^2 < .62.$$

A more detailed treatment of these problems, along with a way to find approximate confidence intervals for both σ_{bet}^2 and σ_e^2, can be found in Scheffé (1959, Chapter 7). In most cases, however, probably only the test of Equation 6-12 can be made with real data. The reasons why are given in the next section.

Importance of Assumptions

The F test on the random-effects model, like that on the fixed-effects model, is robust with respect to most of the requisite assumptions. The random-effects model shares with the fixed-effects model robustness with respect to nonnormality and inequality of variances of the ϵ_{ij}; the two models also share the lack of robustness with respect to nonindependence of the observations. In the random-effects model there is an additional assumption that the α_i are normally distributed. The assumption of normal α_i does not apply when the null hypothesis is Equation 6-12, since this equation is equivalent to the assumption that the α_i are all equal to zero. However, when testing Equation 6-13 or when obtaining a confidence interval for ω^2, the assumption is important. As in Chapter 3, the most important effect here is nonzero kurtosis on the variance of the mean square. In the fixed-effects model the effect was seen to be small when N was relatively large. In the random-effects model nonzero kurtosis of the α_i strongly affects MS_{bet}; these effects can thus be cancelled out only by a large value of I i.e., by sampling from a large number of groups. The difficulty can be in effect eliminated completely by making I infinite. Since the assumption of normally distributed α_i cannot generally be made, the only test that can be performed on most real data with the random model is the test of Equation 6-12.

It should be pointed out that the point estimates (Eqs. 5-9, 5-10 and 5-11) do not require the assumption of normality. They can therefore be made for any data. Suppose, for example, the university decided that it would pay to conduct the survey if ω^2 were larger than .10. Even though the assumption of normality might not be valid, the estimate of .25 (see Table 6-1) suggests that the survey would be worthwhile.

Hypotheses About the Grand Mean

As in the fixed-effects model, it is possible in the random-effects model to test the general hypothesis

$$H_0: \mu = \mu^*. \tag{6-14}$$

However, the test differs radically from that of the fixed-effects model because SS_m is distributed very differently. In this case

$$\overline{X}_{..} = \mu + \bar{\alpha}_{.} + \bar{\epsilon}_{..}, \tag{6-15}$$

where

$$\bar{\alpha}_{.} = (1/I) \ \Sigma_i \ \alpha_i$$

and

$$\bar{\epsilon}_{..} = (1/N) \ \Sigma_i \ \Sigma_j \ \epsilon_{ij}, \tag{6-16}$$

as before. In the fixed-effects model $\bar{\alpha}_{.}$ was assumed to be zero, but this assumption cannot be made in the random-effects model. Although the population mean of the α_i is zero, that is not necessarily true for any finite sample.

From Equations 6-15 and 6-16 we get

$$\overline{X}_{..} \sim N_{(\mu, \sigma_{\text{bet}}^2/I + \sigma_e^2/N)},$$

so that

$$E(\overline{X}_{..} - \mu^*)^2 = V(\overline{X}_{..} - \mu^*) + [E(\overline{X}_{..} - \mu^*)]^2$$
$$= \sigma_e^2/N + \sigma_{\text{bet}}^2/I + (\mu - \mu^*)^2.$$

Finally,

$$E(SS_m^*) = E[N(\overline{X}_{..} - \mu^*)^2] = \sigma_e^2 + n\sigma_{\text{bet}}^2 + N(\mu - \mu^*)^2. \tag{6-17}$$

Suppose now that we attempt to make an F test by dividing SS_m^* by MS_w. Under the null hypothesis (Eq. 6-14) $\mu = \mu^*$, so that

$$SS_m^*/MS_w \sim \frac{(\sigma_e^2 + n\sigma_{bet}^2)\chi_{(1)}^2}{\sigma_e^2\chi_{(N-I)}^2/(N-I)}$$

$$\sim [1 + n\omega^2/(1-\omega^2)]F_{(1,N-I)},$$

which is clearly not distributed as F unless ω^2 is zero, i.e., unless Equation 6-12 is true. However, the F ratio is not intended to be a test of Equation 6-12; a valid test of Equation 6-14 requires a ratio whose distribution is independent of ω^2. From Equation 6-5 we can see that MS_{bet} is the appropriate term to use in the denominator, since under the null hypothesis (Eq. 6-14)

$$SS_m^*/MS_{bet} \sim \frac{(\sigma_e^2 + n\sigma_{bet}^2)\chi_{(1)}^2}{(\sigma_e^2 + n\sigma_{bet}^2)\chi_{(I-1)}^2/(I-1)}$$

$$\sim F_{(1,I-1)}.$$

The appropriateness of MS_{bet} as the denominator for testing Equation 6-14 could also have been determined by noting from Equations 6-6 and 6-17 that when Equation 6-14 is true, MS_{bet} has exactly the same expected value as SS_m^* but MS_w does not.

The test of Equation 6-14, like the other tests discussed in this chapter, is valid only if the α_i are normally distributed or if I is large. Since in most applications neither of these assumptions is met, the test is not often useful in practice. Theoretically, however, it provides a good illustration of a case in which MS_w is not an appropriate error term. Later, more important cases will be studied in which the same problem arises. In addition, the technique (mentioned in the previous paragraph) of comparing Equations 6-6 and 6-17 to determine the validity of using MS_{bet} in the denominator illustrates the general procedure of finding F ratios by comparing expected values of mean squares.

TWO-WAY MODEL

In the two-way design it is possible to choose the levels of either or both factors randomly. If the levels of both factors are chosen randomly, the model is called a *random-effects model;* if the levels of only one factor are chosen randomly, it is a *mixed-effects model.*

We found in Chapter 5 that the two-way analysis of variance is basically a variation on the one-way analysis of variance. That is, the two-way analysis is simply an alternative to the overall one-way F, or a set of planned comparisons. The data from such a study could be as easily analyzed with a one-way as with a two-way analysis of variance. The same is not true of the two-way random and mixed models. For these models the alternative of performing a simple one-way analysis of variance is not open to the experimenter. We can explain the problem most easily with two examples, the first one involving the mixed model and the second involving the random model.

Suppose an experimenter obtains a list of all colleges and universities in the United States containing more than one thousand students; he randomly samples twenty schools from this list. He then samples thirty students from each of the twenty schools and divides the sampled students into three groups of ten each. Each group of ten students is given one of three forms of a proposed college entrance examination. The basic purpose of the study is to determine the extent to which the three forms are interchangeable. The experimenter chose to sample schools randomly rather than to sample students directly because he felt that by sampling schools he was more likely to obtain a representative cross-section of students. A truly random sample of all college students in the United States would be very difficult to obtain, and even if one were obtained, it would be possible for almost all of the students (by pure chance) to come from only a very few schools. By first randomly sampling schools and then randomly sampling students within each school, he makes his random sampling problem simpler and increases the likelihood that he will obtain a more representative sample. In addition, as a sort of side benefit, this design enables him to obtain an estimate of the variability among schools in average student ability.

If the three different forms of the entrance examination are considered to be the three levels of Factor A and the twenty schools are considered to be the levels of Factor B, then we have 3×20 two-way analysis of variance with ten observations per cell (each student's score is counted as one observation). If we regarded this same study as a one-way analysis of variance in which each cell was a group, there would then be sixty (three times twenty) groups. But would it be a fixed-effects or a random-effects design? Clearly, the choice of cells was neither completely random nor completely fixed. Hence, neither the fixed-effects model nor the random-effects model of the one-way analysis of variance would be appropriate for these data. We solve the problem by treating the study as a two-way design with Factor A fixed and Factor B random. Hypothetical data from such a study are shown in Table 6-2.

For an example of the random-effects model, consider a modification of the one-way design in Table 6-1. Suppose that in addition to assessing differences among classes we are also interested in studying different universities. Four universities (Factor A) and six courses (Factor B) are selected; although the universities and the courses are selected randomly, the same six courses are studied at all four universities. Table 6-3 shows hypothetical data from such a study, with scores from two subjects in each course in each university.

Suppose we again regard this as a one-way analysis of variance with twenty-four levels. Now, clearly, the groups have all been chosen randomly. Or have they? From what population are they a random sample?

Table 6-2. Cell means and data analysis for three tests (Factor A) and twenty randomly selected colleges (Factor B), with $n = 10$.

	A_1	A_2	A_3	$\bar{X}$
B_1	22	18	13	17.7
B_2	24	30	19	24.3
B_3	26	21	22	23.0
B_4	21	27	13	20.3
B_5	26	23	17	22.0
B_6	18	21	22	20.3
B_7	16	15	21	17.3
B_8	22	25	23	23.3
B_9	14	22	20	18.7
B_{10}	11	25	17	17.7
B_{11}	24	15	17	18.7
B_{12}	21	19	21	20.3
B_{13}	20	22	20	20.7
B_{14}	22	25	19	22.0
B_{15}	25	19	24	22.7
B_{16}	23	24	23	23.3
B_{17}	22	24	21	22.3
B_{18}	22	14	17	17.7
B_{19}	21	23	12	18.7
B_{20}	17	18	22	19.0
$\bar{X}$	20.85	21.50	19.15	20.50

	RS	SS	df	MS	F	α
m		252,150	1	252,150	1663	< .01
a	252,739	589	2	294.5	1.95	.14
b	255,033	2883	19	151.7	1.76	.03
ab	261,360	5738	38	151.0	1.76	.01
w		46440	540	86.0		
t	307,800	55650	599			

The population must of course be regarded as the collection of all possible combinations of universities and courses. Suppose, for example, that there are 100 universities from which the four were chosen, and that the six courses were chosen from a total set of two hundred. Then the total population of cells would consist of the 20,000 possible ways that the 100 universities could be paired with the 200 courses.

Choosing a random sample from this population would be the same as choosing twenty-four cells from a table with 100 rows and 200 columns. The probability of studying the same course in two universities is the same as the probability of choosing more than one cell from the same column, i.e., less than one in a hundred. The probability of studying the same course

Table 6-3. Hypothetical data from four universities (Factor A) and six courses (Factor B), with $n = 2$.

	B_1	B_2	B_3	B_4	B_5	B_6	Mean
A_1	20 14 $\overline{17.0}$	8 10 $\overline{9.0}$	14 16 $\overline{15.0}$	15 13 $\overline{14.0}$	19 11 $\overline{15.0}$	26 21 $\overline{23.5}$	15.58
A_2	16 8 $\overline{12.0}$	15 6 $\overline{10.5}$	6 2 $\overline{4.0}$	2 10 $\overline{6.0}$	10 11 $\overline{10.5}$	23 14 $\overline{18.5}$	10.25
A_3	4 20 $\overline{12.0}$	7 5 $\overline{6.0}$	8 14 $\overline{11.0}$	7 2 $\overline{4.5}$	14 10 $\overline{12.0}$	18 17 $\overline{17.5}$	10.50
A_4	8 16 $\overline{12.0}$	5 14 $\overline{9.5}$	10 6 $\overline{8.0}$	10 17 $\overline{13.5}$	15 21 $\overline{18.0}$	15 22 $\overline{18.5}$	13.25
Mean	13.25	8.75	9.50	9.50	13.88	19.50	12.40

	RS	SS	df	MS	F	α
m		7375.5	1			
a	7604.6	229.1	3	76.35	5.25	.02
b	8043.1	667.6	5	133.5	9.18	<.01
ab	8490.5	218.3	15	14.55	0.65	—
w		538.5	24	22.44		
t	9029.0	1653.5	47			

in four different universities is infinitesimal. Yet, in this design every course selected is studied in four universities — and the same four universities at that! Obviously, we cannot consider all of the cells to have been randomly sampled from the same population. Instead, we must consider it to be a two-way design in which the individual cells were determined by randomly (and independently) sampling the levels of each factor.

The Model and the Assumptions

The model equation for the two-way random- and mixed-effects models, likê that for the fixed-effects model, is

$$X_{ijk} = \mu + \alpha_i + \beta_j + \alpha\beta_{ij} + \epsilon_{ijk}.$$

The assumptions about the ϵ_{ijk} are the same as for the fixed-effects model, namely,

$$\epsilon_{ijk} \sim N_{(0, \sigma_e^2)}.$$

The assumptions made about the other terms in the model are different, however.

Mixed-effects Model

In the mixed-effects model when Factor A is fixed and Factor B is random, both the β_j and the $\alpha\beta_{ij}$ are sampled randomly. The required assumptions for the analysis of variance are

$$\beta_j \sim N_{(0, \sigma_b{}^2)}$$
$$\alpha\beta_{ij} \sim N_{(0, \sigma_{ab}{}^2)}.$$

In addition, several assumptions of independence must be made. These requirements are usually expressed in abstract terms, but it is not difficult to state them in terms of the observations themselves. For the above requirements to be met, the observations (X_{ijk}) must satisfy four conditions. First, all of the observations within a cell must be sampled randomly and independently. Second, they must all be normally distributed with mean and variance

$$E(X_{ijk}) = \mu + \alpha_i$$
$$V(X_{ijk}) = \sigma_b{}^2 + \sigma_{ab}{}^2 + \sigma_e{}^2.$$

The important points here are that the X_{ijk} are all normally distributed and, although observations in different cells may have different means, they must all have the same variance. Third, the levels of B must be sampled randomly and independently, and all cell means $\overline{X}_{ij.}$ within a given level of A must be normally distributed with the same variance.

The fourth condition is more difficult to explain. The cell means $(\overline{X}_{ij.})$ within the same level (j) of B are usually not independent. That is,

$$E(\overline{X}_{ij.}\overline{X}_{i'j.}) \neq E(\overline{X}_{ij.})E(\overline{X}_{i'j.}), \ (i' \neq i).$$

The expression

$$C(\overline{X}_{ij.}, \overline{X}_{i'j.}) = E(\overline{X}_{ij.}\overline{X}_{i'j.}) - E(\overline{X}_{ij.})E(\overline{X}_{i'j.})$$

is a measure of the degree to which the $\overline{X}_{ij.}$ are dependent. The term, $C(\overline{X}_{ij.}, \overline{X}_{i'j.})$ is called the *covariance* of $\overline{X}_{ij.}$ and $\overline{X}_{i'j.}$. It can be estimated from the data by

$$\hat{C}(\overline{X}_{ij.}, \overline{X}_{i'j.}) = (1/J) \ \Sigma_j \ \overline{X}_{ij.}\overline{X}_{i'j.} - \overline{X}_{i..}\overline{X}_{i'...}$$

A separate $\hat{C}(\overline{X}_{ij.}, \overline{X}_{i'j.})$ can be calculated for each pair of levels of Factor A. The fourth condition that must be met is that all $C(\overline{X}_{ij.}, \overline{X}_{i'j.})$ must be equal.

This condition will perhaps be more familiar to some readers if we point out that the Pearson product-moment correlation between any two random variables is defined to be their covariance divided by the product of their standard deviations:

$$r(x, y) = C(x, y)/(\sigma_x\sigma_y).$$

Since, by the third condition, all of the $\overline{X}_{ij.}$ have the same variance, the fourth condition is equivalent to the condition that all pairs of cell means in the same level of B must have the same correlation.

Estimates of the relevant variances, covariances, and correlations for the data in Table 6-2 are shown in Table 6-4; they are found by treating the cell means in Table 6-2 as though they were a random sample of twenty triplets of observations. The twenty values in column A_1, for example, are the data from which the variance of A_1 is calculated; the twenty pairs of observations in columns A_1 and A_2 are the data from which the covariance of A_1 and A_2, $\hat{C}(\overline{X}_{1j.}, \overline{X}_{2j.})$, is calculated — and so on.

There is no simple method of directly testing the assumption of equal variances and covariances, but the estimates can be examined to determine whether the assumption of approximately equal variances and covariances is reasonable. Generally, the variances and correlations are a better index than are the covariances. In Table 6-4, for example, there appear to be large differences among the covariances. However, these differences are small compared to the size of the variances, so the correlations are very close to zero. Since the correlations are all approximately zero and the variances are no different than we would expect them to be from this small a sample, the assumption of approximately equal variances and covariances is probably reasonable for these data.

The assumption of equal covariances may become clearer if we give an example of its violation. In a study of nonsense syllable learning the experimenter gives each of thirty subjects a list of twenty nonsense syllables to learn by the anticipation method. The number of nonsense syllables correctly anticipated on each of ten trials is recorded. Factor B in this case is subjects, Factor A is trials, and an observation is defined to be the number of syllables correctly anticipated by a given subject on a given trial. It is unlikely, in this case, that condition four will be satisfied, since trials nine and ten, for example, are likely to be much more highly correlated than trials one and ten.

The above list of assumptions is long and complicated; however, we shall see in the section on robustness that only two conditions (in addition to

Table 6-4. Variances, covariances, and correlations for data in Table 6-2 (variances are on the diagonal, covariances above the diagonal, and correlations below the diagonal).

		Covariances			
		A_1	A_2	A_3	σ
	A_1	**15.40**	.71	.34	3.92
rs	A_2	.04	**17.63**	−.97	4.20
	A_3	.02	−.07	**12.34**	3.51

the requirements of independent random sampling) are generally important. The important conditions are that the $\overline{X}_{ij.}$ have equal variances and
equal covariances (correlations).

Random-Effects Model

All of the assumptions made in the mixed-effects model must also be made
for the random-effects model. In addition, the same assumptions made
about the β_j must also be made about the α_i—that is, $\alpha_i \sim N_{(0, \sigma_a^2)}$, and
several additional independence assumptions must also be made. In prac-
tice this means that the same condition of equal covariance (correlation)
of cell means within the same level of B must be extended to cell means
within different levels of B but the same level of A—that is, $C(\overline{X}_{ij.}, \overline{X}_{ij'.}) =$
a constant.

Once again the relevant variances and covariances can be estimated
from the data, but for the random effects design two sets of variances and
covariances must be estimated. To find the first set of estimates, we treat
the cell means at different levels of Factor B as though they were randomly
sampled basic sets of observations, just as we did for the mixed-effects
model of Table 6-2. To find the other set of estimates, we reverse the roles
of Factors A and B, treating the levels of Factor A as randomly sampled.
Both sets of variances, covariances, and correlations are shown in Table
6-5 for the data in Table 6-3. The assumption of equal variances and covariances among levels of Factor A appears reasonable. Among levels
of Factor B, however, the assumption is more doubtful; variance estimates
range from 6.25 to 24.50, and correlation estimates range from $-.50$ to
$+.79$. These estimates are not very accurate, since each is based on only
four observations (the four levels of Factor A) in the case of the variances,
or pairs of observations in the case of the covariances. It is still possible,
therefore, that the population variances and covariances are equal, but the
data indicate otherwise.

Expected Mean Squares and *F* Ratios

In the random- and mixed-effects models the same sums of squares and
mean squares are calculated as in the fixed-effects model; moreover, they
are calculated in exactly the same way (Table 5-5 gives these computational formulas and Tables 6-2 and 6-3 show the calculations for the two
examples in the tables). However, the expected values of the mean squares
in the random and mixed models are different from those in the fixed-effects
model. The expected mean squares for the fixed-, mixed-, and random-
effects models are given in Table 6-6, along with the appropriate *F* ratio
for testing each effect. Table 6-7 shows the variance estimates for the
hypothetical data in Tables 6-2 and 6-3.

The τ^2 values in Table 6-7 are related to the σ^2 by a general formula that
is easily applied. The value of τ^2 for any given effect is found by multiplying σ^2 (the variance of the effect) by as many coefficients as there are
factors in the effect being considered. The coefficient for each factor is
the number of levels in the *total population* of that factor divided by the

Table 6-5. Variances, covariances, and correlations for the data in Table 6-3, (variances are on the diagonal, covariances above the diagonal, and correlations below the diagonal).

		Covariances			
	A_1	A_2	A_3	A_4	σ
A_1	**22.24**	15.28	19.10	12.87	4.72
A_2	.64	**25.68**	16.40	14.62	5.07
A_3	.86	.69	**22.00**	10.95	4.69
A_4	.63	.67	.54	**18.68**	4.32

(*rs* labels the correlations below the diagonal)

			Covariances				
	B_1	B_2	B_3	B_4	B_5	B_6	σ
B_1	**6.25**	.42	9.17	7.50	1.88	6.67	2.50
B_2	.09	**3.75**	−4.50	3.92	.88	1.33	1.94
B_3	.79	−.50	**21.67**	10.17	5.25	8.67	4.65
B_4	.61	.41	.44	**24.50**	14.25	9.17	4.95
B_5	.23	.14	.34	.86	**11.06**	2.50	3.33
B_6	.98	.25	.69	.68	.28	**7.33**	2.71

(*rs* labels the correlations below the diagonal)

same number of levels minus one. Thus, if we let N_a be the total number of levels in the population of Factor A—i.e., if Factor A is fixed, $N_a = I$, and if A is random, N_a is infinite—and N_b be the total number of levels of Factor B, then

$$\tau_a{}^2 = \left(\frac{N_a}{N_a - 1}\right) \sigma_a{}^2$$

$$\tau_b{}^2 = \left(\frac{N_b}{N_b - 1}\right) \sigma_b{}^2$$

$$\tau_{ab}{}^2 = \left(\frac{N_a}{N_a - 1}\right)\left(\frac{N_b}{N_b - 1}\right) \sigma_{ab}{}^2.$$

Since the values of N_a and N_b differ for random and fixed factors, the exact relationships between the τ^2 and the σ^2 depend on the model being considered. In the random model both N_a and N_b are infinite, so that the coefficients are all equal to one:

$$\tau_a{}^2 = \sigma_a{}^2, \quad \tau_b{}^2 = \sigma_b{}^2, \quad \tau_{ab}{}^2 = \sigma_{ab}{}^2 \text{ (random model).} \tag{6-18}$$

Table 6-6. Expected mean squares in the mixed- (A fixed, B random) and random-effects models of the two-way design.

	Mixed Effects		
	Expected Mean Square	F ratio	df
m	$\sigma_e^2 + nI\tau_b^2 + N\mu^2$	SS_m/MS_b	$1, J-1$
a	$\sigma_e^2 + n\tau_{ab}^2 + nJ\tau_a^2$	MS_a/MS_{ab}	$I-1, (I-1)(J-1)$
b	$\sigma_e^2 + nI\tau_b^2$	MS_b/MS_w	$J-1, N-IJ$
ab	$\sigma_e^2 + n\tau_{ab}^2$	MS_{ab}/MS_w	$(I-1)(J-1), N-IJ$
w	σ_e^2		

	Random Effects		
	Expected Mean Square	F ratio	df
m	$\sigma_e^2 + n\tau_{ab}^2 + nJ\tau_a^2 + nI\tau_b^2 + N\mu^2$	[a]	[a]
a	$\sigma_e^2 + n\tau_{ab}^2 + nJ\tau_a^2$	MS_a/MS_{ab}	$(I-1), (I-1)(J-1)$
b	$\sigma_e^2 + n\tau_{ab}^2 + nI\tau_b^2$	MS_b/MS_{ab}	$(J-1), (I-1)(J-1)$
ab	$\sigma_e^2 + n\tau_{ab}^2$	MS_{ab}/MS_w	$(I-1)(J-1), N-IJ$
w	σ_e^2		

[a]No standard F ratio can be computed for this test.

In the mixed model N_b is infinite but $N_a = I$:

$$\tau_a^2 = \left(\frac{I}{I-1}\right)\sigma_a^2, \quad \tau_b^2 = \sigma_b^2 \tag{6-19}$$

$$\tau_{ab}^2 = \left(\frac{I}{I-1}\right)\sigma_{ab}^2 \text{ (mixed model)}.$$

The same principle applies to the fixed-effects model as well (pp. 102–104); for this model, however, $N_a = I$ and $N_b = J$.

The derivations of the expected mean squares are straightforward but too long and involved to present here. It may seem puzzling, however, that when Factor B is changed from fixed to random, the expected mean square

Table 6-7. Estimates of τ^2, σ^2, and ω^2 for data in Tables 6-2 and 6-3.

	For Table 6-2			For Table 6-3		
	τ^2	σ^2	ω^2	τ^2	σ^2	ω^2
a	0.72	0.48	.005	5.15	5.15	.12
b	2.19	2.19	.024	14.87	14.87	.35
ab	6.50	4.33	.047	0.00	0.00	.00
w	86.00	86.00		22.44	22.44	
t		93.01	.076		42.46	.47

of Factor A changes whereas the expected mean square of Factor B remains the same. A partial derivation will give an idea why this is so. From the basic model (Eq. 5-15) for the two-way analysis of variance we can obtain the estimates

$$\overline{X}_{i..} = (1/nJ) \, \Sigma_j \, \Sigma_k \, (\mu + \alpha_i + \beta_j + \alpha\beta_{ij} + \epsilon_{ijk})$$
$$= \mu + \alpha_i + \overline{\beta}_. + \overline{\alpha\beta}_{i.} + \overline{\epsilon}_{i..},$$
$$\overline{X}_{...} = (1/I) \, \Sigma_i \, \overline{X}_{i..} = \mu + \overline{\beta}_. + \overline{\epsilon}_{...},$$

where, as before, a dot replacing a subscript indicates that an average has been taken over the missing subscript. Note that the α_i drop out in the equation for $\overline{X}_{...}$ because, since A is a fixed effect, they sum to zero. The β_j do not drop out, however; being random effects, they have an expected value of zero for the population but they do not necessarily sum to zero for any finite sample. Similarly, the $\alpha\beta_{ij}$ do not drop out of the equation for $\overline{X}_{i..}$, although they do drop out of the equation for $\overline{X}_{...}$, since in that case they are summed over the subscript of a fixed factor. We can now find

$$\hat{\alpha}_i = \overline{X}_{i..} - \overline{X}_{...} = \alpha_i + \overline{\alpha\beta}_{i.} + (\overline{\epsilon}_{i..} - \overline{\epsilon}_{...}).$$

Here, μ and $\overline{\beta}_.$ cancel out of the difference, but $\overline{\alpha\beta}_{i.}$ does not. Since MS_a is based on the sum of the squares of these estimates (Eq. 5-8), and the estimates depend partly on the interaction, it is reasonable that MS_a should also depend on the interaction.

The logic behind the choices of F ratios in Table 6-6 is easily given. In general, the appropriate denominator term for testing a given effect is that term which would have the same expected value as the effect mean square if the effect did not exist, i.e., if τ^2 for the effect were zero. In the mixed-effects model, for example, if there were no A main effect, τ_a^2 would be zero and the expected value of MS_a would be exactly the same as that of MS_{ab}. Accordingly, MS_{ab} is the appropriate term to use in the denominator of the F ratio. The difference between these models and the fixed-effects model should be obvious. In the fixed-effects model the use of MS_{ab} in the denominator of the F ratio was recommended only as a last resort when there was only one observation per cell. In the mixed and random models MS_{ab} is the appropriate denominator for testing some effects no matter how many observations there are per cell. Of course, if there is good reason to assume that there is no interaction, SS_{ab} and SS_w may be pooled, just as in the fixed model; in that case the pooled mean square is the appropriate denominator for testing all of the effects except the grand mean.

A practical illustration of the effect of an interaction on MS_a, and the reason for testing MS_a against MS_{ab} instead of MS_w, can be found in the highly idealized example in Table 6-8. This table shows true mean scores for a hypothetical population of 500 colleges from which the data in Table 6-2 might have been drawn. Note that for half the colleges students score more highly on test A_1 and for the other half they score more highly on test A_2, i.e., there is a large interaction; there are no main effects, however.

Table 6-8. Hypothetical population means for a population of 500 colleges, illustrating the effect of an interaction on the test of the fixed-effect.

College	A_1	A_2	A_3	$\mu_{.j}$
1–250	28	12	20	20
251–500	12	28	20	20
$\mu_{i.}$	20	20	20	20

In drawing a sample of 20 colleges, it could happen by pure chance that say 15 are from the first half of the population and only five are from the second half. Our obtained means would probably then be approximately

$$\overline{X}_{1..} = [(15)(28)+(5)(12)]/20 = 24$$
$$\overline{X}_{2..} = [(15)(12)+(5)(28)]/20 = 16$$
$$\overline{X}_{3..} = [(15)(20)+(5)(20)]/20 = 20.$$

We would obtain a large MS_a and thus be inclined to conclude (erroneously) that there was a large main effect. We would also have a large MS_{ab}, however, so that the F ratio would probably be close to one. If we had used MS_w as our denominator, we would not have had that protection against falsely rejecting the null hypothesis. An example with no interaction can easily be constructed to show that the problem arises only when an interaction is present.

Testing for the grand mean in the random model presents a special problem. To test for it, we should use a denominator mean square whose expected value is $\sigma_e^2 + nJ\tau_a^2 + nI\tau_b^2 + n\tau_{ab}^2$. However, there is no mean square with this expected value and no method of pooling sums of squares will solve the problem. The linear combination of mean squares, $MS_a + MS_b - MS_{ab}$, has the right expected value but it does not have the right distribution for use in the F ratio. (This problem seldom arises in practice with the two-way design, because the grand mean is seldom tested. However, as we shall see, it can also arise in more complicated designs. It arises regularly when there are two or more random factors in the experiment.) The solution to this problem is an approximate F ratio, F^*, sometimes called a *quasi-F*, which has approximately the F distribution. It can be shown that $F^* = SS_m/(MS_a + MS_b - MS_{ab})$ has approximately the F distribution (under the null hypothesis $\mu = 0$), with numerator and denominator degrees of freedom respectively,

$$\nu_1 = 1 \tag{6-20}$$

$$\nu_2 = \frac{[E(MS_a + MS_b - MS_{ab})]^2}{[E(MS_a)]^2/(I-1) + [E(MS_b)]^2/(J-1) + [E(MS_{ab})]^2/(I-1)(J-1)}.$$

The denominator degrees of freedom cannot of course be known exactly because the expected mean squares cannot be known exactly. They can be estimated from the obtained mean squares:

$$\hat{\nu}_2 = \frac{(MS_a + MS_b - MS_{ab})^2}{MS_a^2/(I-1) + MS_b^2/(J-1) + MS_{ab}^2/(I-1)(J-1)}.$$

These conclusions can be generalized in a way that will be of use later. In general, any combination of sums and differences of mean squares,

$$MS_{combined} = \Sigma_{eff}\, k_{eff} MS_{eff},$$

is itself distributed approximately as though it were an ordinary mean square with degrees of freedom,

$$\nu = \frac{[E(MS_{combined})]^2}{\Sigma_{eff}\, k_{eff}^2 E(MS_{eff})^2/\nu_{eff}}. \tag{6-21}$$

The meaning of Equation 6-21 can be clarified by a comparison with Equation 6-20. The degrees of freedom can then be approximated by

$$\hat{\nu} = \frac{(MS_{combined})^2}{\Sigma_{eff}\, k_{eff}^2 MS_{eff}^2/\nu_{eff}}. \tag{6-22}$$

For the data in Table 6-3 we have

$$\hat{\nu} = 6.9 \simeq 7$$
$$F^* = 37.76$$
$$\alpha < .01.$$

Unfortunately, F^* may be a poor approximation in some cases. More specifically, $MS_a + MS_b - MS_{ab}$ may sometimes be smaller than zero. Although the probability of this is small, it is a danger, and F^* would then be negative and uninterpretable. We can avoid the problem by noting (from Table 6-6) that, under the null hypothesis, $SS_m + MS_{ab}$ and $MS_a + MS_b$ both have the same expected value. This suggests using the quasi-F,

$$F^* = (SS_m + MS_{ab})/(MS_a + MS_b).$$

Since both the numerator and the denominator of this quasi-F are sums of positive values, the ratio can never be negative. For this ratio both the numerator and the denominator degrees of freedom must be estimated. The estimates can be derived easily from Equation 6-22:

$$\hat{\nu}_1 = \frac{(SS_m + MS_{ab})^2}{SS_m^2 + MS_{ab}^2/(I-1)(J-1)}$$

$$\hat{\nu}_2 = \frac{(MS_a + MS_b)^2}{MS_a^2/(I-1) + MS_b^2/(J-1)}$$

For the data in Table 6-3 we obtain:

$$\hat{\nu}_1 = 1, \ \hat{\nu}_2 = 8$$
$$F^* = 35.21, \ \alpha < .01.$$

These results are essentially the same (as they should be) as those obtained with the other quasi-F above. The F^* in this case is slightly smaller, but the degrees of freedom are slightly greater.

Note on Terminology

The most commonly used mixed-effects design is the *repeated measures design,* in which each subject receives every treatment. In this design the treatments are the levels of the fixed Factor A and the subjects are the levels of the random Factor B. Some textbooks treat the repeated measures design as a special design, referring to MS_b as the "between subjects" mean square and to MS_{ab} as the "within subjects" mean square. It is more parsimonious to treat such a design as a special case of the mixed-effects model. However, you should be familiar with the alternate terminology, since it is used frequently in experimental reports.

Robustness

The problem of robustness in the random and mixed models is more serious than in the fixed model, simply because there are more assumptions to be met. Concerning the assumptions shared by all three models, about the same things can be said as were said about the fixed model. In general, the random and mixed models are robust with respect to the assumptions of normality and equal variance of the errors so long as each cell contains the same number of observations (n). They are not robust with respect to the assumption of random sampling. The random and mixed models require the additional assumption that the random effects themselves are normally distributed, and about the same things can be said about these assumptions as were said about the one-way random model. In general, if one is testing the null hypothesis that the variance of a random effect is some specified value *different from zero,* the test is not robust with respect to the normality assumption. For all other tests (including any null hypothesis that the variance of a random effect is zero) the assumptions of normality are not critical.

In the fixed-effects models we assumed that the observations in different cells were statistically independent. In random and mixed models this assumption is replaced by the assumption that the cell means have equal variances and are equally correlated. Just as the fixed-effects model is not robust with respect to the assumption of independence, the random and mixed-effects models are not robust with respect to the assumption of equal variances and correlations. It can be shown that, in general, the effect of violating this assumption is to reject the null hypothesis at a higher level of significance than is warranted by the data. This tendency to overestimate the significance of the test will be present in any F ratio for which MS_{ab} is the denominator. Thus, it is present when testing the A main effect in the mixed-effects model.

Fortunately, a partial correction for this exists. Box (1953) has shown that an approximate test can be devised by dividing both the numerator and the denominator degrees of freedom of the F ratio by the same constant K. The exact value of K cannot be known because it depends on the degree to which the assumptions are violated, but the limits on K are known. It is always true that $1 \leq K \leq \nu_1$, where ν_1 are the numerator degrees of freedom as usual. If we test each hypothesis twice, letting K be as small as possible for one test and as large as possible for the other, we can at least find upper and lower limits for the significance of the test. If K is set to 1, its lower limit, we have the standard F test with ν_1 and ν_2 degrees of freedom. If K is set to ν_1, its upper limit, we have an F whose degrees of freedom are 1 and ν_2/ν_1. The significance levels obtained from these two tests are the upper and lower limits on the actual significance of the data for the hypothesis in question. With respect to the A main effect in Table 6-3, for example, the lower limit on α, as determined by the conventional F test with 3 and 15 degrees of freedom, is .02. The upper limit is .08, the significance level that the same value of F has when its degrees of freedom are 1 and 5. We can thus conclude that the significance level of the A main effect, even though the assumptions of equal variances and correlations may be violated, is between .02 and .08. If the violations are small, the significance level is closer to .02; if they are large, it is closer to .08.

An exact test for these effects also exists. It is called the T-squared test; it is a generalization of the matched-sample t test to more than two matched groups, and it does not require the assumptions of equal variances and covariances. It is probably used much less often than it should be, but its complexity and lack of power relative to the F test has kept it from being used more generally. Its explanation is beyond the scope of this book, but an excellent discussion of it can be found in Anderson (1958).

Variance Estimates

The variances of the effects are estimated most easily by estimating the τ^2 from appropriate linear combinations of the mean squares determined by Table 6-7 and then using Equations 6-18 and 6-19 to solve for the σ^2. As an example, we can see from Table 6-6 that $\hat{\tau}_a^2 = (1/nJ)(MS_a - MS_{ab})$ is an unbiased estimate of τ_a^2 in both the random- and mixed-effects models, and in the random-effects model, where $\sigma_a^2 = \tau_a^2$, it is also an unbiased estimate of σ_a^2. In the mixed-effects model, with Factor A fixed, the estimate is

$$\hat{\sigma}_a^2 = \left(\frac{I-1}{I}\right)\hat{\tau}_a^2.$$

The other estimates are found in the same way, and the total variance of the X_{ijk} can be estimated as the sum of variances of the effects and the error: $\hat{\sigma}_t^2 = \hat{\sigma}_a^2 + \hat{\sigma}_b^2 + \hat{\sigma}_{ab}^2 + \hat{\sigma}_e^2$.

The proportion of variance accounted for by any given effect is the variance of that effect divided by the total variance. The proportion of variance accounted for by the AB interaction, for example, is $\omega_{ab}^2 = \sigma_{ab}^2/\sigma_t^2$, and it can be estimated by $\hat{\omega}_{ab}^2 = \hat{\sigma}_{ab}^2/\hat{\sigma}_t^2$.

The complete set of $\hat{\sigma}^2$ and $\hat{\omega}^2$ is given in Table 6-7 for the data in Tables 6-2 and 6-3.

Power Calculations

Power formulas are equally easy once some general principles are understood. If the effect being tested is a random effect, the distribution of the F ratio is

$$MS_{num}/MS_{den} \sim \frac{E(MS_{num})}{E(MS_{den})} F_{(\nu_1, \nu_2)},$$

where MS_{num} is the mean square in the numerator of the F ratio and MS_{den} is the mean square in the denominator of the F ratio. For the B main effect in the random-effects model, for example,

$$MS_b/MS_{ab} \sim \frac{(\sigma_e^2 + n\tau_{ab}^2 + nI\tau_b^2)}{(\sigma_e^2 + n\tau_{ab}^2)} F_{(J-1, IJ-I-J+1)}.$$

Power calculations are then made using the ordinary central F distribution, just as for the one-way random-effects model.

If the effect being tested is a fixed-effect, the F ratio is distributed as noncentral F, with ϕ^2 found by substituting $E(MS_{den})$ for σ_e^2 in Equation 5-22 to obtain

$$\phi^2 = \left(\frac{\nu_1}{\nu_1 + 1}\right) [E(MS_{num})/E(MS_{den}) - 1].$$

For the A main effect in the mixed-effects model, for example,

$$\phi_a^2 = \left(\frac{I-1}{I}\right) [(\sigma_e^2 + n\tau_{ab}^2 + nJ\tau_a^2)/(\sigma_e^2 + n\tau_{ab}^2) - 1].$$

The value of ϕ'^2 is then found by Equation 5-23.

The final principle is that if an effect varies over one or more random factors, it should be regarded as a random effect. Thus, MS_{ab}, since it varies over the random Factor B, should be treated as a random effect in both the random and the mixed model.

Pooling Sums of Squares

Like in the fixed-effects model, sums of squares in random- and mixed-effects models can be pooled to increase the power of statistical tests. In Table 6-3, for example, the AB interaction is not significant. If there is no AB interaction, then, according to Table 6-6, the expected values of MS_{ab} and MS_w are the same. In this case SS_w and SS_{ab} can be added to make a pooled sum of squares with 39 (= 15 + 24) degrees of freedom. The new denominator mean square would then be $MS_{pooled} = (218.3 + 538.5)/39 = 19.41$.

Its expected value would simply be σ_e^2. The expected value of MS_a and MS_b (assuming no AB interaction) would be

$$E(MS_a) = \sigma_e^2 + nJ\tau_a^2$$
$$E(MS_b) = \sigma_e^2 + nI\tau_b^2,$$

so that MS_{pooled} would be an appropriate denominator. The tests on the A and B main effects would be

$$F_{(3,39)} = 76.35/19.41 = 3.93, \alpha = .02 \text{ (A main effect)},$$
$$F_{(5,39)} = 133.5/19.41 = 6.88, \alpha < .001 \text{ (B main effect)}.$$

For this example, the tests are not materially improved by pooling the sums of squares. If MS_{ab} had had fewer degrees of freedom or MS_{w} had had more degrees of freedom, there would probably have been some improvement.

The general rule for pooling sums of squares is as follows: If, when the τ^2 for a given effect is assumed to be zero, two mean squares have the same expected value, those mean squares can be pooled. They are pooled by adding their sums of squares and dividing by the sum of the degrees of freedom. In the mixed-effects design if τ_{ab}^2 is assumed to be zero, it can be pooled with MS_{w} just as in the random effects design. The pooled mean square would then be used to test both the A and the B main effect. Remember again, however, that there should be good a priori reasons for assuming that τ_{ab}^2 is zero.

Planned and Post Hoc Comparisons

Planned and post hoc comparisons are meaningful only when fixed effects are involved. They are therefore not meaningful in the random-effects model, and in the mixed-effects model they are meaningful only when testing the A main effect. That is, any planned comparison must be of the form

$$H_0: \Sigma_i c_i\mu_i. = 0. \tag{6-23}$$

In theory it is possible to test any planned comparison of this form, but special methods are required if the test is not a contrast. For that reason we shall limit our discussion first to tests of contrasts.

Contrasts

It can be shown (although the derivations are somewhat complicated because the $\overline{X}_{i..}$ are not independent) that the same theory given in Chapter 4 can be applied to tests of contrasts in the mixed model. If we let

$$\psi = \Sigma_i c_i\mu_i.$$
$$C = \Sigma_i c_i\overline{X}_{i..}$$
$$(C')^2 = nJC^2/(\Sigma_i c_i^2),$$

then Equation 6-23 can be tested by

$$F_{(1, IJ-I-J+1)} = (C')^2/MS_{\text{ab}},$$
$$t_{(IJ-I-J+1)} = C'/\sqrt{MS_{\text{ab}}}.$$

Basically, the only differences between these equations and those in Chapter 4 are that nJ replaces n wherever it appears and MS_{ab} replaces MS_{e}. All of the theory relating to contrasts in Chapter 4, including post hoc tests (by the S method), confidence intervals (for both planned and post hoc tests), power calculations, proportions of variance accounted for, and combinations of planned and post hoc tests, can be applied to contrasts of the form of Equation 6-23. It is only necessary to replace n by nJ and MS_e by MS_{ab} in the appropriate formula in Chapter 4. Post hoc tests by the T method cannot be made because the means being tested are not independent.

Considerable caution must be exercised in testing contrasts by this method; its robustness is not well known, but it is probably not very robust with respect to violations of the assumption that the $\overline{X}_{ij.}$ have equal variances and covariances.

More General Tests

There is another, more general method for making planned comparisons. It can be used for any planned comparisons, including those that are not contrasts, and it does not require the assumption of equal variances and covariances. It thus has a very general applicability. It pays for this generality, however, by a decrease in power and an increase in the amount of work required to calculate the statistic. The method depends on the fact that any linear combination of normal random variables is also normal.

With a little algebra on the summations, we can show that

$$C = \Sigma_i \, c_i \overline{X}_{i..} = (1/J) \, \Sigma_j \, (\Sigma_i \, c_i \overline{X}_{ij.}).$$

Therefore, if we let

$$Y_j = \Sigma_i \, c_i \overline{X}_{ij.},$$

then C is the mean of the Y_j. Furthermore, the Y_j, being a linear combination of normal random variables, must also be normally distributed. Basically, we simply make the planned comparison by calculating Y_j for each level (B_j) of Factor B and making a t test of the null hypothesis that the mean of the Y_j is zero. The t test will have $(J - 1)$ degrees of freedom, since we will have a sample of J values of Y_j.

However, if we have available the estimated means, variances, and covariances among the levels of A (e.g., the values in Table 6-4), it is not necessary to actually calculate the Y_j. The t test requires only that we know the mean and variance of the Y_j, and these can be found from the formulas:

$$\overline{Y}_. = \Sigma_i \, c_i \overline{X}_{i..} \tag{6-24}$$

$$V(Y_j) = \Sigma_i \, c_i^2 V(\overline{X}_{ij.}) + 2 \, \Sigma_i \, \Sigma_{i'} \, c_i c_{i'} C(\overline{X}_{ij.}, \overline{X}_{i'j.}). \tag{6-25}$$

Here, $\overline{Y}_.$ and $V(Y_j)$ are the mean and variance of the Y_j, and $V(\overline{X}_{ij.})$ and $C(\overline{X}_{ij.}, \overline{X}_{i'j.})$ are the estimates of variances and covariances found as in Table 6-4.

We will illustrate this method by testing the null hypothesis H_0: $\mu_1 + \mu_2 - 2\mu_3 = 0$. In principle, we will test this null hypothesis by finding $Y_j = \overline{X}_{1j.} + \overline{X}_{2j.} - 2\overline{X}_{3j.}$ for each level of Factor B and then doing a t test on the null hypothesis that the mean of the Y_j is zero. Since the only information needed for the t test is the mean and variance of the Y_j, we can find them from Equations 6-24 and 6-25:

$$\bar{Y}_. = \bar{X}_{1..} + \bar{X}_{2..} - 2\bar{X}_{3..} = 20.8 + 21.5 - (2)(19.2)$$
$$= 3.9,$$
$$V(Y_j) = [(1)(15.40) + (1)(17.63) + (4)(12.34)]$$
$$+ [(2)(1)(1)(.71) + (2)(1)(-2)(.34)$$
$$+ (2)(1)(-2)(-.97)] = 86.33.$$

Since the total number of observations for this test is considered to be $J = 20$, the standard error of the mean is $\sqrt{86.33/(19)} = 2.13$, and

$$t_{(19)} = 3.9/2.13 = 1.83,$$
$$F_{(1, 19)} = 1.83^2 = 3.35,$$
$$\alpha = .09.$$

By contrast, if we had made this test using the method described in the previous section, the F would still have been 3.35, but the degrees of freedom would have been 57, and the significance level would have been .08. (It is mostly coincidental that both tests give exactly the same F value in this example; the values will be similar when the estimated variances and covariances are approximately equal but they will seldom be identical.) Of course, the test described in the previous section depends on the assumption of equal variances and covariances, whereas the test described here does not. In a sense, the added assumptions "buy" us 38 more degrees of freedom. Since the assumptions are seldom known to be satisfied, however, the test described here is probably better for routine use. Moreover, this method can be extended to planned comparisons that are not contrasts, while the method described in the previous section cannot.

The method described here cannot be used when one wishes to combine two planned comparisons into a single test with two degrees of freedom in the numerator. The problem is that no two tests of planned comparisons have the same denominator term when this method is used. The method described in the previous section can be used if the assumptions are met; if the assumptions are not met, a variation of the T^2 test can be used.

Exercises

1. In the study of IQs of first grade school children from slum areas in Chicago the following procedure was used: Seven schools were chosen at random from among all of the schools located in what were commonly regarded as slum areas. A list of all children whose parents finished high school but did not attend college was obtained for each school (subjects were limited to this subpopulation to control for economic factors). From among these children in each school six were chosen at random ($n = 6$). The following table gives the mean measured IQs of the sampled first graders from each school ($MS_w = 200$):

School

	A$_1$	A$_2$	A$_3$	A$_4$	A$_5$	A$_6$	A$_7$
Mean IQ	96	99	89	115	97	99	101

(a) Test the null hypothesis that there are no differences in average IQ among the seven schools.

(b) Estimate σ_{bet}^2 and ω^2.

(c) How much power does the test in question (a) have against the alternative $\sigma_{bet}^2 = 2\sigma_e^2$, if $\alpha = .01$ is needed for significance?

(d) Test the null hypothesis that $\sigma_{bet}^2 = 2\sigma_e^2$, using a two-tailed test.

(e) Test the null hypothesis that the average IQ of all these children is no different from the average for the nation, i.e., H_0: $\mu = 100$.

(f) How much power does the test in question (d) have against the alternative $\sigma_{bet}^2 = 0$, if $\alpha = .01$?

(General hint: The calculations are much simpler if, instead of operating on the means given above, you first subtract 100 from each mean and perform the analysis on the differences.)

2. An experimenter has done an experiment with four groups of subjects and obtained the following results:

G$_1$	G$_2$	G$_3$	G$_4$
8	−1	4	−13
12	−2	−4	−1
1	5	7	−3
4	−5	1	−12
6	3	5	3

His results were significant beyond the .05 level. A second experimenter replicated the study using subjects from exactly the same population. His results were:

G$_1$	G$_2$	G$_3$	G$_4$
−6	2	3	9
−6	−8	−1	2
5	−4	1	7
−3	3	10	−1
4	−3	−2	1

His results were not significant.

Since both experimenters drew their subjects from the same population, the results from the two experiments can be combined. Analyze the combined data, assuming that the second experimeter replicated the experiment in the most appropriate manner and assuming that the original experiment was (a) fixed-effects, (b) random-effects. Explain the difference in the results of parts a and b.

3. The following data are correct responses, out of a possible 25, on four difficult concept learning tasks, involving a total of 20, 40, 60, and 80 trials, respectively:

		Total Trials			
		20	40	60	80
Subject	S_1	11	10	15	23
	S_2	2	5	9	13
	S_3	6	3	12	13
	S_4	7	4	13	15
	S_5	11	12	9	19
	S_6	7	5	12	20

(a) Analyze these data as a two-way, mixed-effects design.

(b) Do post hoc tests on all possible pairs of means.

4. An experimenter was interested in comparing the effects of repeated vs. unique words in a verbal learning situation. His subjects were five randomly selected college students. Each S saw 12 different 15-word lists during the hour, and, after seeing each list, was to recall as many words as possible. There were three lists each of four types: Type 1 contained an item followed by an immediate repetition of the same item in positions 8 and 9; Type 2 contained a starred item in position 8; Type 3 contained an item printed in red in position 8; and Type 4 was a control list, with no unique items. The data for lists of Types 1 through 3 were a one if the subject recalled the unique word, and a zero if he did not. The data for Type 4 lists were the proportion of the middle three words (7, 8, and 9) recalled. The data were

	List Type			
	A_1	A_2	A_3	A_4
S_1	1	0	1	.000
	1	1	0	.333
	1	1	0	.000
S_2	1	1	0	.667
	0	1	0	.333
	1	0	1	.000
S_3	1	0	1	.000
	1	1	1	.000
	1	1	1	.333
S_4	1	1	1	.000
	0	1	1	.000
	0	1	1	.000
S_5	0	0	1	.667
	1	0	0	.000
	0	0	1	.333

(a) By summing over certain sets of values in the above table, this can be turned into a simple two-way mixed-effects design with subjects as one factor. Sum the values and test the null hypothesis of no difference among types of lists.

(b) Find the variances and covariances of the data analyzed in part a and comment on the applicability of the assumptions for that analysis.

(c) Analyze the data in the original table as a two-way, mixed-effects design, in which Factor A (fixed) is the type of list, with four levels, and Factor B (random) has 15 levels, corresponding to the fifteen rows in the table. Compare this analysis with the one in part a and comment on the validity of analyzing the data in this way.

(d) Using the analysis in part a, test the two null hypotheses:

$$H_0(1): \mu_{1.} + \mu_{2.} + \mu_{3.} = 3\mu_4.$$
$$H_0(2): \mu_{1.} = \mu_{2.} = \mu_{3.}$$

5. Reanalyze the data in Problem 2, Chapter 5, assuming that:

(a) Factor B is random and Factor A is fixed;

(b) Factor A is random and Factor B is fixed.

(c) Comment on the relationship of the power of the test on the fixed factor to the number of levels of the random factor.

6. Phil Psyche decided to perform an experiment in which rats learned to run down a straight runway to receive either 3, 6, or 9 pellets of food. Thirty-six rats, twelve in each group, were trained for 50 trials each, with each rat in a group receiving the same level of food reward on all 50 trials. After training, he divided each group into three, giving four of the rats in each group three pellets on trial 51, giving another four rats six pellets, etc. Finally, he ran each rat one more time, measuring its running speed in the runway. The data were the running speeds of the rats on this last trial. The design is thus a conventional 3×3 design.
However, Phil had some specific hypotheses to test:

H_1: Among rats receiving the same amount of food in all 51 trials, running speed is a function of the number of pellets received.

H_2: On the average, rats receiving a different number of pellets on the 51st trial from the number they had received on the first 50 trials will run more slowly than rats receiving the same number of pellets on the 51st trial.

H_3: The difference in running speed predicted by H_2 will be greater if the absolute value of the difference between the rewards obtained is six pellets rather than three.

H_4: The difference predicted by H_2 will be greater if there is a decrease rather than an increase in the number of pellets received on the 51st trial.

Phil has come to you for help. First, formulate the above experimental hypotheses as null hypotheses appropriate for testing; then tell *briefly* how you would recommend that he test the null hypotheses.

7. Suppose an experimenter had good reasons to believe that there was no A main effect for a two-way design and that the data supported this belief. He wants to test the grand mean, the B main effect, and the AB interaction. How might he use the absence of an A main effect to increase his power of pooling sums of squares, assuming

(a) that it is a fixed-effects model;

(b) that it is a random-effects model;

(c) that it is a mixed-effects model with Factor A random;

(d) that it is a mixed-effects model with Factor B random.

8. Consider a one-way fixed-effects design with repeated measures (actually, a two-way design where the second factor is subjects), with $I = 2$. Suppose a third level of Factor A (the fixed factor) is added to the analysis. Tell whether each condition listed below would tend to increase the F ratio for three levels over that obtained with only two, whether it would tend to decrease it, or to leave it the same:

(a) σ_e^2 is larger for level A_3 than for levels A_1 and A_2.

(b) $\mu_3 = (\mu_1 + \mu_2)/2$

(c) The correlation between the scores for levels A_3 and A_2 is higher than that between A_1 and A_2.

7

Higher-way Designs

The principles discussed so far with respect to one-way and two-way designs can be extended to designs having any arbitrary number of factors. We need only extend the basic model to include the additional factors; the extension involves additional interaction effects, however. Consider, for example, a three-way analysis of variance ($2 \times 2 \times 3$), such as the one in Table 7-1. The cells in this design vary on three dimensions (factors). We can label these factors A, B, and C, and let ABC_{ijk} be the cell at the ith level of Factor A, the jth level of Factor B, and the kth level of Factor C. In keeping with our previous notation, we will let μ_{ijk} and $\overline{X}_{ijk\cdot}$ be the true and obtained means, respectively, of cell ABC_{ijk}, and we will use dots to replace subscripts over which we have averaged. We will let X_{ijkl} be the lth observation in cell ABC_{ijk}.

As in previous chapters, X_{ijkl} can be expressed as the sum of a number of fixed and random effects plus an error term. In this case, however, a new set of effects must be introduced. The grand mean (μ) will be included in the equation as usual, as will the main effects (α_i, β_j, and γ_k) of Factors A, B, and C. The main effects are defined as

$$\alpha_i = \mu_{i\cdot\cdot} - \mu,$$
$$\beta_j = \mu_{\cdot j\cdot} - \mu,$$
$$\gamma_k = \mu_{\cdot\cdot k} - \mu.$$

In the three-way model there are three two-way interactions, one for each pair of factors, defined as

$$\alpha\beta_{ij} = \mu_{ij.} - \mu_{i..} - \mu_{.j.} + \mu$$
$$= (\mu_{ij.} - \mu) - \alpha_i - \beta_j,$$
$$\alpha\gamma_{ik} = \mu_{i.k} - \mu_{i..} - \mu_{..k} + \mu$$
$$= (\mu_{i.k} - \mu) - \alpha_i - \gamma_k,$$
$$\beta\gamma_{jk} = \mu_{.jk} - \mu_{.j.} - \mu_{..k} + \mu$$
$$= (\mu_{.jk} - \mu) - \beta_j - \gamma_k.$$

All of these terms are still not sufficient to express X_{ijkl}. The complete expression is

$$X_{ijkl} = \mu_{ijk} + \epsilon_{ijkl}$$
$$= \mu + \alpha_i + \beta_j + \gamma_k + \alpha\beta_{ij} + \alpha\gamma_{ik} + \beta\gamma_{jk} + \alpha\beta\gamma_{ijk} + \epsilon_{ijkl},$$

where $\alpha\beta\gamma_{ijk}$ is the *three-way interaction*

$$\alpha\beta\gamma_{ijk} = \mu_{ijk} - \mu - \alpha_i - \beta_j - \gamma_k - \alpha\beta_{ij} - \alpha\gamma_{ik} - \beta\gamma_{jk}$$
$$= \mu_{ijk} - \mu_{ij.} - \mu_{i.k} - \mu_{.jk} + \mu_{i..} + \mu_{.j.} + \mu_{..k} - \mu.$$

The three-way interaction is the extent to which the deviation of the cell mean from the grand mean cannot be expressed as the sum of the main effects and two-way interactions.

Similarly, if we extend the design to a four-way analysis of variance, the model becomes

$$X_{ijklm} = \mu + (\alpha_i + \beta_j + \gamma_k + \delta_l) +$$
$$(\alpha\beta_{ij} + \alpha\gamma_{ik} + \alpha\delta_{il} + \beta\gamma_{jk} + \beta\delta_{jl} + \gamma\delta_{kl}) +$$
$$(\alpha\beta\gamma_{ijk} + \alpha\beta\delta_{ijl} + \alpha\gamma\delta_{ikl} + \beta\gamma\delta_{jkl}) +$$
$$\alpha\beta\gamma\delta_{ijkl} + \epsilon_{ijklm}.$$

The parentheses in the above equation are used to mark off groups of similar terms. As this equation shows, in the four-way design there are four main effects, six two-way interactions, four three-way interactions, and one four-way interaction.

Table 7-1. Hypothetical three-way analysis of variance.

		A₁			A₂	
	C_1	C_2	C_3	C_1	C_2	C_3
B_1	2	8	14	10	8	4
	3	5	11	5	9	7
	0	5	9	3	14	3
	2	11	10	7	12	6
	1.75	7.25	11.00	6.25	10.75	5.00
	9	4	3	3	5	5
	3	11	9	5	5	7
	12	9	10	9	7	5
	8	8	8	5	6	6
	8.00	8.00	7.50	5.50	5.75	5.75

In an m-way analysis of variance there will be m main effects, $\binom{m}{2}$ two-way interactions, $\binom{m}{3}$ three-way interactions, and in general $\binom{m}{r}$ r-way interactions, for each value of r from two through m. The interpretation of a higher-order interaction is usually difficult, and sometimes all but impossible. The interpretation may be somewhat simpler, however, if one keeps in mind the general definition of an interaction. In general, the r-way interaction is defined as follows. Suppose we average over all of the observations in each cell and then average over all factors *except* the r factors involved in the interaction being considered. The result will look like the cell means of an r-way analysis of variance. The r-way interaction is the extent to which the expected values of the cell means of that r-way table cannot be expressed as the sums of the main effects and lower-order interactions of the r factors. (The r-way tables produced by this procedure, for Table 7-1, are shown in Table 7-2.) The following selected examples

Table 7-2. Cell totals (upper values) and means (lower values) for data in Table 7-1.

Three-way Table

	A_1			A_2		
	C_1	C_2	C_3	C_1	C_2	C_3
B_1	7	29	44	25	43	20
	1.75	7.25	11.00	6.25	10.75	5.00
B_2	32	32	30	22	23	23
	8.00	8.00	7.50	5.50	5.75	5.75

Two-way Tables

	A_1	A_2
B_1	80	88
	6.67	7.33
B_2	94	68
	7.83	5.67

	C_1	C_2	C_3
A_1	39	61	74
	4.88	7.62	9.25
A_2	47	66	43
	5.88	8.25	5.38

	C_1	C_2	C_3
B_1	32	72	64
	4.00	9.00	8.00
B_2	54	55	53
	6.75	6.88	6.62

One-way Tables

A_1	A_2
174	156
7.25	6.50

B_1	B_2
168	162
7.00	6.75

C_1	C_2	C_3
86	127	117
5.38	7.94	7.31

$\overline{X}$

330
6.88

of two-, three-, and four-way interactions taken from the four-way analysis of variance illustrate this definition:

$$\beta\delta_{jl} = \mu_{.j.l} - \mu - \beta_j - \delta_l,$$
$$\alpha\gamma\delta_{ikl} = \mu_{i.kl} - \mu - \alpha_i - \gamma_k - \delta_l - \alpha\gamma_{ik} - \alpha\delta_{il} - \gamma\delta_{kl},$$
$$\alpha\beta\gamma\delta_{ijkl} = (\mu_{ijkl} - \mu) - (\alpha_i + \beta_j + \gamma_k + \delta_l) -$$
$$(\alpha\beta_{ij} + \alpha\gamma_{ik} + \alpha\delta_{il} + \beta\gamma_{jk} + \beta\delta_{jl} + \gamma\delta_{kl}) -$$
$$(\alpha\beta\gamma_{ijk} + \alpha\beta\delta_{ijl} + \alpha\gamma\delta_{ikl} + \beta\gamma\delta_{jkl}).$$

Again, the parentheses in the last equation serve to mark off groups of similar terms.

Interpretation of a higher-order interaction depends on the pattern of the data for the particular experiment, and the task of interpretation is not easy. We saw in Chapter 5 that a two-way interaction was due to a lack of parallelism when the data were plotted against the levels of one factor. A three-way interaction is a difference in the degree of parallelism when the curves for two factors are plotted against the third. This difference is most easily visualized in Figure 7-1, which shows the data from Table 7-1. In Figure 7-1 the data for the four combinations of A and B are plotted against the levels of C. The curves for levels AB_{12} and AB_{22} are nearly parallel—if the experiment had included only level B_2, there would have been no AC interaction. The curves for levels AB_{11} and AB_{21}, however, are definitely not parallel. This difference in degree of parallelism accounts for the three-way interaction.

With the four-way interaction, the problem is even more complex. Basically, the four-way interaction is—hold your breath—the difference, over one factor, in the differences in parallelism for the other three factors.

Fortunately, most higher-way interactions result from a single commonly occurring type of case. Most higher-way interactions can be interpreted by looking in the r-way table for one cell (or perhaps two or three scattered cells) for that interaction that has a mean that is greatly different (either smaller or larger) than the means in the other cells. This phenomenon is likely to result not only in an r-way interaction but in large main effects

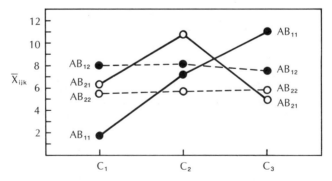

Figure 7-1. Data from Table 7-1 plotted against Factor C.

and lower-order interactions as well. Its greatest effect, however, will tend to be in the r-way table itself. It is often the main cause of a large interaction. For the data in Table 7-1, $\overline{X}_{111.}$ is considerably smaller than the other $\overline{X}_{ijk.}$, whereas $\overline{X}_{113.}$ and $\overline{X}_{212.}$ are somewhat larger.

KINDS OF MODELS

In the one-way analysis of variance we saw that we could have either a fixed or random model. In the two-way analysis of variance three distinct models—fixed, random, and mixed—were possible. In higher-way analyses of variance there are even more possibilities; each factor in the design may be either fixed or random. Thus, in the three-way design we may have zero, one, two, or three random factors with the remaining factors fixed; in the four-way design we add to this the possibility of four random factors —and so on. Usually, the number of random factors in a design is either zero or one (the single random factor is usually subjects), but designs are occasionally constructed with two or more random factors. The general model we will consider in this chapter applies to models with any arbitrary number of fixed and random factors.

ASSUMPTIONS

The assumptions needed in a higher-way analyses of variance are easily stated in terms of the individual effects, but translating these assumptions into statements about the observations themselves is not easy.

Some implications are clear: the assumptions that the observations are normally distributed with constant variance and that they are sampled randomly are implied; the assumption that the random factors are in fact randomly and independently sampled is also implied; the assumptions of equal covariances are similar to those for the mixed-model two-way analysis of variance but not exactly the same. A comprehensive statement of the conditions for the m-way design would be too complicated to give here, but the general approach should be clear from Tables 7-3 and 7-4.

Table 7-3. Assumptions in three-way analysis of variance with Factor C random.

---

1. All X_{ijkl} *within a cell* are randomly sampled and normally distributed with the same mean and variance.
2. The levels of C are randomly sampled.
3. The $\overline{X}_{ijk.}$ are normally distributed over Factor C (within each combination of levels of A and B), with equal variances.
4. For all $i, j, k, i' \neq i$, and $j' \neq j$:
   (a) $C(\overline{X}_{ijk.}, \overline{X}_{i'jk.}) =$ a constant;
   (b) $C(\overline{X}_{ijk.}, \overline{X}_{ij'k.}) =$ a constant;
   (c) $C(\overline{X}_{ijk.}, \overline{X}_{i'j'k.}) =$ a constant.

Table 7-4. Assumptions in three-way analysis of variance with Factors B and C random.

---

1. All X_{ijkl} *within a cell* are randomly sampled and normally distributed with the same mean and variance.

2(a) The levels of B are sampled randomly.

  (b) The levels of C are sampled randomly and independently of the levels of B.

3(a) The $\overline{X}_{ijk\cdot}$ are normally distributed over Factor B (within each combination of levels of A and C).

  (b) The $\overline{X}_{ijk\cdot}$ are normally distributed over Factor C (within each combination of levels of A and B).

4. For all $i, j, k, i' \neq i, j' \neq j, k' \neq k$:

  (a) $C(\overline{X}_{ijk\cdot}, \overline{X}_{i'jk\cdot}) = $ a constant;

  (b) $C(\overline{X}_{ijk\cdot}, \overline{X}_{ij'k\cdot}) = $ a constant;

  (c) $C(\overline{X}_{ijk\cdot}, \overline{X}_{ijk'\cdot}) = $ a constant;

  (d) $C(\overline{X}_{ijk\cdot}, \overline{X}_{i'jk'\cdot}) = $ a constant;

  (e) $C(\overline{X}_{ijk\cdot}, \overline{X}_{i'j'k\cdot}) = $ a constant.

Tables 7-3 and 7-4 list the assumptions required in the three-way design with one and two random factors, respectively. It is sufficient for most practical applications to note that in most experiments the conditions are probably not too well satisfied. A later section on computing F ratios will give a generalization of Box's correction for designs in which the variances and covariances are not equal. In addition, the T^2 test can be used for data for which the assumptions are not met.

Two further points must be made, however, because they are important later in the chapter. The first involves models in which all of the factors are fixed. We saw previously that the assumptions in a fixed-effects model were essentially the same whether there were one or two factors. The same is true no matter how many factors there are so long as all of the factors are fixed. The necessary assumptions are that the observations within the cells are independently sampled from normal populations with the same variance. The only important assumption is that the sampling is independent. All of the tests for main effects and interactions in the m-way fixed-effects model can be expressed as combinations of planned comparisons on a one-way design.

The other point is that this same principle does not extend to designs with one or more random factors unless additional assumptions are met. A three-way design with two factors fixed and one random, for example, cannot simply be thought of as a two-way design with one fixed and one random factor. This is because the assumptions in the three-way design are much less restrictive than they are in the two-way design. For the two fixed factors to be legitimately combined into one, highly restrictive additional assumptions would have to be satisfied.

MEAN SQUARES

Like the two-way analysis of variance, the calculation of sums of squares and mean squares in the m-way analysis of variance does not depend on the particular model. In theory, when calculating the sum of squares for an effect, estimates of the values of the effect are first obtained. These estimates are then squared and summed, and the sum is multiplied by the total number of observations over which a sum was taken to estimate the effect. In the three-way design, for example,

$$SS_a = nJK \, \Sigma_i \, \hat{\alpha}_i^2,$$
$$SS_{ab} = nK \, \Sigma_i \, \Sigma_j \, \widehat{\alpha\beta}_{ij}^2,$$
$$SS_{abc} = n \, \Sigma_i \, \Sigma_j \, \Sigma_k \, \widehat{\alpha\beta\gamma}_{ijk}^2,$$

where I, J, and K are the numbers of levels of Factors A, B, and C respectively. The other sums of squares are found similarly.

An easy way to determine the multiplier for each sum of squares is to note that it is equal to the total number of observations in the experiment divided by the total number of estimates of that effect. Thus, the total number of observations in the experiment is $N = nIJK$, and there are I estimated values of α_i, so the multiplier for SS_a is $nIJK/I = nJK$. For SS_{ab}, IJ values of $\alpha\beta_{ij}$ are estimated, so the multiplier is $nIJK/IJ = nK$. The same principle holds for SS_{abc} and all of the other sums of squares.

As in the one and two-way analyses of variance, the value of SS_t is found as the sum of the squared deviations of all of the observations from the grand mean, and SS_w is the sum of the squared deviations of the observations from their cell means.

Estimating all of the effects and then squaring and summing those estimates is a tedious procedure, and special computing formulas, like those in Tables 3-4, and 6-4, can be given for each higher-way analysis of variance. However, these formulas are generated by a small number of simple rules that apply to any m-way design. So, instead of the individual formulas, only the general rules for calculating sums of squares, mean squares, and degrees of freedom in the m-way design are given here. The application of the rules for the data in Table 7-1 is illustrated in Tables 7-2 and 7-5.

Rules for Calculating Sums of Squares

(1) Find the sum of the squared values of all of the observations; this is RS_t, the raw sum of squares total. For the data in Table 7-1, $RS_t = 2771$.

(2) Find the sum of the observations in each cell. Sum the squares of these values and divide by n, the number of values over which each sum was taken. This is the raw sum of squares (RS) for the m-way interaction. For Table 7-1, this is the three-way table in Table 7-2, and $RS_{abc} = (1/4)$ $(7^2 + 29^2 + 44^2 + \ldots) = 2547.5$.

(3) Sum the cell totals over each factor in turn to obtain m different tables, each of which has $(m - 1)$ dimensions. For the example of Table 7-1, these are the three two-way tables in Table 7-2.

Table 7-5. Summary table for data in Tables 7-1 and 7-2.

	RS	SS	T	df	MS	F	α
m		2268.750		1	2268.75		
a	2275.500	6.750	2	1	6.75	1.08	—
b	2269.500	0.750	2	1	0.75	0.12	—
c	2325.875	57.125	3	2	28.56	4.58	.01
ab	2300.333	24.083	4	1	24.08	3.86	.06
ac	2391.500	58.875	6	2	29.44	4.72	.03
bc	2381.750	55.125	6	2	27.56	4.42	.04
abc	2547.500	76.042	12	2	38.02	6.10	.003
w		224.500		36	6.24		
t	2772.000	503.250	48	47			

(4) For each table separately, find the sum of the squared cell totals and divide this by the *total* number of observations summed over to obtain each cell total. In general, this divisor will be N divided by the number of cell totals that were squared and summed. The results are the m raw sums of squares for the $(m-1)$-way interactions. For the AB interaction in Table 7-2, for example, $RS_{ab} = (1/12)(80^2 + 88^2 + 94^2 + 68^2) = 2300.333$.

(5) Sum the cell totals of each table obtained in Step (3) separately over each factor in the table to obtain $(m^2 - m)/2$ *different* $(m-2)$-dimensional tables. (Actually, $m^2 - m$ tables will be obtained, but one-half of these will be duplicates of the others.) These are the one-way tables in Table 7-2. Then follow the procedure in Step 4 to obtain the $(m^2 - m)/2$ raw sums of squares for the $(m-2)$-way interactions. For the data in Tables 7-1 and 7-2, these are actually the raw sums of squares for the main effects. As an example, $RS_a = (1/24)(174^2 + 156^2) = 2275.5$.

(6) Continue as in Steps 3, 4, and 5 until the total of all of the observations is found by summing over any of the m one-dimensional tables eventually obtained. Square this grand total and divide the squared total by N to obtain SS_m. For Tables 7-1 and 7-2, $SS_m = (1/48)(330)^2 = 2268.75$.

(7) List the raw sums of squares in a table such as Table 7-5 as an aid in calculating the sums of squares.

(8) The sum of squares for each main effect is the RS for that main effect minus SS_m. For example, $SS_a = RS_a - SS_m$. The sum of squares for each effect should be listed in the second column of the table.

(9) The sum of squares for any interaction is the raw sum of squares minus SS_m and minus the sums of squares for *all* main effects and lower-order interactions involving *only* the factors involved in the interaction whose sum of squares is being found, i.e., involving no *other* factors. For the three-way table, for example,

$$SS_{ab} = RS_{ab} - SS_m - SS_a - SS_b$$
$$SS_{abc} = RS_{abc} - SS_m - SS_a - SS_b - SS_c - SS_{ab} - SS_{ac} - SS_{bc}.$$

The value SS_{ac} is not subtracted in finding SS_{ab} because it involves a factor (C) that is not involved in the AB interaction. As another example, for the four-way analysis of variance,

$$SS_{acd} = RS_{acd} - SS_m - SS_a - SS_c - SS_d - SS_{ac} - SS_{ad} - SS_{cd}.$$

The terms SS_b, SS_{ab}, SS_{bc}, and SS_{bd} are not subtracted because they involve Factor B, and the ACD interaction does not involve Factor B.

(10) To find SS_w, subtract the RS for the m-way interaction from RS_t. For Table 7-5, $SS_w = RS_t - RS_{abc} = 2771 - 2547.5 = 233.5$.

(11) $SS_t = RS_t - SS_m$.

The tabular method of summarizing shown in Table 7-5 is especially useful in calculating sums of squares, since one need only scan the list of row headings to determine which sums of squares should be subtracted. In the table, main effects should precede interactions, and lower-order interactions should precede higher-order interactions; SS_m should be at the top, and SS_w and SS_t should be at the bottom.

Rules for Calculating Degrees of Freedom and Mean Squares

(1) For each raw sum of squares, $RS_{abc...}$, find the total number of means (T_{abc}) that were squared and summed to obtain the $RS_{abc...}$. For example, in the three-way design, $T_a = I$; $T_{ab} = IJ$; $T_{abc} = IJK$; and $T_t = N$. For the data of Table 7-1, these values are listed in the third column of Table 7-5.

(2) From each T_{abc}, subtract the number of degrees of freedom (DF) of each SS that was subtracted from the RS in Steps 8–11 above (including SS_m). Again, for the three-way design (note that $DF_m = 1$),

$$DF_a = T_a - 1$$
$$DF_{ab} = T_{ab} - 1 - DF_a - DF_b$$
$$DF_{abc} = T_{abc} - 1 - DF_a - DF_b - DF_c - DF_{ab} - DF_{ac} - DF_{bc}.$$

Similarly, in the four-way design,

$$DF_{acd} = T_{acd} - 1 - DF_a - DF_c - DF_d - DF_{ac} - DF_{ad} - DF_{cd}.$$

(3) The degrees of freedom for SS_w are the total number of observations (N) minus the total number of cells in the original table. In the three-way design, $DF_w = N - IJK = T_t - T_{abc}$.

(4) The degrees of freedom for SS_t are the total number of observations minus one: $DF_t = N - 1 = T_t - 1$. The degrees of freedom are shown in the fourth column of Table 7-5.

(5) Each mean square is then found by dividing the sum of squares by its degrees of freedom.

EXPECTED MEAN SQUARES

As we have seen in previous chapters, estimation of variances and proportions of variance accounted for, F tests, and power calculations all depend on finding the expected mean square of each effect. Fortunately, a relatively simple set of rules exists for finding these expected mean squares.

(1) Write the appropriate model equation for the design: for the three-way analysis of variance, for example,

$$X_{ijkl} = \mu + \alpha_i + \beta_j + \gamma_k + \alpha\beta_{ij} + \alpha\gamma_{ik} + \beta\gamma_{jk} + \alpha\beta\gamma_{ijk} + \epsilon_{ijkl}.$$

(2) Each expected mean square is the sum of a set of terms. Each term in the sum is a constant times the τ^2 for an effect. The constant multiplier of a given τ^2 is the divisor used in finding the RS for that effect. That is, it is the total number of observations divided by the number of cells in the table of sums for that effect. For the A main effect, for example, there are I cells; in the three-way design, there are $nIJK$ observations. The multiplier of τ_a^2, then is $nIJK/I = nJK$, the number of observations summed over in finding the table for the A main effect. The complete set of terms in the three-way model is

$$N\mu^2, \; nJK\tau_a^2, \; nIK\tau_b^2, \; nIJ\tau_c^2, \; nK\tau_{ab}^2, \; nJ\tau_{ac}^2, \; nI\tau_{bc}^2, \; n\tau_{abc}^2, \; \sigma_e^2.$$

These terms are found by referring to the model equation. They should be written out completely for the experiment being analyzed to simplify the calculation of the mean squares.

(3) Not all values enter into the sum for a given expected mean square. The τ^2 values in the expected mean square for a given effect are all values that satisfy two criteria:

> (a) their subscripts include all of the subscripts in the effect whose expected mean square is being calculated;
>
> (b) they have no other subscripts of *fixed* factors (they may have subscripts of random factors).

Finally, σ_e^2 is always a term in the sum. As an example, consider the three-way design with Factors A and B fixed and Factor C random. We will first find the expected mean square of the A main effect, $E(MS_a)$. We want the sum of all τ^2 that have an a in their subscript and that do not have subscripts of any other *fixed* factors. Since B is the only other fixed factor, we must find all τ^2 that have a but not b in their subscript. There are just two that meet this criterion, so the expected mean square of the A main effect is

$$E(MS_a) = \sigma_e^2 + nJ\tau_{ac}^2 + nJK\tau_a^2.$$

It is a common (and useful) practice in writing an expected mean square to write σ_e^2 first. The τ^2 terms then appear, with the higher-order interactions preceding lower-order interactions in the sum. The order of the terms in the expected mean square is thus the reverse of their order in the list in Step 2. If this practice is followed, the τ^2 for the effect whose expected mean square is being written will always be the last term in the sum, and the finding of appropriate F ratios will be facilitated.

If B and C were both random factors, τ_{ab}^2 and τ_{abc}^2 would also have satisfied the criterion, since b would not then be the subscript of a fixed factor. The expected mean square would then be

$$E(MS_a) = \sigma_e^2 + n\tau_{abc}^2 + nJ\tau_{ac}^2 + nK\tau_{ab}^2 + nJK\tau_a^2.$$

To find the expected mean square of the AB interaction, we must find terms whose subscripts contain both a and b, but no other subscripts of fixed factors. If C is fixed, τ_{ab}^2 is the only term that satisfies this criterion, so

$$E(MS_{ab}) = \sigma_e^2 + nK\tau_{ab}^2.$$

However, if C is random, τ_{abc}^2 also satisfies the requirement, and

$$E(MS_{ab}) = \sigma_e^2 + n\tau_{abc}^2 + nK\tau_{ab}^2.$$

(4) When finding expected mean squares, the grand mean is regarded as an effect without subscripts. Therefore, the terms that enter into its expected mean square are all of those that have no subscripts of fixed factors, i.e., all of their subscripts are of random factors; the term $N\mu^2$ is added, of course. Thus, in the three-way design, if there are no random factors (all factors are fixed),

$$E(SS_m) = \sigma_e^2 + N\mu^2;$$

if Factor C is the only random factor,

$$E(SS_m) = \sigma_e^2 + nIJ\tau_c^2 + N\mu^2;$$

if B and C are both random,

$$E(SS_m) = \sigma_e^2 + nI\tau_{bc}^2 + nIJ\tau_c^2 + nIK\tau_b^2 + N\mu^2,$$

and so on.

(5) The expected value of MS_w is always σ_e^2.

A careful study of Tables 7-6 and 7-7 will further help to clarify the application of these rules to specific designs.

Table 7-6. Expected mean squares in the three-way design with Factor C random.

Effect	$E(MS)$
m	$\sigma_e^2 + nIJ\tau_c^2 + N\mu^2$
a	$\sigma_e^2 + nJ\tau_{ac}^2 + nJK\tau_a^2$
b	$\sigma_e^2 + nI\tau_{bc}^2 + nIK\tau_b^2$
c	$\sigma_e^2 + nIJ\tau_c^2$
ab	$\sigma_e^2 + n\tau_{abc}^2 + nK\tau_{ab}^2$
ac	$\sigma_e^2 + nJ\tau_{ac}^2$
bc	$\sigma_e^2 + nI\tau_{bc}^2$
abc	$\sigma_e^2 + n\tau_{abc}^2$
w	σ_e^2

Table 7-7. Expected mean squares in the three-way design with Factors B and C random.

Effect	$E(MS)$
m	$\sigma_e^2 + nI\tau_{bc}^2 + nIJ\tau_c^2 + nIK\tau_b^2 + N\mu^2$
a	$\sigma_e^2 + n\tau_{abc}^2 + nJ\tau_{ac}^2 + nK\tau_{ab}^2 + nJK\tau_a^2$
b	$\sigma_e^2 + nI\tau_{bc}^2 + nIK\tau_b^2$
c	$\sigma_e^2 + nI\tau_{bc}^2 + nIJ\tau_c^2$
ab	$\sigma_e^2 + n\tau_{abc}^2 + nK\tau_{ab}^2$
ac	$\sigma_e^2 + n\tau_{abc}^2 + nJ\tau_{ac}^2$
bc	$\sigma_e^2 + nI\tau_{bc}^2$
abc	$\sigma_e^2 + n\tau_{abc}^2$
w	σ_e^2

F RATIOS AND VARIANCE ESTIMATES

The appropriate denominator terms for calculating F ratios are found in the way described in Chapter 6. The denominator mean square for testing each effect is the mean square whose expected value would be the same as that of the effect tested if the null hypothesis were true.

Estimates of variances of the effects are obtained from the expected mean squares, as in Chapter 6. Estimates of the τ^2 are first found from appropriate linear combinations of the mean squares. The equation defining τ^2 in terms of σ^2 is then used to solve for $\hat{\sigma}^2$. (For a review of this procedure see pp. 127–129 and 134–135.)

To estimate proportions of variance accounted for, we first calculate $\hat{\sigma}_t^2$ as the sum of the variance estimates for all of the effects (including $\hat{\sigma}_e^2$). We then divide $\hat{\sigma}_t^2$ into each variance estimate to obtain $\hat{\omega}^2$.

Just as with two-way designs, power calculations in three-way and higher designs follow the general formulas given in Chapter 6 (p. 135).

Pooling Sums of Squares

Sums of squares can be pooled in much the same way as described in Chapter 6 (pp. 135–136). Sums of squares can be pooled whenever a main effect or interaction can legitimately be assumed to have a true value of zero, i.e., whenever a τ^2 or, equivalently, a σ^2 can be assumed to be zero. To determine which sums of squares can be pooled, we once again refer to expected mean squares. We first cross out from each expected mean square the τ^2 of each effect assumed to have a true value of zero. We can then pool the sums of squares of any effects whose expected mean squares are equal. The pooling is accomplished by adding the associated sums of squares together to obtain SS_{pooled}. The degrees of freedom of SS_{pooled} are the sum of the degrees of freedom of the sums of squares that were added.

Suppose, for example, that the true value of τ_{abc}^2 in the design illustrated in Table 7-6 is assumed to be zero. Two expected mean squares, $E(MS_{ab})$

and $E(MS_{abc})$, would be affected by this assumption. Under the assumption of no ABC interaction $(\tau_{abc}^2 = 0)$, their expected mean squares become

$$E(MS_{ab}) = \sigma_e^2 + nK\tau_{ab}^2, \ E(MS_{abc}) = \sigma_e^2.$$

In this case SS_{abc} and SS_w can be pooled because their associated expected mean squares are equal; MS_{pooled} would then be the denominator for testing the AB interaction as well as the C main effect and the AC and BC interactions.

Theoretically, numerator sums of squares can also be pooled, although this rarely occurs in practice because the necessary assumptions are seldom reasonable. Suppose, however, that we assume there is no A main effect. In that case

$$E(MS_a) = E(MS_{ac}) = \sigma_e^2 + nJ\tau_{ac}^2.$$

In such a case, $SS_{pooled} = SS_a + SS_{ac}$ could be used to increase numerator degrees of freedom in testing for the AC interaction. Of course, the assumption of no A main effect will seldom be encountered in practice, but it does illustrate the range of possibilities for pooling sums of squares.

One Observation Per Cell

As with the two-way design, special assumptions may be necessary when there is only one observation per cell. To find an error term for testing some or all of the effects, at least one of the effects must be assumed to have a true value of zero. Usually, the effect chosen is the highest interaction in the design, i.e., the three-way interaction in the three-way design, the four-way interaction in the four-way design, and so on. Although there is no mathematical reason for choosing the highest interaction, it is usually the most difficult effect to interpret, and most experimenters tend to feel that it is least likely to be very large. The mean square for the chosen effect becomes the denominator mean square for testing other effects for which MS_w would normally be appropriate.

If the effect chosen for the denominator does not have a true value of zero, i.e., if the assumptions necessary for using the effect as an error estimate are not met, the direction of the bias can be determined. If the true value of the effect is not zero, the expected value of its mean square will tend to be larger than it should be. The corresponding F ratios will tend to be smaller, with the result that obtained significance levels will not be as high as they should be. Whenever a null hypothesis is rejected with this type of test, one can be confident that there is, in fact, an effect. When designing an experiment with only one observation per cell, however, one runs the risk that otherwise significant differences will be masked by too large a denominator mean square.

Quasi-F Ratios

If no single mean square can be used, mean squares can be combined as in Chapter 6 (pp. 131–133) to obtain quasi-F ratios. The same principles for constructing F ratios and degrees of freedom can be applied. As in chapter

6, either the denominator alone or both the numerator and the denominator may be linear combinations of mean squares.

As a final example, note that when Factors B and C are both random (Table 7-7), the A main effect has to be tested with a quasi-F. We might use

$$F^* = MS_a / (MS_{ab} + MS_{ac} - MS_{abc}).$$

The numerator degrees of freedom for this F^* are $(I - 1)$, and the denominator degrees of freedom are estimated to be

$$\hat{\nu}_2 = \frac{(MS_{ab} + MS_{ac} - MS_{abc})^2}{\dfrac{MS_{ab}^2}{(I-1)(J-1)} + \dfrac{MS_{ac}^2}{(I-1)(K-1)} + \dfrac{MS_{abc}^2}{(I-1)(J-1)(K-1)}}$$

For the data in Table 7-1 (see Table 7-5 for the mean squares) we have $\hat{\nu}_2 < .2$. A better quasi-F in this case would be $F^* = (MS_a + MS_{abc})/(MS_{ab} + MS_{ac})$. For the data in Table 7-1 this quasi-F would be $F^*_{(2.6, 2.8)} = .84$. For this example the latter quasi-F obviously produces more reasonable degrees of freedom.

Box's Correction

If there is some doubt that the assumptions of equal variances and covariances are satisfied, Box's correction can again be used. It is appropriate whenever the denominator of the F ratio is an interaction mean square. (The one exception to this rule occurs when the appearance of an interaction term in the denominator is due to pooling; Box's correction need not be used.) Limits can be found for the significance level by first finding the level of significance with the degrees of freedom (ν_1 and ν_2) that would be appropriate if the assumptions were met. The level of significance is then found again, with one and ν_2/ν_1 degrees of freedom. The "true" significance level is between these two limits.

COMPARISONS

Standard Method

Standard planned and post hoc comparisons can be carried out in higher-way analyses of variance just as in the one-way and two-way designs. However, unless all of the factors are fixed, some complications may arise. To begin with, the lack of robustness of such comparisons against violations of the equal variance and covariance assumptions hold for all higher-way models with one or more random factors.

In the second place, if there are more than two fixed factors, not all planned comparisons can be readily tested. The problem is that different denominator mean squares may be used for testing different effects. For some planned comparisons none of the mean squares are appropriate for the denominator of the F ratio. The following three examples of orthogonal contrasts will make the problem more clear. Appropriate denominator mean squares can be found for the first two contrasts. For the third there is no appropriate denominator mean square. All three examples involve the three-way, $2 \times 3 \times 5$ design with Factor C random and $n = 1$ shown in Table 7-8.

Table 7-8. Three-way design with Factor C random, illustrating tests of comparisons.

| | A_1 | | | | | A_2 | | | | |
	C_1	C_2	C_3	C_4	C_5	C_1	C_2	C_3	C_4	C_5
B_1	11	10	10	14	4	1	7	2	−1	0
B_2	6	−2	3	8	3	−1	1	4	6	−7
B_3	−7	−7	−7	−3	−1	−3	−1	2	4	1

	B_1	B_2	B_3
A_1	49 / 9.8	18 / 3.6	−25 / −5.0
A_2	9 / 1.8	3 / 0.6	3 / 0.6

	C_1	C_2	C_3	C_4	C_5
A_1	10 / 3.33	1 / 0.33	6 / 2.00	19 / 6.33	6 / 2.00
A_2	−3 / −1.00	7 / 2.33	8 / 2.67	9 / 3.00	−6 / −2.00

	C_1	C_2	C_3	C_4	C_5
B_1	12 / 6.0	17 / 8.5	12 / 6.0	13 / 6.5	4 / 2.0
B_2	5 / 2.5	−1 / −0.5	7 / 3.5	14 / 7.0	−4 / −2.0
B_3	−10 / −5.0	−8 / −4.0	−5 / −2.5	1 / 0.5	0 / 0.0

A_1	A_2
42 / 2.8	15 / 1.0

B_1	B_2	B_3
58 / 5.8	21 / 2.1	−22 / −2.2

$\overline{X}$: 57 / 1.9

C_1	C_2	C_3	C_4	C_5
7 / 1.17	8 / 1.33	14 / 2.33	28 / 4.67	0 / 0.00

	RS	SS	T	df	MS	Den	F	α
m		108.30	1	1	108.30	c	5.86	
a	132.60	24.30	2	1	24.3	ac	1.90	
b	428.90	320.60	3	2	160.30	bc	10.99	<.01
c	182.17	73.87	5	4	18.47			
ab	689.80	236.60	6	2	118.30	abc	13.64	<.01
ac	257.67	51.20	10	4	12.80			
bc	619.5	116.73	15	8	14.59			
abc	1001.00	69.74	30	8	8.68			

The first example is a test of the null hypothesis

$$H_0(1): \mu_{11.} + \mu_{21.} - \mu_{12.} - \mu_{22.} = 0. \qquad (7\text{-}1)$$

This hypothesis can be expressed more simply by noting that

$$\mu_{11.} + \mu_{21.} = 2\mu_{.1.},$$
$$\mu_{12.} + \mu_{22.} = 2\mu_{.2.}.$$

Remembering that multiplication or division of the coefficients by a constant does not change the null hypothesis itself, $H_0(1)$ can be rewritten

$$H_0(1): \mu_{.1.} - \mu_{.2.} = 0. \qquad (7\text{-}2)$$

Equation 7-2 expresses $H_0(1)$ entirely in terms of the B main effect. In fact, $H_0(1)$ is one of two orthogonal contrasts making up the B main effect, the other contrast being

$$H_0(1'): \mu_{.1.} + \mu_{.2.} - 2\mu_{.3.} = 0.$$

The appropriate denominator term for testing the B main effect is MS_{bc}; it is also the appropriate denominator for testing $H_0(1)$.

The test of $H_0(1)$ can be performed equally well on the $\overline{X}_{ijk.}$, the $\overline{X}_{ij..}$, or the $\overline{X}_{.j..}$. It is only necessary to modify the multiplier when finding $(C')^2$. In general, the multiplier is the number of observations over which the means used in computing C were taken. This is best explained by an illustration. We will test $H_0(1)$ first by using Equation 7-1 and then by Equation 7-2. Using Equation 7-1, we calculate

$$C_1 = 9.8 + 1.8 - 3.6 - 0.6 = 7.4,$$
$$(C_1')^2 = 5(7.4)^2/4 = 68.45,$$
$$F_{(1, 8)} = 68.45/14.59 = 4.69,$$
$$\alpha = .07.$$

In the equation for $(C_1')^2$ the multiplier was 5 because it was necessary to average over 5 observations to obtain each $\overline{X}_{ij..}$ used in the equation for C_1. If there had been n observations per cell (n larger than one), the multiplier would have been $5n$.

With the null hypothesis formulated as in Equation 7-2, the test would be

$$C_1 = 5.8 - 2.1 = 3.7,$$
$$(C_1')^2 = 10(3.7)^2/2 = 68.45.$$

This example makes it clear that the value of C depends on the way in which the null hypothesis is formulated and tested, but the value of $(C')^2$ (and hence the test itself) does not, so long as the correct multiplier is used in calculating $(C')^2$. The multiplier in this second case is 10 because C_1 is a linear combination of the $m_{.j.}$, each of which is the average of 10 observations. If the null hypothesis had been calculated from the entire ABC table, as

$$H_0(1): (\mu_{111} + \mu_{112} + \mu_{113} + \mu_{114} + \mu_{115}) +$$
$$(\mu_{211} + \mu_{212} + \mu_{213} + \mu_{214} + \mu_{215}) -$$
$$(\mu_{121} + \mu_{122} + \mu_{123} + \mu_{124} + \mu_{125}) -$$
$$(\mu_{221} + \mu_{222} + \mu_{223} + \mu_{224} + \mu_{225}) = 0,$$

then C_1 should have been calculated as a linear combination of the $\overline{X}_{ijk.}$, but $(C_1')^2$ would have had the same value (verify this for yourself). In general, if the comparison can be expressed entirely in terms of a main effect, the appropriate denominator is the mean square for testing that main effect.

The second example is the null hypothesis

$$H_0(2): \mu_{11.} + \mu_{22.} - \mu_{12.} - \mu_{21.} = 0.$$

This is one of two orthogonal contrasts making up the AB interaction. The other is

$$H_0(2'): \mu_{11.} + \mu_{12.} + 2\mu_{23.} - \mu_{21.} - \mu_{22.} - 2\mu_{13.} = 0.$$

To see this, we add the $(C')^2$ for these two null hypotheses:

$$C_2 = 9.8 + 0.6 - 3.6 - 1.8 = 5.0$$
$$(C_2')^2 = 5(5)^2/4 = 31.25.$$

$$C_{2'} = 9.8 + 3.6 + (2)(0.6) - 1.8 - 0.6 - (2)(-5.0) = 22.2$$
$$(C_{2'}')^2 = 5(22.2)^2/12 = 205.35.$$

$$(C_2')^2 + (C_{2'}')^2 = 31.25 + 205.35 = 236.60 = SS_{ab}$$

The last of these equations can be verified from Table 7-8. Since $H_0(2)$ is part of the AB interaction, the correct denominator for testing it is MS_{abc}, the denominator used in testing the AB interaction. The result is

$$F_{(1,8)} = 31.25/8.68 = 3.60; \alpha = .10.$$

The third example is the contrast

$$H_0(3): \mu_{13.} - \mu_{23.} = 0. \tag{7-3}$$

This test is orthogonal to Factor B since it involves a difference only between two levels of A. The null hypothesis may be false, however, if there is *either* an A main effect *or* an AB interaction. Since the test is on the difference between two levels of Factor A, the relevance of the A main effect should be clear. Even if there were no A main effect, however, there could still be a difference between these two means; the difference might be counteracted (in the A main effect) by another difference (in the opposite direction) at a different level of B. If that were to occur, it would indicate the presence of an AB interaction. Consequently, both the A main effect and the AB interaction are involved in this contrast. Which denominator should we use to test it, MS_{ac} or MS_{abc}? The answer is neither. The null hypothesis $H_0(3)$ cannot be tested by the standard method. In general, to be testable by the standard method, a planned comparison must be independent of all but one effect. The denominator for testing it is then the denominator that would be used in testing that effect.

One obvious conclusion from this discussion is that, aside from the test of the grand mean, all planned comparisons must be contrasts if they are to be testable by the standard method. Two points should be remembered here. First, if the assumptions necessary for combining Factors A and B into a single factor can be met, the design can be treated as a two-way analysis of variance and any planned comparison not involving the C main effect and its interactions can be tested. Second, even though the assumptions for combining the A and B factors cannot be met, methods to be described later can be used to test planned comparisons that are not testable by the standard method.

The generalization of the above discussion to S-method post hoc comparisons should be clear. If a planned comparison can be tested by the standard method, it is transformed to a post hoc comparison by merely treating $(C')^2$ as if it has more than one degree of freedom. The appropriate number of degrees of freedom is determined by the considerations discussed in Chapter 4. In most cases it will be the sum of the degrees of freedom of all fixed effects in the design (the post hoc comparisons will be independent of all random effects). In the three-way design with only A fixed, for example, it would be $(I-1)$; in the same design with both A and B fixed it would be $(I-1)+(J-1)+(I-1)(J-1)=IJ-1$. However, if the actual post hoc comparisons are chosen from a more limited set of potential comparisons (as discussed in Chapter 4), fewer degrees of freedom might be appropriate.

Power calculations for planned comparisons follow the formulas given on pp. 58 and 135.

Analyzing Planned Comparisons

We saw in the last section that a planned comparison may be part of a single effect only or it may involve two or more effects. To test the comparison appropriately, we first have to know which effects are involved. This is a relatively simple problem with many planned comparisons; with others it can be much more difficult unless a relatively simple method for analyzing planned comparisons is available. The following describes such a method.

To determine the effects involved in a planned comparison and the extent to which these effects are involved, we need only do an analysis of variance on the coefficients of the planned comparison, treating them as data with one observation per cell. Any effect with a nonzero sum of squares in this analysis is involved in the planned comparison. Any effect with a zero sum of squares is independent of the planned comparison.

We will demonstrate this on the contrast $H_0(3)$, Equation 7-3. The first step is to put the coefficients into the cells of an AB table:

	B_1	B_2	B_3
A_1	0	0	1
A_2	0	0	−1

The analysis of these coefficients is shown in Table 7-9. The coefficients are treated just as though they were the data in a two-way analysis of variance with one observation per cell. The analysis follows the procedure outlined on pp. 149–151. Unlike an ordinary analysis of variance, however, this analysis need not be done on all three factors. Instead, it need only be done on those factors that may be involved in the planned comparison. In this example Factor C is clearly not involved in the planned comparison, so it is omitted from the analysis in Table 7-9. The omission of factors clearly not involved in the comparison is not essential, but it does considerably simplify the analysis.

   In Table 7-9 the analysis proceeds in the usual way up to the calculation of sums of squares. The symbols rs and ss (note lower case) indicate that the values have been calculated on the coefficients of the planned comparison instead of on data. Since ss_a and ss_b are the only nonzero sums of squares, we conclude that this particular comparison involves both the A main effect and the AB interaction, but no other effect. Unfortunately, in this design the A main effect and the AB interaction are tested with different denominator mean squares. Consequently, the ordinary method of planned comparisons cannot be used. However, the next section describes a method that may be used for any planned comparison.

Table 7-9. Approximate test of $H_0(3)$ (Eq. 7-3). Data from Table 7-8.

	Coefficients					Analysis			
	B_1	B_2	B_3	Σ		rs	ss	ss^*	Den
A_1	0	0	+1	1	m		0	0	c
					a	2/3	2/3	1/3	ac
A_2	0	0	−1	−1	b	0	0	0	bc
					ab	2	4/3	2/3	abc
Σ	0	0	0	0					

$$F^* = \frac{78.40}{(1/3)(12.80) + (2/3)(8.68)} = \frac{78.40}{10.05} = 7.80$$

$$\hat{\nu} = \frac{(10.05)^2}{\left(\frac{1}{9}\right)\left(\frac{12.80^2}{4}\right) + \left(\frac{4}{9}\right)\left(\frac{8.68^2}{8}\right)} \cong 11.6, \ \alpha = .025$$

Approximate Planned Comparisons

When there is no appropriate denominator mean square for a planned comparison, an approximate denominator can be created. The approximate denominator is a linear combination of mean squares, following exactly the same principles as those for a quasi-F. Moreover, both the mean squares and their coefficients in the linear combination are found from the analysis described in the last section. The mean squares are those that would be used to test the effects involved in the planned comparison. To find the coefficients, we normalize the ss values found in the above analysis, dividing them by their sum, so that the normalized ss values add to one.

The procedure is simpler than it sounds. The step-by-step procedure is described below. Each step is illustrated using the planned comparison in Table 7-9 on the data in Table 7-8.

(1) Normalize the ss values. First, sum the ss values, then divide each ss by the sum. The resulting values (labeled ss^* in Table 7-9) will sum to one. In Table 7-9 the ss values add to 2; consequently, we normalize these by dividing each ss by 2.

(2) The denominator for the quasi-F is then

$$\Sigma_{\text{eff}} \, ss^*_{\text{eff}} MS_{\text{den(eff)}},$$

where $MS_{\text{den(eff)}}$ symbolizes the mean square that would appear in the denominator when testing the effect. These denominator terms (taken from Table 7-8) are in the last column of Table 7-9. Basically, the above sum is the sum of the cross-products of the values in the last two columns of Table 7-9; for those data the sum would be $(1/3)MS_{\text{ac}} + (2/3)MS_{\text{abc}}$, and the quasi-$F$ would be

$$F^* = \frac{(C')^2}{(1/3)MS_{\text{ac}} + (2/3)MS_{\text{abc}}}.$$

From Table 7-8 we see that the denominator is $(1/3)(12.80) + (2/3)(8.68) = 10.05$. The value of $(C')^2$ is 78.40, so the F ratio is 7.02.

(3) As usual, the numerator degrees of freedom are one. The denominator degrees of freedom are estimated from Equation 6-22, letting $k_{\text{eff}} = ss^*_{\text{eff}}$:

$$\hat{\nu} = \frac{(MS_{\text{combined}})^2}{\Sigma_{\text{eff}} \, ss^{*2}_{\text{eff}} MS_{\text{eff}}^2 / \nu_{\text{eff}}}.$$

For the current problem,

$$\hat{\nu} = \frac{(10.05)^2}{[(1/3)^2(12.80)^2/4] + (2/3)^2(8.68)^2/8]} \approx 11.6.$$

With 1 and 11.6 degrees of freedom, an F^* of 7.02 is significant at about the .025 level.

If more terms had been involved in the planned comparison, there would have been more terms in the denominator, but the principle would have been the same.

Table 7-10. Test of H_0: $\mu_{1..} = 0$, design of Table 7-8.

Coefficients					Analysis	
A_1	A_2	Σ		rs	ss	ss^*
1	0	1	m		1/2	1/2
			a	1	1/2	1/2

This approach can be used even when the effects involved are themselves testable only by quasi-F ratios. Suppose, for example, that in the design in Table 7-8 we wanted to test H_0: $\mu_{1..} = 0$. This is not a contrast, and the relative weights of the grand mean and the A main effect are in Table 7-10. Suppose, furthermore, that B and C are both random. The denominator for testing the grand mean would then be $(MS_b + MS_c - MS_{bc})$, and the denominator for testing the A main effect would be $(MS_{ab} + MS_{ac} - MS_{abc})$. The F ratio would then be

$$F^* = \frac{(C')^2}{(1/2)(MS_b + MS_c - MS_{bc}) + (1/2)(MS_{ab} + MS_{ac} - MS_{abc})}$$

$$= \frac{(C')^2}{(1/2)[MS_b + MS_c - MS_{bc} + MS_{ab} + MS_{ac} - MS_{abc}]}$$

$$= \frac{117.6}{143.3} = .82.$$

The denominator degrees of freedom would be

$$\hat{\nu} = \frac{(143.3)^2}{\left(\frac{1}{4}\right)\left[\frac{(160.3)^2}{2} + \frac{(18.47)^2}{4} + \frac{(14.59)^2}{8} + \frac{(118.3)^2}{2} + \frac{(12.8)^2}{4} + \frac{(8.68)^2}{8}\right]}$$

$$= \frac{20535}{5002} \approx 4.$$

The F is of course not significant.

As in our previous examples of quasi-F ratios, we could eliminate the negative terms from the denominator by adding them to the numerator. Of course, we would then have to use Equation 6-22 to estimate both the numerator and the denominator degrees of freedom.

Exact Planned Comparisons — Special Cases

The method described here for testing planned comparisons applies to any comparison made on fixed factors, and the only assumption needed is that the random factor is normally distributed. The method is limited, however, to mixed-effects models for which only one factor is random. Fortunately, the most commonly used type of mixed-model, multifactor design has only one random factor, although some designs involve two or more random factors. The method given here is not applicable to designs with more than one random factor.

Moreover, this method is usually less powerful than the method described in the last section—as a general rule, the fewer assumptions one makes, the less power the test has. However, the fact that it is an exact test using minimal assumptions is a strong argument for its use whenever appropriate.

The method is simplicity itself, though some of the calculations are a little tedious. Instead of averaging over the levels of the random factor and then calculating the value of the planned comparison, C, from these averages, we calculate C separately for each level of the random factor. The result is as many different values of C as there are levels of the random factor. The average of these separate C values is equal to the C that would normally be calculated for the planned comparison. More to the point, however, is that under the null hypothesis the separate C values are a random sample from a normally distributed population with a mean of zero. To test the planned comparison, we simply do a one-sample t test on the C values. We will illustrate this method by testing the three null hypotheses on pp. 158–159.

To test $H_0(1)$, on the B main effect, we use the cell means in the BC table, calculating $\overline{X}_{.1k} - \overline{X}_{.2k}$ for each level of the random factor (C) to obtain the five values, 3.5, 9.0, 2.5, −0.5, 4.0. An ordinary t test on the null hypothesis that the mean of these five values is zero results in a t of 2.41, with 4 d.f. and $\alpha = .06$.

To test $H_0(2)$ and $H_0(3)$, we must use the complete ABC table. For $H_0(2)$ we calculate $X_{11k} + X_{22k} - X_{12k} - X_{21k}$ for each level of Factor C, giving the five values, 3, 6, 9, 13, −6. The t value is 1.56, which is not significant.

Finally, for $H_0(3)$ we calculate the five values of $X_{13k} - X_{23k}$. The values are −4, −6, −9, −7, −2. The value of t is −4.63 and $\alpha = .006$.

The loss of power is not evident from a comparison of these results with those on pp. 158–159 and 161. Usually, however, the obtained significance level will not be as high with this exact test as with either the standard or approximate method of testing planned comparisons.

Exercises

1. As part of a program testing athletic ability, six randomly selected students were asked to take a basketball-shooting test. Each subject was asked to take two shots from each of three distances: five, ten, and twenty feet, and then to repeat the set of shots. Each student then took 100 practice shots, after which the tests were repeated. The following table shows the number of baskets made by each student under each condition:

		5 ft						10 ft				
		S_1	S_2	S_3	S_4	S_5	S_6	S_1	S_2	S_3	S_4	S_5	S_6
Practice	Set 1	2	1	2	1	1	2	2	0	1	2	2	1
	Set 2	2	2	0	2	2	1	2	1	0	1	2	0
No	Set 1	0	2	2	1	1	2	0	1	1	0	0	0
practice	Set 2	0	1	2	2	1	1	0	1	2	0	1	0

			20 ft			
S_1	S_2	S_3	S_4	S_5	S_6
0	0	0	1	0	1
1	1	0	0	1	0
0	1	0	0	1	0
0	1	1	0	0	0

(a) Analyze these data as a 4-way, distance × practice × repetition × subjects design, finding cell means and significance levels for each effect.

(b) Reanalyze the data, assuming that the scores on the two repetitions are independent of each other, i.e., that they are the two observations in a single cell.

(c) Comparing the results of parts a and b, comment on the validity of the assumption in part b. If it is not valid, how might it affect conclusions about the other three factors?

(d) Reanalyze the data, keeping all four factors as in part a, but assume that the three distances used were selected randomly from the set of all distances between $\frac{1}{2}$ foot and $49\frac{1}{2}$ feet.

2. Reanalyze the data in Problem 4, Chapter 6, treating the three lists of each type as the levels of a third factor. Assume (a) that this third factor is fixed, and (b) that it is random.

8

Nested Designs

In each of the designs discussed so far all cells contained observations. Sometimes, however, it is either impossible, impractical, or otherwise undesirable to obtain observations in every cell. The next two chapters will discuss some designs in which not all cells contain observations. These *incomplete designs* are convenient for some purposes, but they also have disadvantages. In choosing a design, the advantages and the disadvantages must be weighed against each other.

The designs in this chapter are probably the most commonly used type of incomplete design. Characteristically each level of some factor is paired with one and only one level of some other. Consider, for example, the

Table 8-1. Hypothetical data comparing teachers (B) and schools (A).

		A_1		A_2		A_3	
		B_1	B_2	B_3	B_4	B_5	B_6
		20	19	14	12	13	9
Scores		18	20	18	12	16	4
		14	20	14	9	13	4
	B	52	59	46	33	42	17
		17.33	19.67	15.33	11.00	14.00	5.67
Totals and means	A	111		79		59	
		18.50		13.17		9.83	
	$\overline{X}$			249			
				13.83			

following experiment. An educator was interested in the extent to which high school students' knowledge of algebra depended on the school and the teacher. He got the cooperation of the three high schools in his town, each of which had two algebra teachers. Three randomly-selected students were assigned to each teacher's class, making a total of eighteen students in all (three students in each of two classes in each of three schools). At the end of the semester each student was given a standardized test in algebra; the scores are shown in Table 8-1.

If we treat the schools as the levels of Factor A and the teachers as the levels of Factor B, we have a 3×6 design with three observations in each cell containing observations. Not all cells contain observations, however. This is illustrated in Table 8-2. The experimenter wanted to have each teacher teach one class in each of the three schools, but both schools and the teachers felt that that would be stretching cooperation a little too far. Consequently, the experimenter chose the design illustrated here.

Table 8-2. Data from Table 8-1 arranged as a 3×6 design.

	A_1	A_2	A_3	Sum / Mean
B_1	20 18 14			52 17.33
B_2	19 20 20			59 19.67
B_3		14 18 14		46 15.33
B_4		12 12 9		33 11.00
B_5			13 16 13	42 14.00
B_6			9 4 4	17 5.67
Sum / Mean	111 18.50	79 13.17	59 9.83	249 13.83

The important point in Table 8-2 is that each level of Factor B is paired with one and only one level of Factor A, whereas each level of Factor A is paired with two levels of Factor B. Whenever each level of one factor is paired with one and only one level of another factor, the former is said to be *nested* in the latter. In our example Factor B is nested in Factor A. Notice that nesting is strictly a "one-way" relationship; Factor A is not nested in Factor B, since each level of Factor A is paired with two levels of Factor B. When one factor is nested in another, the design is called a *nested design*. When each level of one factor is paired with each level of the other, the factors are said to be *crossed,* and if all factors are crossed with each other, the design is called *completely crossed.* All of the designs up until now have been completely crossed.

TWO-WAY MODELS

Now let us consider the design of Tables 8-1 and 8-2 in terms of the model developed previously. For two-way designs this model is

$$X_{ijk} = \mu + \alpha_i + \beta_j + \alpha\beta_{ij} + \epsilon_{ijk}. \tag{8-1}$$

The average of the observations in each cell is

$$\overline{X}_{ij.} = (1/n) \, \Sigma_k \, X_{ijk} = \mu + \alpha_i + \beta_j + \alpha\beta_{ij} + \bar{\epsilon}_{ij.}.$$

To take other averages, we must have some new definitions. First, instead of letting J equal the total number of levels of Factor B, we will let it be the number of levels of B that are paired with each level of A. In the example of Tables 8-1 and 8-2, J equals 2 (instead of 6, as it would if it were the total number of levels of B).

Second, we have to define a new kind of summation index. The average over the levels of B, within a level of A, can only be taken over the levels of B that are paired with that level of A. We will use the symbol $j(i)$ to indicate that level j of Factor B is paired with level i of Factor A. The average over the levels of B, for a given level of A, is then

$$\overline{X}_{i..} = (1/J) \, \Sigma_{j(i)} \, \overline{X}_{ij.} = \mu + \alpha_i + (1/J) \, \Sigma_{j(i)} \, (\beta_j + \alpha\beta_{ij}) + \bar{\epsilon}_{i..}. \tag{8-2}$$

The term $\bar{\epsilon}_{i..}$ in this case is of course an average only over the levels of Factor B that are paired with level A_i. In order to be able to estimate the α_i and test for the A main effect, it is necessary to make the assumption

$$\Sigma_{j(i)} \, (\beta_j + \alpha\beta_{ij}) = 0. \tag{8-3}$$

Notice that this assumption is different from the assumptions made in the completely crossed two-way design. In the completely crossed design the terms for both the B main effect and the interaction summed to zero, but the summation in each case was over all of the levels of B. In the nested design the terms need not sum to zero separately, but their combined sum must be zero *when taken only over those levels of B that are paired with A_i.*

Notice also that this assumption, like the similar assumptions with the completely crossed designs, is one that is *imposed* on the model; that is, by defining the terms of the model appropriately, we can *require* that Equation 8-3 hold. The assumption does not depend on the nature of the data themselves—we will have more to say on this later. Assuming Equation 8-3, Equation 8-2 becomes

$$\overline{X}_{i..} = \mu + \alpha_i + \bar{\epsilon}_{i..}. \tag{8-4}$$

Finally, we can average over the levels of A to obtain

$$\overline{X}_{...} = \mu + \bar{\epsilon}_{...}. \tag{8-5}$$

The α_i cancel out of this equation by the assumption (identical to that in the completely crossed models) that they sum to zero. The $\overline{X}_{ij.}$ for the example presented above are shown in the row labeled B in Table 8-1 and in the last column of Table 8-2; the $\overline{X}_{i..}$ are shown in the row labeled A in Table 8-1 and in the last row of Table 8-2.

From Equations 8-4 and 8-5 we can easily derive the estimates

$$\hat{\mu} = \overline{X}_{...}$$
$$\hat{\alpha}_i = \overline{X}_{i..} - \overline{X}_{...}.$$

Estimating β_j and $\alpha\beta_{ij}$ is a different problem. Ordinarily, we would use the row means in Table 8-2 to estimate the β_j and then use the means of the individual cells to estimate the $\alpha\beta_{ij}$. However, Table 8-2 clearly shows that the means of the rows and the means of the individual cells are identical. Each row mean is the mean of the single cell in that row. Consequently, we cannot obtain separate estimates of β_j and $\alpha\beta_{ij}$; the best we can do is estimate their sum for the cells containing data. If we let

$$\beta_{j(i)} = \beta_j + \alpha\beta_{ij}, \tag{8-6}$$

then

$$\hat{\beta}_{j(i)} = \overline{X}_{ij.} - \overline{X}_{i..},$$

for each cell (AB_{ij}) containing observations. The parentheses around the subscript i merely indicate that B is nested in A rather than vice versa.

To make it clear that the β_j and $\alpha\beta_{ij}$ cannot be estimated separately, we also modify the model by inserting Equation 8-6 into Equation 8-1 to obtain

$$X_{ijk} = \mu + \alpha_i + \beta_{j(i)} + \epsilon_{ijk}. \tag{8-7}$$

Equation 8-3 now becomes

$$\Sigma_{j(i)} \beta_{j(i)} = 0.$$

The estimates of μ, α_i, and $\beta_{j(i)}$, for the example of Table 8-1, are shown in Table 8-3. There appear to be large differences among schools, and the size of the difference between teachers in the same school appears to vary from school to school. Note, however, that this design can only compare

Table 8-3. Estimates of effects for data in Table 8-1.

| | A_1 | | A_2 | | A_3 | |
	B_1	B_2	B_3	B_4	B_5	B_6
$\hat{\beta}_{j(i)}$	−1.17	1.17	2.17	−2.17	4.17	−4.17
$\hat{\alpha}_i$		4.67		−0.67		−4.00
$\hat{\mu}$				13.87		

teachers in the same school. Because the $\beta_{j(i)}$ sum to zero within each school, we cannot directly compare $\hat{\beta}_{j(i)}$ in different schools. We cannot say, for example, that teacher B_5 is a better teacher than teacher B_2, even though

$$\hat{\beta}_{5(3)} = 4.17, \text{ and } \hat{\beta}_{2(1)} = 1.17.$$

In fact, the *average* score for teacher B_2, shown in Table 8-1, is *higher* than the average for teacher B_5. Of course, the difference in *average* scores may be due to the difference between schools rather than teachers, but we cannot tell from this experiment. The next section discusses this problem in more detail.

Sums of Squares and Degrees of Freedom

Statistical tests follow the same pattern as in the crossed model. A statistical test exists for each effect in Equation 8-7. The sum of squares for each effect is found by squaring and summing the estimates of the effect and multiplying that sum by the number of observations averaged over to obtain each estimate (a simpler computational method will be given in a later section).

$$SS_a = nJ \, \Sigma_i \hat{\alpha}_i^2$$
$$SS_{b(a)} = n \, \Sigma_{j(i)} \, \hat{\beta}_{j(i)}^2$$
$$SS_t = \Sigma_i \, \Sigma_{j(i)} \, \Sigma_k \, (X_{ijk} - \overline{X}...)^2$$
$$SS_m = N\overline{X}...^2$$

The grand mean, as usual, has one degree of freedom, and the A main effect has $(I - 1)$ degrees of freedom. The degrees of freedom for error are equal to $(n - 1)$ times the total number of cells containing observations, or $(n - 1)IJ = N - IJ$. The number of degrees of freedom remaining to test the B main effect are $N - 1 - (I - 1) - (N - IJ) = I(J - 1)$. This can be seen also by noting that because the J estimates of $\beta_{j(i)}$ within a given level of A must sum to zero, only $(J - 1)$ of them are free to vary. Since there are $(J - 1)$ "free" estimates in each of the I levels of A, the total degrees of freedom are $I(J - 1)$. We might also note that in a completely crossed $I \times J$ design the sum of the degrees of freedom for the B main effect and the AB interaction is $(J - 1) + (I - 1)(J - 1) = I(J - 1)$. We saw earlier that $\beta_{j(i)}$ was the sum of the B main effect and the AB interaction; correspondingly, the test of the B main effect is really a combined test on both the B main effect and the AB interaction. The B main effect and the AB interaction are *confounded*. This confounding is one price that must be paid for the practical advantages of a nested design.

There is a sense in which a similar problem occurs with respect to the A main effect, although the problem in this case is in the interpretation rather than in the statistical test itself. One is tempted, in a model such as this one, to attribute the A main effect to factors independent of the levels of B. Thus, in the schools example a significant effect due to schools is likely to be interpreted as due to differences in physical facilities, administration, and other factors that are independent of the teaching abilities of the teachers themselves. The temptation to interpret the A main effect in that way rests on the arguement that the test on differences between teachers is independent of the test on differences between schools. In fact, however, the test on the B main effect is only a test on differences between teachers in the *same school*. Differences between teachers in *different* schools are part of the A main effect, and the observed differences between schools could be due entirely to the fact that some schools have better teachers than others. In the design described here the differences between schools could even be due to the fact that some schools have smarter children attending them. Both of these possibilities could have been ruled out by randomly assigning both the teachers and the students to schools as well as classrooms; however, in the described design this could not be done.

The preceding considerations could help to clarify the nature of the confounding between the B main effect and the AB interaction as well. The B main effect in this design is the extent to which each teacher is an inherently better (or worse) teacher than a colleague in the *same* school. The AB interaction is the extent to which the teaching abilities of the teachers are differentially affected by the school in which they teach. Thus, teacher B_5 appears to be much better than B_6 in the example of Tables 8-1 and 8-3. It may be, however, that B_6 is having difficulties with the administration of school A_3 and that her teaching would greatly improve if she were transferred to school A_1. The teaching of B_5, on the other hand, might deteriorate if she were transferred to A_1. Thus, it is entirely possible that in school A_1, B_6 might turn out to be a better teacher than B_5. The existence of a significant B main effect does not tell us whether the cell mean of one level of B is generally higher than another or whether it is higher only when paired with that particular level of A. It tells us only that there are differences among cells that lie within the same level of A. It tells us that the variance of the true means of the J cells *within each level of* A is not zero.

The sums of squares and mean squares for the data in Table 8-1 are given in Table 8-4.

Table 8-4. Sum of squares and mean squares from Table 8-1.

	SS	df	MS
m	3444.50	1	3444.5
a	229.33	2	114.7
b(a)	140.50	3	46.83
e	58.67	12	4.889

Models

We saw that three distinct models were possible in the completely crossed two-way analysis of variance. The models corresponded to zero, one, or two random factors. In the nested design there are four different models in all. The four models are listed in Table 8-5 with the expected values of the mean squares for each model (the derivation of the expected mean squares is straight-forward but tedious). In Table 8-5, $\tau_a^2 = \sigma_a^2$ when A is random and, similarly, $\tau_{b(a)}^2 = \sigma_{b(a)}^2$ when B is random. The corresponding equations when the factors are fixed are

$$\tau_a^2 = \left(\frac{I}{I-1} \right) \sigma_a^2$$

$$\tau_{b(a)}^2 = \left(\frac{J}{J-1} \right) \sigma_{b(a)}^2.$$

Whether Factor B is considered to be random or fixed depends on the way that the levels of B within each level of A are chosen. Accordingly, whether B is random or fixed does not depend on the nature of Factor A.

Table 8-5. Expected mean squares in a nested two-way design.

Model	Effect	$E(MS)$
A and B fixed	μ	$\sigma_e^2 + nIJ\mu^2$
	α_i	$\sigma_e^2 + nJ\tau_a^2$
	$\beta_{j(i)}$	$\sigma_e^2 + n\tau_{b(a)}^2$
A fixed, B random	μ	$\sigma_e^2 + n\tau_{b(a)}^2 + nIJ\mu^2$
	α_i	$\sigma_e^2 + n\tau_{b(a)}^2 + nJ\tau_a^2$
	$\beta_{j(i)}$	$\sigma_e^2 + n\tau_{b(a)}^2$
A random, B fixed	μ	$\sigma_e^2 + nJ\tau_a^2 + nIJ\mu^2$
	α_i	$\sigma_e^2 + nJ\tau_a^2$
	$\beta_{j(i)}$	$\sigma_e^2 + n\tau_{b(a)}^2$
A and B random	μ	$\sigma_e^2 + n\tau_{b(a)}^2 + nJ\tau_a^2 + nIJ\mu^2$
	α_i	$\sigma_e^2 + n\tau_{b(a)}^2 + nJ\tau_a^2$
	$\beta_{j(i)}$	$\sigma_e^2 + n\tau_{b(a)}^2$

Suppose, for example, that the schools were chosen randomly from a large population of potential schools, each school having only two algebra teachers. Factor B would then be fixed, even though Factor A was random. Conversely, if each school had a large number of teachers from which two had been chosen randomly, Factor B would be random no matter how the schools themselves had been chosen.

The information presented in Table 8-5, together with that given above, can be used in exactly the same way as outlined in previous chapters to estimate the variances of the effects and the proportions of variance accounted for, as well as to make the usual statistical tests.

Assumptions

As with the previous models, the required assumptions can be most easily given in terms of the effects, but they can be most easily *applied* in terms of the data themselves. Accordingly, we will give the assumptions here in terms of the data.

First, for all models the assumption must be made that the observations within each cell are randomly sampled from a normal distribution with constant variance. Once again, the assumptions of normality and constant variance are generally unimportant, but the assumption of independent random sampling is crucial.

When Factor A is random, it is likewise necessary that the levels of A be an independent random sample. The same assumption must be made of B if it is random. Technically, the assumption must also be made that any random factor has a normal distribution, although the tests are relatively robust with respect to this assumption.

HIGHER-WAY MODELS

We can generalize directly to higher-way models. The number and complexity of the possible models increases considerably as a result, but any model, no matter how complex, can be broken down into relatively simple components to which specific rules apply in the calculation of mean squares and expected mean squares.

The complexity arises from the fact that in the general m-way design each factor may be either fixed or random, and each pair of factors may be either crossed or nested. Whether a given pair of factors is crossed or nested can be determined most easily by a consideration of the two-way table containing the results of summing or averaging over all other factors. If every cell in this table contains a sum or average, the factors are crossed. If each level of one factor is paired with only one level of the other, the former is nested in the latter. The only exception to this rule is the case in which two factors are both nested in a third. An example of this is given below, and the nature of the exception is explained there. The following are a number of examples to help clarify the points made here.

A × B × C(B) Design. Suppose subjects are required to learn four lists of nonsense syllables under three different sets of instructions. Each subject learns all four lists of nonsense syllables but is given only one set of instructions. The levels of Factor A are the lists of nonsense syllables, the levels of Factor B are the sets of instructions, and the levels of Factor C are the subjects. Since each level of Factor C (each individual subject) is paired with only one level of Factor B, C is nested in B. However, since all four lists of nonsense syllables are learned under each condition, A and B are crossed. Similarly, since every subject learns all four lists of nonsense syllables, A and C are crossed. A set of hypothetical data from this design are shown in Table 8-6. The data from this table will be used later to illustrate the general rules for analyzing data from nested designs. This type of design is sometimes referred to as a two-way design with repeated measures on Factor A. It may also be referred to as a "partially replicated" design. Still another term that may be used is "split-plot" design, the name being derived from its use in agricultural experiments. Frequently, each of a number of different plots of ground will be divided up into groups with a different level of Factor B applied to the plots in each group. Each

Table 8-6.   A × B × C(B) Design.

	B_1			B_2			B_3		
	C_1	C_2	C_3	C_4	C_5	C_6	C_7	C_8	C_9
A_1	8	20	14	21	23	26	15	6	9
A_2	15	24	20	18	21	29	12	11	18
A_3	12	16	19	17	17	26	13	10	9
A_4	17	20	20	28	31	30	18	12	23

	A_1	A_2	A_3	A_4
B_1	42 / 14.0	59 / 19.7	47 / 15.7	57 / 19.0
B_2	70 / 23.3	68 / 22.7	60 / 20.0	89 / 29.7
B_3	30 / 10.0	41 / 13.7	32 / 10.7	53 / 17.7

	A_1	A_2	A_3	A_4
	142 / 15.8	168 / 18.7	139 / 15.4	199 / 22.1

	B_1	B_2	B_3
	205 / 17.1	287 / 23.9	156 / 13.00

	B_1			B_2			B_3			$\overline{X}$
	C_1	C_2	C_3	C_4	C_5	C_6	C_7	C_8	C_9	
	52 / 13.0	80 / 20.0	73 / 18.2	84 / 21.0	92 / 23.0	111 / 27.8	58 / 14.5	39 / 9.8	59 / 14.8	648 / 18.0

plot is then divided ("split") into a number of subplots, with a different level of Factor A applied to each subplot. For this design the plots are regarded as Factor C, and, since each plot is given only one level of B, C is nested in B. All of these names appear to imply that designs of this type are somehow basically different from other designs that might have nested factors. As a result, special terms such as "within subjects" and "between subjects" variance have to be invented. Both the terminology and the analysis are simplified (as will be seen later) by merely treating it as a three-way design with a nested factor.

A × B × C(AB). Two different textbooks are to be tested. Three teachers are recruited to do the teaching, and each takes four classes. Two of the classes are taught using textbook A_1 and the other two using textbook A_2. Thus, each teacher teaches two classes from each textbook for a total of twelve classes in all. The levels of Factor A are the textbooks, the levels of Factor B are the teachers, and the levels of Factor C are the classes. Factors A and B are crossed, since each teacher uses both textbooks. Factor C, however, is nested in both A and B, since each class has only one teacher and one textbook. If the classes were randomly assigned both to teachers and textbooks, they would be regarded as a random factor. This is true even though there may have been a total of only twelve potential classes to begin with. (Scheffé (1959) has argued that in such a case a "permutation" model is actually most appropriate; a good approximation to the permutation model can be had, however, by regarding Factor C as random.)

A × B(A) × C(A). Two kinds of intelligence tests are to be tested on first-grade children. Both kinds are paper-and-pencil tests designed for group administration; one kind requires a minimal amount of reading ability while the other does not. Two forms of each type of test are to be administered, making a total of four tests. The tests are given to twenty randomly selected children in each of six randomly selected first-grade classes. All the children in a single class receive the same *type* of test, but half (10) of the children in the class receive one *form* of the test and the other half receive the other form. In this design the levels of Factor A are the two types of tests and the levels of Factor B are the two tests of each type. Since each form can be of only one type, Factor B is nested in Factor A. The levels of Factor C are the classes, and, since each class is given only one type of test, C is nested in A. The scores of the individual children represent the observations within the cells ($n = 10$).

This design is diagrammed in Table 8-7, where an X in a cell means that that cell contains data; empty cells contain no data. Notice that neither B nor C is nested in the other but, at the same time, the levels of B are not paired with all levels of C e.g., B_1 is not paired with C_4. Each level of B within level A_1, however, is paired with every level of C that is also within level A_1, i.e., B_1 and B_2 are paired with C_1, C_2, and C_3, and each level of B within A_2 is paired with every level of C that is also within A_2. In this case, B and C are considered to be crossed. In general, when two factors

Table 8-7. A × B(A) × C(A) design.

	A₁						A₂					
	C_1	C_2	C_3	C_4	C_5	C_6	C_1	C_2	C_3	C_4	C_5	C_6
B_1	X	X	X									
B_2	X	X	X									
B_3										X	X	X
B_4										X	X	X

are both nested in a third, they are considered to be crossed if they are crossed within each level of the third factor. Hypothetical data from this design are shown in Table 8-8. They will be used later as another illustration of the procedures for finding mean squares and expected mean squares.

A × B(A) × C(AB). An experiment is conducted to assess the effects of personality variables on teacher performance. Three teachers with high scores and three with low scores on an "extroversion scale" each teach two different sections of an introductory psychology class (making twelve sections in all). The two levels of Factor A are the high and low extroversion scores, the levels of Factor B are the teachers, and the levels of Factor C are the classes. Since no teacher will have both a high and a low score, Factor B is nested in Factor A. Also, since each class is taught by only one teacher, Factor C is nested in Factor B. This also means, however, that no class is taught by teachers having both high and low scores. That is, C is also nested in A. In general, whenever C is nested in B and B is nested in A, C is also nested in A.

The Model Equations

As we have seen, writing out the model equation for a given design can help to specify the effects that are involved. In nested designs the model equation specifies clearly which main effects and interactions are confounded. It is indispensable for finding expected mean squares. For the A x B x C(B) design described above, for example, the equation is

$$X_{ijkl} = \mu + \alpha_i + \beta_j + \gamma_{k(j)} + \alpha\beta_{ij} + \alpha\gamma_{ik(j)} + \epsilon_{ijkl}. \qquad (8\text{-}8)$$

This equation tells us that six different effects (in addition to the grand mean) can be estimated and tested, that the C main effect is confounded with the BC interaction, and that the AC interaction is confounded with the ABC interaction. The latter information is obtained by comparing the Greek letter indicating an effect with its subscripts. Thus, γ indicates the C main effect, but the subscripts k and j are those of the BC interaction. Similarly, $\alpha\gamma$ indicates the AC interaction, but the subscripts i, k, and j are those of the ABC interaction.

Table 8-8. Hypothetical cell totals (upper values) and means (lower values) from A × B(A) × C(A) design, assuming 10 observations per cell.

	A_1					A_2		
	C_1	C_2	C_3			C_4	C_5	C_6
B_1	147	199	197	B_3		144	108	126
	14.7	19.9	19.7			14.4	10.8	12.6
B_2	192	156	149	B_4		156	101	154
	19.2	15.6	14.9			15.6	10.1	15.4

	A_1		A_2				A_1			A_2		
	B_1	B_2	B_3	B_4			C_1	C_2	C_3	C_4	C_5	C_6
	543	497	378	411			339	355	346	300	209	280
	18.1	16.6	12.6	13.7			17.0	17.8	17.3	15.0	10.4	14.0

A_1	A_2		$\overline{X}$
1040	789		1829
17.3	13.2		15.2

Once written, the model equation appears complicated, but is really very easy to write. We will illustrate the procedure by deriving Equation 8-8. The first step is to write the equation that would obtain if all of the factors were completely crossed. For the three-way design this is

$$X_{ijkl} = \mu + \alpha_i + \beta_j + \gamma_k + \alpha\beta_{ij} + \alpha\gamma_{ik} + \beta\gamma_{jk} + \alpha\beta\gamma_{ijk} + \epsilon_{ijkl}.$$

Next, the subscripts of all effects (except error) involving nested factors are modified by adding the subscripts of the factors in which they are nested in parentheses. In the A × B × C(B) design, for example, any effect containing k as a subscript is modified by appending (j). The equation then becomes

$$X_{ijkl} = \mu + \alpha_i + \beta_j + \gamma_{k(j)} + \alpha\beta_{ij} + \alpha\gamma_{ik(j)} + \beta\gamma_{jk(j)} + \alpha\beta\gamma_{ijk(j)} + \epsilon_{ijkl}.$$

Finally, each term in the equation that contains the same subscript both inside and outside parentheses is struck out. In our example there are two such terms: $\beta\gamma_{jk(j)}$ and $\alpha\beta\gamma_{ijk(j)}$. Eliminating these gives Equation 8-8, the final form of the model equation. In the same way, the equations for the other designs described can be shown to be

$$A \times B \times C(AB): X_{ijkl} = \mu + \alpha_i + \beta_j + \gamma_{k(ij)} + \alpha\beta_{ij} + \epsilon_{ijkl}$$
$$A \times B(A) \times C(A): X_{ijkl} = \mu + \alpha_i + \beta_{j(i)} + \gamma_{k(i)} + \beta\gamma_{jk(i)} + \epsilon_{ijkl}$$
$$A \times B(A) \times C(AB): X_{ijkl} = \mu + \alpha_i + \beta_{j(i)} + \gamma_{k(ij)} + \epsilon_{ijkl}.$$

The reader should derive each of these equations, using the rules given above, to test his understanding of the procedures. Note that no letter ever appears more than once in any subscript. For example, in the term $\beta\gamma_{jk(i)}$ in the $A \times B(A) \times C(A)$ design the i appears only once even though j and k both represent factors nested in A.

Sums of Squares and Degrees of Freedom

The rules given in Chapter 7 for finding sums of squares and degrees of freedom in completely crossed designs can be extended to nested and partially nested designs with little difficulty. The rules are extended by making the following changes.

The number of different tables obtained by summing over factors in a nested design is smaller than in a completely crossed design. This is because any attempt to sum over the levels of a factor before first summing over the levels of all factors nested in it would result in "summing" over only one cell – the result would of course be simply the original table. This can be seen most clearly by referring again to Table 8-2, in which any attempt to sum over Factor A before summing over Factor B would simply reproduce the six original cell totals. Tables 8-7 and 8-8 also show that in the $A \times B(A) \times C(A)$ design one cannot sum over the levels of A without first summing over the levels of B and C. The total number of different tables obtained will be exactly equal to the number of terms in the model equation, and each table will correspond to one of the terms. For the $A \times B \times C(B)$

Table 8-9. Summary table and expected mean squares for $A \times B \times C(B)$ design in Table 8-6.

	RS	SS	T	df	MS	F	α
m		11664.00	1	1	11664.0		
a	11923.33	259.33	4	3	86.44	10.14	$<.001$
b	12394.17	730.17	3	2	365.1	8.24	.02
c(b)	12660.00	265.83	9	6	44.30		
ab	12720.67	67.17	12	6	11.20	1.31	—
ac(b)	140.00	123.50	36	18	8.528		

	E(MS)
m	$\sigma_e^2 + nI\tau_{c(b)}^2 + N\mu^2$
a	$\sigma_e^2 + n\tau_{ac(b)}^2 + nJK\tau_a^2$
b	$\sigma_e^2 + nI\tau_{c(b)}^2 + nIK\tau_b^2$
c(b)	$\sigma_e^2 + nI\tau_{c(b)}^2$
ab	$\sigma_e^2 + n\tau_{ac(b)}^2 + nK\tau_{ab}^2$
ac(b)	$\sigma_e^2 + n\tau_{ac(b)}^2$

design, for example, Equation 8-8 shows that six tables, one for the grand mean and one for each of the five effects, will be obtained. If each cell contains more than one observation, there will be a seventh table (containing the raw data) corresponding to the error term.

An RS is obtained for each term in the model equation, i.e., for each table, by the usual procedure of squaring each sum in the table, summing the squares, and dividing the result by the number of observations over which the original sums were taken. The number of observations over which they were taken will always be the total number of observations (N) divided by the number of cells in the table. For the design of Table 8-6, for example, the divisor for $RS_{c(b)}$ is $36/9 = 4$. The complete equation is

$$RS_{c(b)} = (52^2 + 80^2 + 73^2 + 84^2 + 92^2 + 111^2 + 58^2 + 39^2 + 59^2)/4.$$

The raw sums of squares are then placed in a summary table with the same form as Tables 8-9 and 8-10, which show the summaries for the data in Tables 8-6 and 8-8 respectively. It is important, in the summary table, to indicate nesting with appropriate parentheses as was done in Tables 8-9 and 8-10. For calculating sums of squares it is essential that factors in which an effect is nested be listed with the effect. The parentheses themselves, while they are ignored in the actual calculations, make it clear that the effect is nested.

The sums of squares are then calculated from the raw sums of squares in exactly the same way as with the completely crossed design. That is, from each RS we subtract SS_m and the SS for each effect that does not involve any factors other than those in the effect whose mean square is being found. For this purpose, any parentheses are ignored. In Table 8-9,

Table 8-10. Summary table and expected mean squares for A × B(A) × C(A) design in Table 8-8 (assuming $n = 10$, with $RS_T = 36,000$).

	RS	SS	T	df	MS	F	α
m		27,877.00	1	1	27877.00	474	.01
a	28,402.02	525.02	2	1	525.02	8.93	.05
b(a)	28,455.43	53.41	4	2	26.70	0.35	
c(a)	28,637.15	235.13	6	4	58.78	0.91	
bc(a)	28,994.90	304.34	12	4	76.08	1.17	
w		7005.10		108	64.86		
t	36,000.00	8123.00	120	119			

	$E(MS)$
m	$\sigma_e^2 + nJ\tau_{c(a)}^2 + N\mu^2$
a	$\sigma_e^2 + nJ\tau_{c(a)}^2 + nJK\tau_a^2$
b(a)	$\sigma_e^2 + n\tau_{bc(a)}^2 + nK\tau_{b(a)}^2$
c(a)	$\sigma_e^2 + nJ\tau_{c(a)}^2$
bc(a)	$\sigma_e^2 + n\tau_{bc(a)}^2$

for example, the C main effect, because it is nested in B, is treated as the BC interaction. Its SS is thus

$$SS_{c(b)} = RS_{c(b)} - SS_m - SS_b,$$

subtracting, in this case, the SSs for all effects that do not involve Factor A. Similarly, in Table 8-10 the BC interaction is treated as the ABC interaction for purposes of calculating $SS_{bc(a)}$:

$$SS_{bc(a)} = RS_{bc(a)} - SS_m - SS_a - SS_{b(a)} - SS_{c(a)}.$$

If there is only one observation per cell in the design, SS_w cannot be found. If there is more than one observation, however, SS_w is found as follows. In every design there will be one effect that has the subscripts (either in or out of parentheses) of all of the factors in the design. The value of SS_w is found by subtracting the RS for this highest interaction from RS_t. The value of SS_t can then be found as usual by subtracting SS_m from RS_t. The sums of squares for the data in Tables 8-6 and 8-8 are shown in the columns labeled SS of Tables 8-9 and 8-10.

Degrees of freedom are found by basically the same procedure as that given in Chapter 7. First, in the column headed "T" are placed the number of cells in each table. Then, the degrees of freedom are found by subtracting from each number the degrees of freedom for each effect (including SS_m) that was subtracted from RS to find the sum of squares for the effect in question. In Table 8-9, for example,

$$df_{c(b)} = T_{c(b)} - 1 - df_b.$$

(Note, again, that $DF_m = 1$.) In Table 8-10

$$df_{bc(a)} = T_{bc(a)} - 1 - df_a - df_{b(a)} - df_{c(a)}.$$

Mean squares are then found by dividing each sum of squares by its degrees of freedom.

Expected Mean Squares

The rest of the information necessary for estimating and testing effects is found, as usual, from the expected values of the mean squares. The rules for finding expected mean squares in nested designs can be obtained from some relatively simple modifications of the rules given in Chapter 7 for completely crossed designs. We begin, as in Chapter 7, by listing all of the terms that can enter into the expected mean squares. As in Chapter 7, the appropriate terms are found from the model equation; there is a term for each effect, and each term is a τ^2 multiplied by the divisor used in finding the RS for the associated effect. For the data in Table 8-6 the terms are

$$N\mu^2, \ nJK\tau_a^2, \ nIK\tau_b^2, \ nI\tau_{c(b)}^2, \ nK\tau_{ab}^2, \ n\tau_{ac(b)}^2.$$

(Remember that for this purpose the number of levels of a nested factor is regarded as the number of levels over which one must sum in summing over that factor. Factor C in Table 8-6 has nine different levels, but one sums over only three levels of C to obtain each cell of the A $\times$ B table, so that $K = 3$.) For the data in Table 8-8 the terms are

$$N\mu^2, \quad nJK\tau_a^2, \quad nK\tau_{b(a)}^2, \quad nJ\tau_{c(a)}^2, \quad n\tau_{bc(a)}^2.$$

In Chapter 7 the terms included in the expected mean squares for a given effect had to meet two criteria: they had to have subscripts of all of the factors involved in the effect whose expected mean square was being considered, and they could have no other fixed factors as subscripts. For nested designs the first criterion remains the same; each term entering into an expected mean square must contain subscripts of all of the factors involved in the effect whose expected mean square is being considered. It does not matter whether the subscripts are inside or outside of the parentheses, so long as they are present in the subscript of the term. For the data in Table 8-9, for example, $\tau_{c(b)}^2$ is part of the expected mean square of the B main effect even though the b subscript is found inside the parentheses. Similarly, every term entering into $E(MS_{c(b)})$ must have both b and c as subscripts.

The second criterion in Chapter 7 is modified slightly for nested designs. The subscript of each term in an expected mean square must contain no other symbols for fixed factors *outside of the parentheses*. Fixed factors unrelated to the effect whose expected mean square is found may appear inside but not outside the parentheses. This point is best illustrated by finding $E(MS_a)$ for the A $\times$ B $\times$ C(AB) design with Factor C random and Factors A and B fixed. The appropriate model for this design is

$$X_{ijkl} = \mu + \alpha_i + \beta_j + \alpha\beta_{ij} + \gamma_{k(ij)} + \epsilon_{ijkl}.$$

The terms entering into the expected mean squares are, accordingly,

$$N\mu^2. \quad nJK\tau_a^2, \quad nIK\tau_b^2, \quad nK\tau_{ab}^2, \quad n\tau_{c(ab)}^2.$$

In finding the expected mean square of the A main effect, the terms $N\mu^2$ and $nIK\tau_b^2$ are eliminated because they do not contain an a in their subscripts. The term $nK\tau_{ab}^2$ is also eliminated because it contains b (a fixed factor) in its subscript. The term $n\tau_{c(ab)}^2$, however, is retained; the b in its subscript does not eliminate it because, unlike τ_{ab}^2, the b appears inside the parentheses. (The presence of a c in the subscript does not disqualify the term because c refers to a *random* rather than a fixed factor.) The expected mean square for the A main effect is thus

$$E(MS_a) = \sigma_e^2 + n\tau_{c(ab)}^2 + nJK\tau_a^2.$$

In summary, the two qualifications that must be met for a term to be included in an expected mean square are (1) that it include all of the subscripts (including any in parentheses) of the effect whose expected mean square is being found, and (2) that it include no *other* subscripts of *fixed* factors *outside* the parentheses. Additional examples of expected mean squares are found in Tables 8-5, 8-9, and 8-10. These tables can be used most profitably by first finding for yourself the expected mean squares and then checking the results against those in the tables.

F Ratios

F ratios are formed in the usual way. To test for a given effect, another effect must be found whose expected mean square would be exactly equal to the expected mean square of the effect tested if the null hypothesis were true. If no mean square exists, a quasi-F must be used.

The following rules are usually not useful for *finding* the correct F ratios, although they can serve as a check on the F ratios after the ratios have been formulated. If these rules are violated by any of the F ratios, there is an error somewhere in the original calculations.

(1) The subscript of the denominator mean square (if it is not MS_w) must contain symbols (inside or outside of the parentheses) for all of the factors in the numerator effect.

(2) The subscript of the denominator mean square (if it is not MS_w) must contain the subscript of at least one *random* factor, *outside of the parentheses*, that is not also in the subscript of the numerator effect.

(3) The subscript of the denominator mean square must contain no symbols for fixed factors, *outside of the parentheses*, that are not *also* in the numerator effect.

(4) An effect must be tested with a quasi-F *if and only if* it is crossed with *at least two random* factors. Otherwise, it can always be tested with an ordinary F ratio.

Variance Estimates

The τ^2 for each effect is found in exactly the same way as with completely crossed designs. It is found most easily through the F ratio. The τ^2 for an effect can be estimated by first subtracting the denominator from the numerator of the F ratio used to test for that effect. The resulting difference is then divided by the multiplier of the τ^2 for that effect in the equation of expected mean squares. Earlier (p. 180) it was stated that the multiplier of τ^2 was the divisor used in finding the associated RS.

As an example, we will derive the equation for the B main effect in the $A \times B \times C(B)$ design of Table 8-6. According to Table 8-9, $MS_{c(b)}$ is the appropriate denominator for testing the B main effect, and

$$E(MS_b) = \sigma_e^2 + nI\tau_{c(b)}^2 + nIK\tau_b^2,$$

$$E(MS_{c(b)}) = \sigma_e^2 + nI\tau_{c(b)}^2,$$

$$E(MS_b - MS_{c(b)}) = E(MS_b) - E(MS_{c(b)}) = nIK\tau_b^2,$$

$$\hat{\tau}_b^2 = \frac{MS_b - MS_{c(b)}}{nIK}.$$

In studies with only one observation per cell, σ_e^2 is conventionally assumed to be zero for purposes of estimating the τ^2. Thus, in the same $A \times B \times C(B)$ design, if $n = 1$ so that σ_e^2 cannot be estimated,

$$\hat{\tau}_{c(b)}^2 = MS_{c(b)}/nI = MS_{c(b)}/I,$$
$$\hat{\tau}_{ac(b)}^2 = MS_{ac(b)}/n = MS_{ac(b)}.$$

To find the variance (σ^2) for any effect, the τ^2 for that effect must be multiplied by as many constants as there are *fixed factors outside the parentheses* in the subscript of the τ^2 (subscripts in parentheses are treated as if they were not there). The appropriate multiplier for each factor is the number of levels of that factor divided into the number of levels minus one (compare with pp. 127–129, 134–135, 154, 171–173). For this purpose the number of levels of each nested factor is defined as the number of levels over which one must sum when summing over that factor. In Table 8-6, for example, Factor C is considered to have only three levels instead of nine, and in Table 8-8 Factor C is considered to have three levels instead of six. Table 8-11 illustrates these principles for the data in Tables 8-6 and 8-8, and Table 8-12 shows the estimates for the same data.

The method of estimating ω^2 for each effect is the same as on pp. 134–135 and 154; the estimated variances are totaled to obtain an estimate of σ_t^2, and this is divided into each variance estimate to obtain $\hat{\omega}^2$.

Assumptions and Box's Correction

The necessary assumptions are even more difficult to spell out in detail for multifactor nested designs than for multifactor crossed designs. Some assumptions are basically the same as in previous chapters. The observations within a cell are assumed to be normally distributed with constant variance, but, as usual, this assumption is not critical if each cell contains the same number of observations. In addition, the cell means for random factors must be normally distributed and covariances must be assumed to be equal.

It is not necessary, however, to spell out the assumptions in detail in order to tell when Box's correction (or, alternatively, the more exact T^2 test) may be appropriate. For each effect to be tested, compare the subscripts *outside of the parentheses* on both the numerator and the denominator mean square. If the subscripts outside of the parentheses on the denominator mean square include all of those on the numerator mean square, Box's correction is applicable. If not, the test is robust and Box's correction need not be used. In Table 8-9, for example, the denominator for testing the A main effect is $MS_{ac(b)}$. The letter a is the only subscript on MS_a and that letter also appears outside the parentheses in the subscript of $MS_{ac(b)}$. Hence, for the test Box's correction may be appropriate. However, the appropriate denominator for the B main effect is $MS_{c(b)}$. The numerator MS_b contains only the subscript b; $MS_{c(b)}$ also contains b, but the b in this case is within the parentheses. Therefore, Box's correction does not apply.

In general, a factor may be thought of as *directly* involved in an effect if its symbol appears as a subscript *outside of the parentheses* on the mean square for that effect. It is *indirectly* involved in the effect if its symbol appears inside the parentheses. With these terms, the principle above can be restated as follows. Box's correction is applicable whenever all of the factors *directly* involved in the numerator mean square are also *directly* involved in the denominator mean square.

Table 8-11. Formulas relating σ^2 to τ^2 in Tables 8-9 and 8-10.

For Table 8-9:

$$\sigma_a^2 = \left(\frac{I-1}{I}\right)\tau_a^2 = (3/4)\tau_a^2$$

$$\sigma_b^2 = \left(\frac{J-1}{J}\right)\tau_b^2 = (2/3)\tau_b^2$$

$$\sigma_{c(b)}^2 = \tau_{c(b)}^2$$

$$\sigma_{ab}^2 = \left(\frac{I-1}{I}\right)\left(\frac{J-1}{J}\right)\tau_{ab}^2 = (1/2)\tau_{ab}^2$$

$$\sigma_{ac(b)}^2 = \left(\frac{I-1}{I}\right)\tau_{ac(b)}^2 = (3/4)\,\tau_{ac(b)}^2$$

For Table 8-10:

$$\sigma_a^2 = \left(\frac{I-1}{I}\right)\tau_a^2 = (1/2)\tau_a^2$$

$$\sigma_{b(a)}^2 = \left(\frac{J-1}{J}\right)\tau_{b(a)}^2 = (1/2)\tau_{b(a)}^2$$

$$\sigma_{c(a)}^2 = \tau_{c(a)}^2$$

$$\sigma_{bc(a)}^2 = \left(\frac{J-1}{J}\right)\tau_{bc(a)}^2 = (1/2)\tau_{bc(a)}^2$$

Table 8-12. Variance estimates for data in Tables 8-9 and 8-10.

For Table 8-9:

	$\hat{\tau}^2$	$\hat{\sigma}^2$	$\hat{\omega}^2$
a	8.66	6.50	.15
b	26.73	17.82	.42
c(b)	11.08	11.08	.26
ab	0.89	0.44	.01
ac(b)	8.53	6.40	.15
t		42.24	

For Table 8-10:

	$\hat{\tau}^2$	$\hat{\sigma}^2$	$\hat{\omega}^2$
a	7.77	3.88	.06
b(a)	(−1.65)	(−0.82)	(−.01)
c(a)	(−0.30)	(−0.30)	.00
bc(a)	1.12	0.56	.01
w	64.86	64.86	.95
t		68.18	

Two special cases should also be considered: effects tested by quasi-F ratios, and effects tested by a pooled mean square. Both of these are, in at least some sense, only approximate tests anyway, and the appropriateness of Box's correction in general has not been determined for them. For these cases it is best just to make the usual tests but to interpret the results with some caution.

As stated previously, Box's correction is always applied by treating the obtained F ratio as if it had one and v_2/v_1 degrees of freedom, where v_1 are the numerator degrees of freedom and v_2 the denominator degrees of freedom without Box's correction. In the example of Table 8-9 the F ratio for the A main effect would have 1 and 6 degrees of freedom ($\alpha = .02$) with Box's correction, where it would have 3 and 18 degrees of freedom ($\alpha < .001$) without Box's correction. It should be remembered, however, that Box's correction, employed in this way, only sets a lower bound on the significance of the result; in practice, the test should be made both with and without Box's correction, and the two results should be treated as upper and lower bounds on the actual significance level.

Power

The rules for calculating the power of a test are also easily adapted to nested designs. In calculating power, an effect is considered to be fixed if all of the letters outside parentheses in its subscript represent fixed factors; if one or more of the letters correspond to random factors, the effect is considered to be random.

The F distribution for a fixed effect when the null hypothesis is false is noncentral F, with parameters ϕ^2 and ϕ'^2 defined in the same way as in Chapter 6 (p. 135); the F distribution for a random effect is a constant times an ordinary central F (as described on p. 135).

Comparisons

The basic principles for testing comparisons are the same as those on pp. 156–164; there are some differences, however. First, the exact method described on pp. 163–164 is more complicated and more difficult to apply — it will not be discussed here.

Second, there are times when a planned comparison may involve more than one effect and still be testable by the standard method. We will illustrate by testing $\mu_{43.}$ against the average of the other three values in the same row of the AB table in Table 8-6. Our null hypothesis is

$$H_0: 3\mu_{43.} - \mu_{13.} - \mu_{23.} - \mu_{33.} = 0. \tag{8-9}$$

Just as in Chapter 7, our first step is to put the coefficients into the cells of the AB table and then do an analysis of variance on them to determine the effects involved. This analysis, shown in Table 8-13, reveals that the A main effect and AB interaction are involved. Table 8-9 shows, however, that these effects are both tested against $MS_{ac(b)}$. Consequently, $MS_{ac(b)}$ is the appropriate denominator for testing Equation 8-9:

Table 8-13. Analysis of coefficients in Equation 8-9.

	Coefficients						Analysis		
	A_1	A_2	A_3	A_4	Σ		rs	ss	ss^*
B_1	0	0	0	0	0	m	0	0	0
						a	4	4	1/3
B_2	0	0	0	0	0	b	0	0	0
						ab	12	8	2/3
B_3	−1	−1	−1	3	0				
Σ	−1	−1	−1	3	0				

$$F_{(1,18)} = \frac{(C')^2}{MS_{c(ab)}} = \frac{87.42}{8.528} = 10.2,$$

$$\alpha = .005.$$

Note that the analysis of variance on the coefficients must take exactly the same form as on the original data. That is, any nesting relationships in the original data must be preserved in the analysis on the coefficients.

A third difference is that care is sometimes needed when estimating degrees of freedom for a quasi-F. We will illustrate this with the contrast

$$H_0: \mu_{13.} - \mu_{41.} = 0 \tag{8-10}$$

on the data in Table 8-6. From the analysis of these coefficients (Table 8-14), both main effects and the interaction are involved. Comparing these results with the analysis in Table 8-9, we see that the appropriate F ratio is

$$F^* = \frac{(C')^2}{(1/4)MS_{ac(b)} + (1/3)MS_{c(b)} + (5/12)MS_{ac(b)}}.$$

Our estimate of v_2 may be in error, however, unless we combine the two terms involving $MS_{ac(b)}$ to obtain

$$F^* = \frac{(C')^2}{(1/4 + 5/12)MS_{ac(b)} + (1/3)MS_{c(b)}}$$

$$= \frac{(C')^2}{(2/3)MS_{ac(b)} + (1/3)MS_{c(b)}}$$

$$= \frac{121.5}{20.45} = 5.94.$$

The estimated degrees of freedom are then

$$\hat{v}_2 = \frac{(20.45)^2}{\left(\frac{4}{9}\right)\dfrac{(8.528)^2}{18} + \left(\frac{1}{9}\right)\dfrac{(44.30)^2}{6}} \cong 11$$

Table 8-14. Analysis of coefficients in Equation 8-10.

| | Coefficients | | | | | | Analysis | | |
	A_1	A_2	A_3	A_4	Σ		rs	ss	ss^*
B_1	0	0	0	−1	−1	m	0	0	0
						a	2/3	2/3	1/3
B_2	0	0	0	0	0	b	1/2	1/2	1/4
						ab	2	5/6	5/12
B_3	1	0	0	0	1				
Σ	1	0	0	−1	0				

In general, before using Equation 7-4 to estimate degrees of freedom, all terms involving the same mean square in the F ratio must be combined.

Treating Error as a Random Effect

By definition, a random effect is a random component of the differences among scores, attributable to a specific factor whose levels have been randomly sampled. By this definition it is possible to treat random error itself as a random effect. Like the levels of a random factor, the random errors in a design are always due to a specific random-selection procedure. In a typical design, where only one observation is obtained from each subject, differences between scores of subjects in the same cell are treated as due to error. If we modify the design so that two observations are taken from each subject, subjects will be treated as levels of a random factor in the experiment (whenever more than one observation is taken on each subject, subjects should be treated as levels of a random factor). Thus, differences between subjects are treated as random error in the former design and as the subjects main effect in the second.

Since random error is always the result of a specific random sampling of the levels of some variable (e.g., subjects), it is equally logical to treat it either as random error or as the levels of a random factor. The advantage of the latter concept of error is parsimony. Instead of having a design with factors plus error, one has only a design with different factors. Most commonly-used computer programs take advantage of this parsimony to simplify their operation. To use such programs effectively, the user must understand how random error can be treated as a specific factor in the experiment. Moreover, an understanding of this point can lead to a better understanding of the general nature of error in all of the designs that can occur in the analysis of variance.

To see just exactly how random error can be treated as a factor, consider, for example, a two-way fixed-effects design with n subjects per cell. Ordinarily, differences between subjects in the same cell are treated as due to random error. If we had obtained more than one observation from each subject, however, we would have had to treat subjects as a random effect. If, for example, each subject had been given two trials, we would have had an $A \times B \times C(AB) \times D$ design, where D would have been the trials factor (two levels) and C(AB) the subjects factor (n levels per cell).

Since there is only one trial per subject in the simple two-way design, it would seem logical to think of it as the same kind of design, but with the trials factor omitted, leaving an $A \times B \times C(AB)$ design. Factor C (subjects) is now treated as a random factor, nested in both A and B, whereas in the simple two-way design it would be treated as random error. The following two model equations should then be equivalent:

$$X_{ijk} = \mu + \alpha_i + \beta_j + \alpha\beta_{ij} + \epsilon_{ijk}.$$
$$X_{ijk} = \mu + \alpha_i + \beta_j + \alpha\beta_{ij} + \gamma_{k(ij)}.$$

The first equation is the standard equation for the two-way design; the second is the equation for the same design with error (subjects) treated as a random factor. Notice that no additional error term appears in the second equation; Factor C is the "error."

In general, the error term can be treated equivalently either as random error or as a random effect that is nested in all of the other effects (both random and fixed) in a design. MS_w is then equivalent to the mean square for that random effect.

The point can be clarified more fully if we review the rules for finding sums of squares and expected mean squares (pp. 177–181), treating error as a factor. We will again use the $A \times B \times C(AB)$ design, with C random, as our example. (You are encouraged to refer again to Chapter 6 before reading this explanation.) Since the subscript of $SS_{c(ab)}$ contains symbols for all of the factors in the design, the RS table for the C(AB) main effect must be the table of original observations. Consequently, $RS_{c(ab)} = RS_t$. In addition, $SS_{c(ab)}$ is found (following the method described on pp. 177–180) by subtracting out the sums of squares for all of the other effects in the design (including SS_m). $SS_{c(ab)}$ is thus a residual sum of squares, consisting of the remainder when the sums of squares of all other effects have been subtracted from SS_a. By definition, however, SS_w is the sum of squares that remains after subtracting from SS_t the sums of squares for all systematic effects. Hence, $SS_{c(ab)}$ in the $A \times B \times C(AB)$ design is equivalent to SS_w in the simple two-way design. Similarly, $MS_{c(ab)}$ is equivalent to MS_w, and both have the same number of degrees of freedom. The simple two-way design can be treated either as a two-way design with error or as an $A \times B \times C(AB)$ design with only one observation per cell. The resulting mean squares are identical.

Identical expected mean squares are also obtained. We can see this by reviewing the rules given on pp. 180–181. A small change must be made in these rules; since we are treating error as a random factor, we do not add a separate error term to the expected mean squares. For the simple two-

Table 8-15. Expected mean squares for two-way design regarded as an $A \times B \times C(AB)$ design with Factor C random.

	$E(MS)$
m	$T_{c(ab)}{}^2 + N\mu^2$
a	$T_{c(ab)}{}^2 + nJKT_a{}^2$
b	$T_{c(ab)}{}^2 + nIKT_b{}^2$
ab	$T_{c(ab)}{}^2 + nKT_{ab}{}^2$
c(ab)	$T_{c(ab)}{}^2$

way design with error treated as a random factor, the terms that can enter into the expected mean square are

$$N\mu^2, \ nJ\tau_a{}^2, \ nI\tau_b{}^2, \ n\tau_{ab}{}^2, \ \tau_{c(ab)}{}^2.$$

In these terms we regard n as the number of levels of Factor C. Since $\tau_{c(ab)}{}^2$ contains both of the other factors in its subscript, and the only symbol outside of the parentheses is for a random factor, $\tau_{c(ab)}{}^2$ enters into the expected mean squares of all of the other effects. The complete list of expected mean squares (calculated according to the rules on pp. 180–181) is given in Table 8-15. If we substitute $\sigma_e{}^2$ for $\tau_{c(ab)}{}^2$, these equations are identical to those in Chapter 5, pp. 102–104.

These results can be generalized to any design. In any design the error can be treated as a random factor nested in all other factors in the design. When it is so treated, MS_w is the mean square and $\sigma_e{}^2$ the τ^2 for the main effect of the error random factor. (There cannot be any interactions involving the error factor because it is nested in all other factors.) This point is important primarily for the use of computer analysis of variance programs; it should also help to clarify the nature of the error terms in all of the designs we have discussed.

Exercises

1. *Life* magazine once reported that the California climate was so healthy the women in California were frequently stronger than the average man in the East. A certain researcher decided to test whether or not California males were stronger than Michigan males. He chose three California towns and three Michigan towns at random, and he chose two male high school students at random from each town. He then tested each of these twelve subjects for the number of push-ups they could do (Test 1) and the number of times they could chin themselves (Test 2). He obtained the following data:

	California				Michigan	
Town:	A	B	C	D	E	F
Test:	1 2	1 2	1 2	1 2	1 2	1 2

Subjects	11 7	13 5	10 6	20 15	11 13	18 16
	12 9	12 3	10 3	19 13	13 14	12 16

(a) Analyze the data, testing for main effects and interactions. On the basis of these data, what can you conclude about differences between California and Michigan males?

(b) Can you draw any conclusions about the main effect of subjects or the interaction between subjects and tests? What assumptions would such conclusions require?

2. Suppose that in Problem 5, Chapter 3, each odd answer was the answer of the husband, and each even answer was that of the wife of the person giving the preceding answer. Specify the correct design for this experiment and reanalyze the data.

3. Experimenters were puzzled by an apparent difference in average running times of rats under apparently identical conditions in different laboratories. One investigator speculated that rats run in southern climates were healthier and thus faster. He chose four laboratories at random from each of three ranges of latitude: 30–35°N, 35–40°N, and 40–45°N. Ten randomly selected rats were run in each laboratory. The running times for each rat in each laboratory were as follows:

	Lat. 30–35°				Lat. 35–40°				Lat. 40–45°			
Laboratory	1	2	3	4	5	6	7	8	9	10	11	12
	11	5	3	2	13	11	9	10	13	15	12	12
	15	1	2	4	1	1	7	6	14	21	16	14
	2	3	19	12	5	4	15	4	16	18	6	6
	9	1	10	10	17	5	1	5	17	16	19	18
	2	13	20	5	16	6	22	14	12	10	20	12
	9	2	7	11	10	13	14	5	23	14	13	27
	2	4	8	4	15	3	11	8	8	18	12	16
	13	14	11	14	5	9	18	3	13	18	4	17
	8	10	12	4	2	6	16	11	20	17	16	8
	10	20	18	11	4	1	10	12	20	7	15	13

(a) Analyze these data and interpret the results, estimating the proportion of variance accounted for by each effect.

(b) If the only test of interest is the one comparing different latitudes, how might the data be combined to simplify the analysis?

4. A study was made of the effect of ward conditions in mental hospitals on patients with mild catatonia. The experimenter suspected that subtle differences among conditions in different wards of the same hospital could affect the patients. He randomly chose eight patients in each of two hospitals, placing each patient first in one ward and then in another, with each patient spending three months in each ward. (Each patient spent time in two different wards, with half the patients in one ward first and the other half in the other ward first.) The data (scores on an objective test of mental health) were as follows:

| | Hosp. 1 | | Hosp. 2 | |
	Ward 1A	Ward 1B	Ward 2A	Ward 2B
Patient in	3	2	3	3
Ward A	3	3	2	3
first	1	2	0	1
	3	1	3	2
Patient in	1	0	3	1
Ward B	2	2	1	2
first	1	0	2	0
	0	0	1	1

Analyze these data; do not assume that wards in different hospitals are comparable.

5. Twelve randomly selected college students who had spent two years in the army were compared with twelve who had not. Six students from each group were in engineering school; the other six were in liberal arts. Each student was evaluated in two courses. In each course he received a score of one if his grade was B or better; he received a score of zero otherwise. For the engineering students the courses were freshman math and freshman chemistry; for the liberal arts students the courses were freshman English and American history. The data were as follows:

| | | Army | | | | | | No Army | | | | | |
		S_1	S_2	S_3	S_4	S_5	S_6	S_7	S_8	S_9	S_{10}	S_{11}	S_{12}
L.A.	Eng	0	0	0	0	0	0	1	0	1	1	1	1
	Hist	0	0	1	1	0	0	0	0	0	0	1	1
Engin.	Math	1	1	1	1	1	0	0	0	0	0	0	0
	Chem	0	0	0	0	0	0	0	1	0	0	0	0

Analyze these data:
(a) specify the model exactly;
(b) find the F ratio for testable effects;
(c) comment on the meaning for this experiment of the significant effects (use .05 level);

(d) find the proportion of variance accounted for by each effect;

(e) test the following null hypotheses by planned comparisons, using the modified method whenever no existing mean square can be used, and using both methods whenever an existing mean square can be used:

H_0 (1): There is no difference between army veterans and others in performance in English.

H_0 (2): There is no difference between liberal arts and engineering students in performance in English.

H_0 (3): There is no difference between army veterans and others in overall performance in engineering courses.

(f) How might you extract more information from the data if the engineering courses were math and English instead of math and chemistry?

6. An experimenter studied the effects of early environment on the problem-solving ability of rats raised in normal, "enriched," and "restricted" environments. Each rat, of course, was raised in only one environment. The experimenter measured the number of trials required to learn a visual discrimination task (Task *a*) and a tactile discrimination task (Task *b*). Half the rats were given Task *a* first and the other half were given Task *b* first. The data, in number of trials required to learn, are as follows:

		Task *a* first			Task *b* first	
		Task *a*	Task *b*		Task *a*	Task *b*
	Subject			Subject		
Enriched	E1	45	62	E5	50	68
	E2	20	39	E6	43	63
	E3	20	39	E7	66	84
	E4	33	55	E8	59	82
Normal	N1	28	44	N5	50	71
	N2	53	69	N6	43	59
	N3	36	53	N7	42	66
	N4	52	75	N8	50	62
Restricted	R1	69	86	R5	63	78
	R2	59	77	R6	48	60
	R3	81	107	R7	61	82
	R4	83	101	R8	37	63

(a) Specify each factor, telling how many levels it has and whether it is fixed or random; determine which factors are crossed and which are nested, and write the model equation.

(b) Analyze the data.

7. A psychologist studied the effects of two tranquilizers on extinction of appetitive and aversive responses. Twelve rats were divided into three

groups of four each. Group 1 received Drug 1, Group 2 received Drug 2, and Group 3 received no drug. All rats were trained to push a lever for food in the presence of a green light and to push a lever to avoid shock in the presence of a red light. Each animal was then given five extinction trials to each cue, but the order was counterbalanced. Half the animals in each group were extinguished to the red light first, and the other half were extinguished to the green light first. On each trial the score was 1 if the animal responded, and 0 if it did not. The data are as follows:

		SA first[a]				F first[b]			
		Animal 1		Animal 2		Animal 3		Animal 4	
		SA	F	SA	F	SA	F	SA	F
	Trial								
	1	1	1	0	1	0	1	1	1
	2	0	1	1	1	0	1	0	1
Group 1	3	0	0	0	0	0	0	0	1
(Drug 1)	4	0	1	0	0	0	0	0	0
	5	0	0	0	0	0	0	0	0
	1	0	1	0	1	0	1	0	1
	2	0	1	1	1	0	1	0	0
Group 2	3	0	0	0	1	0	0	0	0
(Drug 2)	4	0	1	0	0	0	0	0	0
	5	0	0	0	0	0	0	0	0
	1	1	1	1	1	1	1	1	1
	2	1	1	1	1	1	1	1	0
Group 3	3	1	0	1	0	0	0	1	0
(No drug)	4	1	0	1	1	0	0	0	0
	5	0	0	1	0	1	0	0	0

[a]SA: Bar press for shock avoidance
SA first: Extinction of shock avoidance measured first
[b]F: Bar press for food
F first: Extinction of bar press for food measured first

(a) Do an analysis of variance on the above data, and estimate proportions of variance accounted for.

(b) Discuss the assumptions required for the analysis of variance and their importance for the conclusions reached in this design.

(c) Test the following null hypotheses, testing each in the way you feel is most appropriate:

H_0 (1): There is no difference between the two drugs, averaging over all other factors.

H_0 (2): On the average, the two drugs do not differ from the control group.

H_0 (3): Averaging over the two drug groups, there is no difference between the first and last trials of extinction to shock avoidance.

(d) Suppose you suspected that there were a main effect and interactions involving the five trials but you had no interest in such effects. Could you combine the data in a way that would simplify the analysis?

(e) Suppose you were interested in trial effects but you were not interested in effects involving the order in which the two responses are extinguished. Assuming that such effects exist but you are not interested in them, can you combine the data in a way that would simplify the analysis?

8. In an $A \times B \times C(A)$ design, with C random, how should each of the following planned comparisons be tested when (a) all assumptions are met, (b) assumptions of equality of variances and covariances are not met?

(1) $\mu_{1..} = \mu_{2..}$

(2) $\mu_{11.} = \mu_{12.}$

(3) $\mu_{11.} = \mu_{21.}$

(In answering, consider each hypothesis by itself; i.e., do not worry about whether or not the tests are orthogonal.)

9. For each of the following designs (1) tell what type of design it is (e.g., $A \times B \times C(A)$); (2) write the model equation; (3) write the expected mean square for each effect; and (4) tell what denominator mean square would be used to test each effect.

(a) Thirty randomly selected subjects were tested in an experiment on the effects of sunglasses on visual acuity. The tests were conducted using all possible combinations of two types of sunglasses and five different lighting conditions. In addition, each subject was tested under three different levels of initial light adaptation. A single visual acuity score was obtained from each subject under each level of adaptation. (The total N is 90.)

(b) A memory experiment used three different lists of 10 words each. List N contained 10 nouns, List V contained 10 verbs, and List A contained 10 adjectives. The words in each list were selected independently of (and hence were not directly comparable to) the words in each of the other two lists (but the words were not selected randomly). In addition, the words in each list were presented in each of four different random orders, so that there was a grand total of 12 lists in all. Each subject received one of the three types of list in one of the four random orders. After hearing the entire list, he was asked to recall as many of the words as he could. If he recalled a word correctly, he received a score of 1 for that word; if not, he received a score of 0. Each subject's scores, then, were a series of 10 ones and zeroes. A total of 144 randomly selected subjects were run – 12 under each of the 12 combinations of type of list and word order. ($N = 1440$).

(c) The design is identical to that in part b except that the words within each list were selected randomly.

(d) An experimenter studying the effects of early environment on the problem-solving ability of rats raised three groups in "normal," "enriched," and "restricted" environments, respectively. He then measured the number of trials required to learn (a) a visual discrimination task, and (b) a tactile discrimination task. Every rat was given both tasks, but half of the rats from each group were given task (a) first, and the other half were given task (b) first. The experimenter expects the order in which the tasks are given to have an effect. The total number of rats is 60.

(e) Six randomly selected first-grade teachers were each assigned 12 randomly selected pupils to teach. Each teacher taught her pupils two subjects, reading and arithmetic. At the end of the year all six classes were given the same standard tests in these two subjects. The data were the test scores of the individual pupils in each of the two subjects.

(f) The design in Problem 4, Chapter 3.

(g) The design in Problem 5, Chapter 3.

(h) The design in Problem 7, Chapter 3.

9

Other Incomplete Designs

We found previously that if Factor B was nested in Factor A, it was impossible to separate the B main effect from the AB interaction. The best we could do was to estimate the sum of these two effects. When two different effects cannot be separated, they are confounded. Whenever the factors in a design are not completely crossed, some effects will be confounded. The factors in a particular experimental design, however, need not necessarily be either completely crossed or completely nested. A number of designs are possible in which the factors are neither crossed nor nested but are partly nested. An example of such a design is the 4×6 design shown in Table 9-1; in this design, it is obvious that the factors are not completely crossed, yet neither is nested within the other. The analysis of most such partially nested designs is extremely difficult, but for certain designs of this kind the analysis is relatively simple.

Partially nested designs like the one shown in Table 9-1 offer a number of advantages, the most obvious of which is a savings in experimental labor. In a completely crossed 4×6 design there would be twenty-four cells, requiring that at least twenty-four observations be taken. In the design of Table 9-1, however, observations are taken in only twelve of the twenty-four cells. For designs with three or more factors the saving may be even greater. Ordinarily, for example, a $4 \times 4 \times 4$ design requires at least sixty-four observations, but a $4 \times 4 \times 4$ latin square (discussed later in this chapter) requires only 16.

In addition, partially nested designs sometimes make it possible to improve precision by adding additional factors to the design. Consider, for example, an experimenter who wants to study the effects of four different drugs on the behavior of mice. He has a number of mice on which to experiment, but he is concerned that the heredities of the mice might affect their susceptibilities to the drugs. If he does not control for them, differences in heredity might increase his error mean square and reduce his chances of obtaining significant results. One way to control for heredity would be to separate the mice according to the litters they came from and choose four animals from each litter to be given the drugs. The litter from which an animal came could then be treated as a second factor in the experiment. However, although the experimenter has six different litters to work with, some of them contain only two animals each. Fortunately, the experimenter may be able to solve his problem by using the design of Table 9-1, with Factor A being drugs and Factor B being litters; as can be seen in Table 9-1, this design requires no more than two animals from each litter.

Balancing the advantages, however, there are a number of disadvantages — in fact, the disadvantages usually outweigh the advantages. To begin with, the number and variety of partially nested designs is limited unless one is willing to accept sizeable difficulties in both calculation and interpretation. If the experimenter of our example had had say only five litters, no computationally simple design would have been available to him. In general, partially nested designs are possible only when the number of levels of the factors meet certain rather rigid requirements.

The most serious problem with partially nested designs is that in most of them all interactions are confounded not only with each other but with main effects as well. Moreover, they are confounded in complicated ways. In a completely nested design the B main effect and the AB interaction are confounded in a simple additive way — the B(A) main effect is simply the sum of the B main effect and the AB interaction. In a partially nested design the problem is much more complicated. In a latin square design, for example, the A main effect is confounded with part of the BC interaction, and the confounding is not simple; because only part of the BC interaction is confounded with the main effect, a large BC interaction may either increase or decrease SS_a. Consequently, a large BC interaction may make MS_a nonsignificant, even though there is a sizeable A main effect. Accordingly, the results of most partially nested experiments are almost completely uninterpretable unless all interactions can be assumed to be negligible. The experimenter of our example must assume that a general susceptibility to drugs is inherited but that mice do not inherit differential susceptibilities to specific drugs. If such an assumption is not plausible, the design of Table 9-1 is a poor choice. A common error in the literature is the use of a partially nested design, such as a latin square, in an experiment in which interactions are almost certainly present. Unless interactions are specifically discussed, all of the designs considered in the following sections require the assumption that there are no nonnegligible interactions.

Table 9-1. Partially nested (balanced incomplete blocks) 4×6 design. The Xs mean the cells contain observations; the empty cells contain no observations.

	B_1	B_2	B_3	B_4	B_5	B_6
A_1	X		X		X	
A_2	X			X		X
A_3		X	X			X
A_4		X		X	X	

TWO-FACTOR DESIGNS – INCOMPLETE BLOCKS

An incomplete blocks design is a two-factor design in which each level of Factor A is paired with only a small number of levels of Factor B. In the drugs experiment described above and in Table 9-1, for example, each drug is administered to animals from only three of the six litters. Letting the drugs be the levels of Factor A and the litters be the levels of Factor B, we have an incomplete blocks design in which each level of Factor A is paired with only three of the six levels of Factor B.

Suppose there are I levels of Factor A, and each is paired with g levels of B (in Table 9-1, $g = 3$). Then the total number of cells containing data must be gI. We can reason about Factor B in the same way; each level of B is paired with the same number (h) of different levels of A. Hence, the total number of cells containing observations must be hJ. We can therefore conclude that $gI = hJ$. In the design of Table 9-1, $I = 4$, $J = 6$, $g = 3$, $h = 2$, and $gI = hJ = 12$.

Frequently, incomplete blocks designs are diagrammed in a table with J columns representing the J levels of B; each column contains the h levels of A that are paired with that level of B. This way of diagramming the design is illustrated in Table 9-2 for the design of Table 9-1.

For the general case, the data from an incomplete blocks design are difficult both to analyze and interpret because it is difficult to separate the

Table 9-2. The design of Table 9-1 diagrammed in the conventional form.

B_1	B_2	B_3	B_4	B_5	B_6
A_1	A_3	A_1	A_2	A_1	A_2
A_2	A_4	A_3	A_4	A_4	A_3

effects of A from the effects of B. For one special case, the *balanced incomplete blocks design,* the task is much simpler.

The term "incomplete blocks" derives from the fact that the incomplete blocks design is generally thought of as an incomplete version of the *randomized blocks design.* Basically, the randomized blocks design is one in which the experimenter is primarily interested in only one factor, but performs the experiment as a crossed two-factor design in order to reduce MS_w. In the design described in Tables 9-1 and 9-2, for example, the experimenter's primary interest is in the drugs, but if he had used a one-way design, differences between litters would have contributed to error and increased MS_w. In this case the introduction of Factor B is equivalent to dividing the observations within the levels of Factor A into distinct groups, or *blocks,* and treating the blocks as levels of a second factor. The entities referred to as blocks in other statistics texts are thus identical to what we here call the *levels* of B.

Balanced Incomplete Blocks

If an incomplete blocks design is represented as in Table 9-2, it is possible to count the number of times that any two levels of Factor A, A_i and A_i', appear together in the same column, i.e., within the same level of B. We can let this number be λ_{ii}'. An incomplete blocks design is said to be *balanced* if $\lambda_{ii}' = \lambda$ (a constant) for all i, $i' \neq i$. That is, a balanced incomplete blocks design is one in which each level of A appears together with each other level of A, within the same level of B, the same number of times. In the design in Table 9-2 levels A_1 and A_2 appear together under B_1 but under no other level of B; similarly, every other pair of levels of A_i occurs together under one and only one level, B_j. The design of Tables 9-1 and 9-2 is thus a balanced incomplete blocks design with $\lambda = 1$.

Another example of a balanced incomplete blocks design is shown in Table 9-3. This table shows a 7×7 design with $g = h = 4$. In Table 9-3 each level of A appears with each other level in two different columns.

Table 9-3. Example of a 7×7 balanced incomplete blocks design.

B_1	B_2	B_3	B_4	B_5	B_6	B_7
A_1	A_1	A_1	A_2	A_2	A_1	A_3
A_2	A_3	A_2	A_3	A_4	A_4	A_5
A_3	A_4	A_5	A_4	A_5	A_6	A_6
A_6	A_5	A_7	A_7	A_6	A_7	A_7

Levels A_2 and A_6, for example, appear together in columns B_1 and B_5 but in no other columns. It is thus a balanced incomplete blocks design with $\lambda = 2$.

The balanced incomplete blocks design places additional restrictions on the number of levels of each factor. With h levels of A paired with each level of B, level A_i will appear with $h - 1$ other levels of A within each level of B with which it is paired. If A_i is paired with g different levels of B, then the number of times A_i appears with all other levels of A combined is $g(h - 1)$. On the other hand, since each level of A is paired with each other level of A λ times, the number of times A_i is paired with all of the other $I - 1$ levels of A combined must be $\lambda(I - 1)$. Therefore, it must be true that $g(h - 1) = \lambda(I - 1)$. These requirements are highly restrictive, since g, h, I, J, and λ must all be integers. However, an additional restriction can also be shown to hold: $I < J$. Not even all of these restrictions are sufficient to guarantee that values of I, J, g, h, and λ that satisfy them will correspond to a possible balanced incomplete blocks design. A detailed list of possible designs can be found in Cochran and Cox (1957, p. 469).

The following list summarizes the definitions of symbols introduced so far.

I = number of levels of Factor A

J = number of levels of Factor B

g = number of levels of B with which each A_i is paired

h = number of levels of A with which each B_j is paired

λ = number of times any two levels of A are paired with the same level of B

We shall now develop the theory for estimating the A and B main effects and testing for their significance for either one observation or more than one observation in each cell in which observations are made.

Randomization Condition

All of the theory to follow is heavily dependent on the particular pairings of AB_{ij} that are chosen. To avoid serious biases it is essential that these pairings be chosen randomly. The pairings often can be made most readily by choosing a particular design and then randomly assigning the labels B_1, B_2, etc., to the levels of Factor B. An equally good alternative is to randomly assign the labels to the levels of Factor A. Still another possibility is to choose a design and randomly permute the columns when the design is represented like that in Tables 9-2 or 9-3. Whatever method is used, it is imperative that the particular levels of B_j with which a given A_i is paired be chosen randomly.

One Observation per Cell

It is easiest to begin by deriving the formulas for designs in which there is only one observation in each cell; in a later section these results will be generalized to designs with more than one observation per cell. For the design with only one observation per cell, the cell total (t_{ij}) and the cell mean $(\overline{X}_{ij})$ are both just the value X_{ij} of the observation in cell AB_{ij}. The values of $t_{i\cdot}$ and $t_{\cdot j}$ are then found by summing over the B and A factors, respectively, as in previous designs. As with nested designs, summation can be only over those cells containing observations. Since neither factor is completely nested in the other, however, both factors can be summed over. This summation is illustrated, for our hypothetical data, in Table 9-4. For example,

$$t_{1\cdot} = 18 + 19 + 23 = 60,$$
$$t_{\cdot 1} = 18 + 20 = 38,$$
$$T = 38 + 14 + 34 + \cdots + 52 = 60 + 73 + 42 + 29 = 204.$$

Subsequent formulas can be simplified if we define $Q_{ij} = 1$ if A_i appears in column B_j, otherwise $Q_{ij} = 0$. That is, $Q_{ij} = 1$ if and only if cell AB_{ij} contains an observation. With this definition, the formulas for $t_{i\cdot}$, $t_{\cdot j}$, and T can be written

$$t_{i\cdot} = \Sigma_j \, Q_{ij} t_{ij} \tag{9-1}$$

$$t_{\cdot j} = \Sigma_i \, Q_{ij} t_{ij} \tag{9-2}$$

$$T = \Sigma_i \, \Sigma_j \, Q_{ij} t_{ij}. \tag{9-3}$$

The model for the balanced incomplete blocks design, assuming there are no interactions, is

$$X_{ij} = \mu + \alpha_i + \beta_j + \epsilon_{ij}.$$

By simple algebra,

$$E(T) = E \, \Sigma_i \, \Sigma_j \, Q_{ij} X_{ij} = \Sigma_i \, \Sigma_j \, Q_{ij} E(\mu + \alpha_i + \beta_j + \epsilon_{ijk})$$
$$= N \mu + \Sigma_i \, \alpha_i \, \Sigma_j Q_{ij} + \Sigma_j \, \beta_j \, \Sigma_i \, Q_{ij}.$$

Table 9-4. Sample data for a 4 × 6 balanced incomplete blocks design with $t = 2$ and $r = 3$.

	B_1	B_2	B_3	B_4	B_5	B_6	t_i	t_i^*	$\hat{\alpha}_i$
A_1	18		19		23		60	52	4.0
A_2	20			21		32	73	62	5.5
A_3		7	15			20	42	50	−4.0
A_4		7		13	9		29	40	−5.5
$t_{\cdot j}$	38	14	34	34	32	52	$T = 204$		
$\overline{X}_{\cdot j}$	19	7	17	17	16	26	$\overline{X}_{\cdot\cdot} = 17$		
$\hat{\beta}_j$	−2.75	−5.25	0.00	0.00	−0.25	8.25			

The last equality is found by distributing both the summations and the expected value over the terms in the parentheses, with N being the total number of observations in the table. Now, a reference to Table 9-1 or Table 9-4 makes it clear that

$$\Sigma_j Q_{ij} = g, \ \Sigma_i Q_{ij} = h.$$

Consequently,

$$E(T) = N\mu + g \ \Sigma_i \ \alpha_i + h \ \Sigma_j \ \beta_j = N\mu.$$

The last equality derives from the usual restriction that the α_i and β_j sum to zero. An unbiased estimate of μ is thus

$$\mu = \overline{X}_{..} = T/N.$$

To obtain unbiased estimates of the α_i, we must first define

$$\overline{X}_{.j} = t_{.k}/h$$

$$t_i^* = \Sigma_j \ Q_{ij}\overline{X}_{.j}. \qquad (9\text{-}4)$$

In other words, $\overline{X}_{.j}$ is the mean of the observations in column B_j, and t_i^* is the sum of the $\overline{X}_{.j}$ of all of the levels of B that are paired with level A_i. It can then be shown, by a derivation too involved to present here, that the least squares estimate of α_i is

$$\alpha_i = (t_{i.} - t_i^*)/(gH),$$

where

$$H = \frac{I(h-1)}{h(I-1)}.$$

The error variance of each estimate $(\hat{\alpha}_i)$ is $\sigma_e^2/(gH)$. If each level of A were replicated g times in a complete design, the error variance would be σ_e^2/g, so that the balanced incomplete blocks design results in an error variance $1/H$ times as large as an equivalent complete design. For this reason, H (which is always smaller than one) is called the *efficiency factor* of the balanced incomplete blocks design. Put differently, $1 - H$ is the proportional loss in accuracy due to pairing only h of the I levels of A with each level of B. In general, the smaller the value of h and the larger the value of I, the less efficient will be the experiment. For the designs in Tables 9-2 and 9-3 the efficiency factors are 0.67 and 0.88, respectively.

The estimates of the β_j are then found by

$$\hat{\beta}_j = \overline{X}_{.j} - \overline{X}_{..} - (\Sigma_i \ Q_{ij}\hat{\alpha}_i)/h$$

These calculations are illustrated in Table 9-4, which shows a set of sample data for the 4×6 experiment of Tables 9-1 and 9-2.

The sum of squares for testing the grand mean is, as usual, $SS_m = T^2/N$, and it has one degree of freedom.

The sum of squares for the A main effect is

$$SS_a = gH \sum_i \hat{\alpha}_i^2 = \sum_i (t_{i.} - t_i^*)^2/(gH)$$

(SS_a has $(I - 1)$ degrees of freedom). The simplest way to find SS_b is first to compute the following:

$$RS_a = \sum_i t_{i.}^2/g \qquad (9\text{-}5)$$

$$RS_b = \sum_j t_{.j}^2/h. \qquad (9\text{-}6)$$

These are analogous to the RS values for completely crossed designs. Then

$$SS_b = RS_b + SS_a - RS_a \qquad (9\text{-}7)$$

(SS_b has $(J - 1)$ degrees of freedom).

The reason for this somewhat unusual formula for SS_b is that in the model we are considering the A and B main effects are not independent of each other. A large A main effect may either increase or decrease the sum of squares for the B main effect, depending on the pairings of the levels of A with the levels of B, and vice versa. This dependence is the reason for the requirement that the levels of A and B should be paired randomly (p. 200). Since both the amount and direction of dependence depend on the pairings, the random pairing makes the degree of dependence itself a random variable that will tend to "average out" over the long run. The dependence can thus be ignored for practical purposes — there is no way to tell how large it is in a single experiment anyway — so long as the requirement of random pairings has been met. None of the above discussion applies to SS_m, which is independent of both the A and the B main effects.

The dependence between the tests for the A and B main effects explains the rather unusual formula given above for calculating SS_b. Similarly, the denominator for testing the grand mean and both main effects is also calculated from a somewhat different formula. The denominator in this case is simply the remainder after the effects of SS_a, SS_b, and SS_m have been subtracted from RS_t. Because of the dependence between SS_a and SS_b, however, we cannot simply subtract their sum. The correct formula is

$$SS_{rem} = RS_t - RS_a - SS_b = RS_t - SS_a - RS_b. \qquad (9\text{-}8)$$

(SS_{rem} has $(N - I - J + 1)$ degrees of freedom). These results are summarized in Table 9-5, which also gives expected mean squares. Table 9-6 shows the summary for the data of Table 9-4.

In general, planned comparisons on the B main effect are difficult to make, but planned comparisons on the A main effect are very simple. The quantity

$$\frac{(C')^2}{MS_{rem}} = \frac{gH(\sum_i c_i\hat{\alpha}_i)^2}{MS_{rem}(\sum_i c_i^2)} \qquad (9\text{-}9)$$

is distributed as F with one degree of freedom in the numerator and $(N - I -$

Table 9-5. Computational summary for balanced incomplete blocks design with one observation per cell.

	RS	SS	df	E(MS)
m		T^2/N	1	$\sigma_e^2 + N\mu^2$
a	$(\Sigma_i\, t_{i.}^2)/g$	$gH\,\Sigma_i\,\hat{\alpha}_i^2$	$I - 1$	$\sigma_e^2 + gH\sigma_a^2$
b	$(\Sigma_j\, t_{.j}^2)/h$	$RS_b + SS_a - RS_a$	$J - 1$	$\sigma_e^2 + [(N - I)/(J - 1)]\sigma_b^2$
rem		$RS_t - RS_b - SS_a$ $= RS_t - RS_a - SS_b$	$N - I - J + 1$	σ_e^2
t	$\Sigma_i\,\Sigma_j\,Q_{ij}X_{ij}^2$	$RS_t - SS_m$	$N - 1$	

$J + 1$) degrees of freedom in the denominator. To make post hoc comparisons by the Scheffé method, the F ratio in Equation 9-9 is divided by $(I - 1)$ and treated as though it had $(I - 1)$ degrees of freedom in the numerator.

More Than One Observation per Cell

With more than one observation per cell, the theory is basically the same as for only one observation per cell, but the power of the test is increased considerably because the estimates of the effects are more accurate and more degrees of freedom are available for estimating the error variance. The basic computational difference is that t_{ij} is now the sum of n observations and $N = ngI = nhJ$; otherwise the theory presented above, particularly Equations 9-1 through 9-3, still holds. The row and column means are

$$\overline{X}_{i..} = t_{i.}/(gn)$$
$$\overline{X}_{.j.} = t_{.j}/(hn)$$

and Equation 9-4 becomes

$$t_i^* = n\,\Sigma_j\,Q_{ij}\overline{X}_{.j.} = \Sigma_j\,Q_{ij}t_{.j}/h.$$

Table 9-6. Summary table for data of Table 9-4.

	RS	SS	df	MS	F	α
m		3468.00	1	3468.00		
a	3844.67	185.00	3	61.67	6.85	.08
b	3840.00	180.33	5	36.07	4.01	.15
rem		27.00	3	9.00		
t	4052.00	584.00	11			

Table 9-7. Computational formulas for balanced incomplete blocks design with n observations per cell.

	RS	SS	df	E(MS)
m		T^2/N	1	$\sigma_e^2 + N\mu^2$
a	$(\Sigma_i\, t_{i.}^2)/(ng)$	$ngH\,\Sigma_i\,\hat{\alpha}_i^2$	$I-1$	$\sigma_e^2 + ngH\sigma_a^2$
b	$(\Sigma_j\, t_{.j}^2)/(nh)$	$RS_b + SS_a - RS_a$	$J-1$	$\sigma_e^2 + n[(N-1)/(J-1)]\sigma_b^2$
rem		$RS_t - RS_b - SS_a$ $= RS_t - RS_a - SS_b$	$N-I-J+1$	σ_e^2
t	$\Sigma_i\,\Sigma_j\,Q_{ijk}\,\Sigma\,X_{ijk}^2$	$RS_t - SS_m$	$N-1$	

Table 9-8. Sample analysis of 4×6 balanced incomplete blocks design with $n = 2$.

	B_1	B_2	B_3	B_4	B_5	B_6	$t_{i.}$	t_i^*	$\hat{\alpha}_i$
A_1	19 17		22 16		25 21		120	104	4.0
A_2	20 20			24 18		35 29	146	124	5.5
A_3		6 8	18 12			23 17	84	100	−4.0
A_4		3 11		10 16	11 7		58	80	−5.5
$t_{.j}$	76	28	68	68	64	104	$T = 408$		
$\overline{X}_{.j.}$	19	7	17	17	16	26	$\overline{X}_{...} = 17$		
β_j	−2.75	−5.25	0.00	0.00	−0.25	8.25			

	RS	SS	df	MS	F	α
m		6936.00	1			
a	7689.33	370.00	3	123.33	8.64	.01
b	7680.00	360.67	5	72.13	5.06	.01
rem		214.00	15	14.27		
t	8264.00	1328.00	23			

The estimates of α_i and β_j are

$$\hat{\alpha}_i = (t_{i.} - t_i^*)/(ngH)$$
$$\hat{\beta}_j = \overline{X}_{.j.} - \overline{X}_{...} - (\Sigma_i\, Q_{ij}\hat{\alpha}_i)/h.$$

Table 9-7 summarizes the formulas for finding sums of squares.

When there are more than one observation per cell, SS_{rem}, which is still defined by Equation 9-8 and has $(N - I - J + 1)$ degrees of freedom, is the sum of two different sums of squares. The first is the usual within-cells sum of squares:

$$SS_w = \Sigma_i \Sigma_j Q_{ij} \Sigma_k (X_{ijk} - \overline{X}_{ij\bullet})^2.$$

The second is the remaining between-cells sum of squares after all of the effects have been parcelled out:

$$SS_{rem(bet)} = RS_{ab} - RS_a - SS_b = RS_{ab} - SS_a - RS_b,$$

where

$$RS_{ab} = (\Sigma_i \Sigma_j t_{ij}^2)/n.$$

It can be shown that $SS_{rem} = SS_w + SS_{rem(bet)}$. Table 9-8 shows the calculations for sample data in which the cell means are the same as those in Table 9-4 but in which there are two observations per cell.

The equation analogous to Equation 9-9 for planned and post hoc comparisons is

$$F = \frac{(C')^2}{MS_{rem}} = \frac{ngH(\Sigma_i c_i \hat{\alpha}_i)^2}{MS_{rem}(\Sigma_i c_i^2)}$$

All of the above theory applies whether Factors A and B are fixed or random. Occasionally, however, it is possible to improve the estimates of α_i and thus to improve the test for the A main effect when Factor B is random. The method for doing this was developed by Yates (1940); it is also presented in Scheffé (1959, p. 170). The method is useful only if the number of levels of B is fairly large (about 15 or more) and the B main effect is not very large.

THREE-FACTOR DESIGNS

The balanced incomplete blocks design discussed in the previous section can be extended in a number of ways to the case in which there are three factors. Incomplete three-factor designs are more commonly used than incomplete two-factor designs because many completely-crossed three-factor designs require a prohibitive number of different treatments. However, the requirement in three-factor designs that there be no interactions is much more restrictive than in two-factor designs. In most three-factor designs available all three of the two-factor interactions as well as the three-factor interaction must be assumed to be zero. Thus, the model for these designs is usually

$$X_{ijkl} = \mu + \alpha_i + \beta_j + \gamma_k + \epsilon_{ijkl}.$$

Youden Squares

When a balanced incomplete blocks design is represented as in Table 9-2, it may be possible to balance the positions of the levels of A so that each appears the same number of times in each row of the table. It is impossible to do this with the designs in Tables 9-2 or 9-3, but it has been done with the 4×4 design shown in Table 9-9. In this design each level of A appears once in each row. In general, if each level of A appears q times in each row, the number of columns (levels of B) must be q times the number of levels of A:

$$J = qI. \tag{9-10}$$

(In Table 9-9 $q = 1$, but other examples exist in which q is larger than one.)

It can be proved that any balanced incomplete blocks design that satisfies the requirement of Equation 9-10, with q an integer, can be arranged so that each level of A appears q times in each row of the table. Such a design is called a *Youden square*.

One common reason for arranging a balanced incomplete blocks design in such a way is to balance out the effects of the position of a treatment in a sequence. Suppose the experiment requires tasting four different sucrose solutions, with a number of subjects each tasting three of the four. In such a case, the order in which the solutions are tasted is likely to be important. With Factor A being sucrose solutions and Factor B being subjects, the three levels of Factor A can be given to each subject in the order in which they appear in that subject's column in Table 9-9. In such a design the order in which the levels of A are presented will be balanced out across subjects, so that the estimates of the main effects of A and B, as well as SS_a and SS_b, can be calculated as for the usual balanced incomplete blocks design. In addition, however, the position effects themselves can be estimated and tested for very easily.

To test for position effects, we regard the rows of Table 9-9 as the levels of a third factor, C. Then, if we define $t_{..k}$ as the sum of the observations in row k,

$$t_{..k} = \Sigma_i \Sigma_j Q_{ijk} t_{ijk},$$

where Q_{ijk} is equal to one if treatment AB_{ij} appears in row k and is equal to zero otherwise. In other words, $Q_{ijk} = 1$ if and only if cell ABC_{ijk} contains observations; it is zero otherwise. In summing across a row of the table, each level of B (each column) will appear once in the summation,

Table 9-9. Example of a 4×4 Youden square with $g = h = 3$, $\lambda = 2$.

	B_1	B_2	B_3	B_4
C_1	A_1	A_2	A_3	A_4
C_2	A_2	A_3	A_4	A_1
C_3	A_3	A_4	A_1	A_2

and, because A is balanced with respect to C, each level of A will appear exactly q times. The expected value of $t_{..k}$ is thus easily shown to be

$$E(t_{..k}) = nJ\mu + q\,\Sigma_i\,\alpha_i + \Sigma_j\,\beta_j + nJ\gamma_k$$
$$= nJ\mu + nJ\gamma_k,$$

because the α_i and β_j both sum to zero. The estimate of γ_k is then

$$\hat{\gamma}_k = t_{..k}/(nJ) - \overline{X}_{....},$$
$$= \overline{X}_{..k.} - \overline{X}_{....}$$

and

$$SS_c = nJ\,\Sigma_k\,\hat{\gamma}_k^{\,2},$$

which can be shown to be

$$SS_c = \Sigma_k\,(t_{..k})^2/(nJ) - SS_m,$$

where $SS_m = T^2/N$, as usual. SS_c has $(K - 1) = (h - 1)$ degrees of freedom because the number of levels of C must be equal to h, the number of rows in the table.

The values of SS_m, SS_a, and SS_b are found just as in the ordinary balanced incomplete blocks design because the effects of C are balanced out across the levels of both A and B. SS_{rem} is the SS_{rem} of the balanced incomplete blocks design (cf. Table 9-7 and Eq. 9-8) with SS_c subtracted out:

$$SS_{rem} = RS_t - SS_a - RS_b - SS_c = RS_t - RS_a - SS_b - SS_c.$$

The degrees of freedom of SS_{rem} are those of Table 9-7 minus the degrees of freedom of SS_c, or $(N - I - J - K + 2)$. Table 9-10 summarizes the calculations for the Youden square design, and Table 9-11 gives a numerical example for the 4×4 design of Table 9-9.

It is not necessary that the levels of Factor C be the positions in a sequence of presentations. They may be the levels of any third factor that

Table 9-10. Computational formulas for the Youden square design.

	RS	SS	df	E(MS)
m	T^2/N		1	$\sigma_e^2 + N\mu^2$
a	$(\Sigma_i\,t_{i..}^2)/(ng)$	$ngH\,\Sigma_i\,\hat{\alpha}_i^2$	$I - 1$	$\sigma_e^2 + ngH\sigma_a^2$
b	$(\Sigma_j\,t_{.j.}^2)/(nh)$	$RS_b + SS_a - RS_a$	$J - 1$	$\sigma_e^2 + n[(N - I)/(J - 1)]\sigma_b^2$
c	$(\Sigma_k\,t_{..k}^2)/(nJ)$	$RS_c - SS_m$	$K - 1$	$\sigma_e^2 + [nJ/(J - 1)]\sigma_c^2$
rem		$RS_t - SS_a - RS_b - SS_c$ $= RS_t - RS_a - SS_b - SS_c$	$N - I - J - K + 2$	σ_e^2
t	$\Sigma_i\,\Sigma_j\,\Sigma_k\,X_{ijk}^2$	$RS_t - SS_m$	$N - 1$	

Table 9-11. Numerical example of 4×4 Youden square design with $g = h = 3$, $\lambda = 2$, and $H = 8/9$.

	B_1	B_2	B_3	B_4	$t_{..k}$	$\overline{X}_{..k}$	$\hat{\gamma}_k$
C_1	A_1 15	A_2 7	A_3 14	A_4 8	44	11.0	-1.0
C_2	A_2 17	A_3 6	A_4 8	A_1 7	38	9.5	-2.5
C_3	A_3 22	A_4 14	A_1 11	A_2 15	62	15.5	$+3.5$
$t_{.j.}$	54	27	33	30	$T = 144$		
$\overline{X}_{.j.}$	18	9	11	10	$\overline{X}_{...} = 12$		
$\hat{\beta}_j$	6.00	-5.25	-0.25	-0.50			

	A_1	A_2	A_3	A_4
$t_{i..}$	33	39	42	30
t_i^*	39	37	38	30
$\hat{\alpha}_i$	-2.25	0.75	1.50	0.00

	RS	SS	df	MS	F	α
m		1728	1			
a	1758	21	3	7	1.00	
b	1878	141	3	47	6.71	.08
c	1806	78	2	39	5.57	.09
rem		21	3	7		
t	1998	270	11			

is relevant to the experiment. Once again, however, it is important that levels be paired in a random manner. In addition to pairing the levels of A with the levels of B randomly, as in the balanced incomplete blocks design, it is also necessary to assign the levels of C randomly to the rows of the table. The designs given in Cochran and Cox (pp. 469–482) that meet the requirement $J = qI$, with q an integer, can be easily modified to generate a Youden square.

Latin Squares

The incomplete blocks design can be extended of course to a "complete" blocks design by making t equal to I so that every level of Factor A is paired with every level of Factor B. The result is then simply a completely crossed two-factor design. When such a design is represented in the same way as the incomplete blocks designs in Tables 9-2, 9-3, and 9-9, it may be possible to add a third factor by balancing the positions of the levels of A

Table 9-12. Computational formulas for the generalized latin square design ($I = K$, $J = ql$).

	RS	SS	df	E(MS)
m		T^2/N	1	$\sigma_e^2 + N\mu^2$
a	$(\Sigma_i\, t_{i..}^2)/(nI)$	$RS_a - SS_m$	$I-1$	$\sigma_e^2 + n[I/(I-1)]\sigma_a^2$
b	$(\Sigma_j\, t_{.j.}^2)/(nJ)$	$RS_b - SS_m$	$J-1$	$\sigma_e^2 + n[J/(J-1)]\sigma_b^2$
c	$(\Sigma_k\, t_{..k}^2)/(nI)$	$RS_c - SS_m$	$I-1$	$\sigma_e^2 + n[I/(I-1)]\sigma_c^2$
rem		$RS_t - SS_m - SS_a - SS_b - SS_c$	$N - 2I - J + 2$	σ_e^2
t	$\Sigma_i \Sigma_j \Sigma_k \Sigma_l\, X_{ijkl}^2$	$RS_t - SS_m$	$N-1$	

Table 9-13. Numerical example of 4×4 latin square design.

	B_1	B_2	B_3	B_4	$t_{..k}$	$\overline{X}_{..k}$	$\hat{\gamma}_k$
C_1	A_1 10	A_2 8	A_3 5	A_4 4	27	6.75	-3.25
C_2	A_2 11	A_4 13	A_1 16	A_3 12	52	13.00	3.00
C_3	A_3 10	A_1 14	A_4 9	A_2 10	43	10.75	0.75
C_4	A_4 8	A_3 6	A_2 11	A_1 13	38	9.50	-0.50
$t_{.j.}$	39	41	41	39	$T = 160$		
$\overline{X}_{.j.}$	9.75	10.25	10.25	9.75	$\overline{X}_{...} = 10$		
$\hat{\beta}_j$	-0.25	0.25	0.25	-0.25			

	A_1	A_2	A_3	A_4
$t_{.j.}$	53	40	33	34
$\overline{X}_{i..}$	13.25	10.00	8.25	8.50
$\hat{\alpha}_i$	3.25	0.00	-1.75	-1.50

	RS	SS	df	MS	F	α
m		1600.0	1			
a	1663.5	63.5	3	21.17	7.94	.02
b	1601.0	1.0	3	0.333	0.12	
c	1681.5	81.5	3	27.17	10.19	.01
rem		16.0	6	2.667		
t	1762.0	162.0	15			

Table 9-14. Numerical example of 3 × 6 generalized latin square design.

	B_1	B_2	B_3	B_4	B_5	B_6	$t_{..k}$	$\overline{X}_{..k.}$	$\hat{\gamma}_k$
C_1	A_1 -1 $\underline{4}$ 3	A_3 -2 $\underline{2}$ 0	A_3 1 $\underline{1}$ 2	A_1 -1 $\underline{1}$ 0	A_2 -3 $\underline{1}$ -2	A_2 -8 $\underline{-8}$ -16	-13	-1.08	-1.50
C_2	A_3 -1 $\underline{2}$ 1	A_2 -2 $\underline{-2}$ -4	A_2 2 $\underline{1}$ 3	A_3 2 $\underline{4}$ 6	A_1 0 $\underline{1}$ 1	A_1 -2 $\underline{-1}$ -3	4	0.33	-0.08
C_3	A_2 1 $\underline{4}$ 5	A_1 6 $\underline{4}$ 10	A_1 -3 $\underline{5}$ 2	A_2 -4 $\underline{-1}$ -5	A_3 6 $\underline{5}$ 11	A_3 -1 $\underline{2}$ 1	24	2.00	1.58
$t_{.j.}$	9	6	7	1	10	-18	$T = 15$		
$\overline{X}_{.j..}$	1.50	1.00	1.17	0.17	1.67	-3.00	$\overline{X}_{....} = .42$		
$\hat{\beta}_j$	1.08	0.58	0.75	-0.25	1.25	-3.42			

	A_1	A_2	A_3
$t_{i..}$	13	-19	21
$\overline{X}_{i...}$	1.08	-1.58	1.75
$\hat{\alpha}_i$	0.67	-2.00	1.33

	RS	SS	df	MS	F	α
m		6.25	1			
a	80.92	74.67	2	37.33	5.82	.01
b	98.50	92.25	5	18.45	2.88	.04
c	63.42	57.17	2	28.58	4.46	.03
rem		166.67	26	6.410		
t	397.00	390.75	36			

so that each appears the same number of times in each row of the table. This is possible of course only if $J = qI$ and $K = I$. The special case in which $q = 1$, i.e., $I = J = K$, is called a *latin square design,* but the general method of analysis for such a design applies to any positive integral value of q.

The analysis of a latin square design, or a generalized latin square with $q > 1$, can be undertaken in exactly the same manner as for a Youden square. Because every level of A is paired with every level of B, however, the calculations can be considerably simplified. For latin square designs,

$$\hat{\alpha}_i = \overline{X}_{i...} - \overline{X}_{....}$$
$$\hat{\beta}_j = \overline{X}_{.j..} - \overline{X}_{....}$$
$$\hat{\gamma}_k = \overline{X}_{..k.} - \overline{X}_{....}.$$

The calculations of sums of squares is similarly simplified. The calculational formulas are given in Table 9-12; a numerical example of a latin square is given in Table 9-13, and a numerical example of a generalized latin square is given in Table 9-14.

Once again, it is important that levels of different factors be paired randomly. Cochran and Cox (pp. 145–147) give a number of plans for latin squares in "standard" form. A latin square is put into standard form by permuting both the rows and columns so that the levels of A appear in numerical order in both the first row and the first column. The latin square in Table 9-13, for example, is in standard form. The requirement of random pairings can be satisfied most readily by choosing a design in standard form and randomly permuting first the rows and then the columns of the matrix.

Repeated Balanced Incomplete Blocks

The latin square and the Youden square are the most commonly used special three-way designs. Nevertheless, other variations on the balanced incomplete blocks design are also possible. The design to be described here is not commonly used, but it can be useful in special situations. It is a relatively simple, logical extension of the balanced incomplete blocks design; it is an example of how any incomplete design can be extended to involve additional factors.

Suppose, in the example of drugs and mice given at the beginning of the chapter, the experimenter had had three different strains of mice, with six litters from each strain, and at least two animals per litter. The experimenter could then have done a separate balanced incomplete blocks experiment, in the manner of Tables 9-1 and 9-2, on each strain of mice. By doing a separate statistical analysis on each strain, however, he loses information on differences between strains, as well as on overall A and B main effects, averaged over strains. If he can reasonably assume that there are no interactions between drugs, litters, or strains, he can treat strains as a third factor, C, and do a three-way analysis that will enable him to recover the information that he would lose if he did three separate analyses.

The three-way analysis is very simple to perform, if the analysis of the balanced incomplete blocks design is understood. Table 9-15 gives a worked example of an experiment in which the design of Table 9-8 is repeated twice. Since we assume that the same balanced incomplete blocks design has been repeated K times over the K levels of Factor C, the estimates of α_i and β_j are averages of the estimates over the separate balanced incomplete blocks designs:

$$\hat{\alpha}_i = (1/K) \ \Sigma_k \ \hat{\alpha}_{i(k)}$$
$$\hat{\beta}_j = (1/K) \ \Sigma_k \ \hat{\beta}_{j(k)},$$

Table 9-15. Repeated 4×6 balanced incomplete blocks design with $k = 2$, $n = 2$.

	B_1	B_2	B_3	B_4	B_5	B_6	$t_{i\bullet(1)}$	$t^*_{i(1)}$	$\hat{\alpha}_{i(1)}$
								C_1	
A_1	19 17		22 16		25 21		120	104	4.0
A_2	20 20			24 18		35 29	146	124	5.5
A_3		6 8	18 12			23 17	84	100	−4.0
A_4		3 11		10 16	11 7		58	80	−5.5
$t_{\bullet j(1)}$	76	28	68	68	64	104	$t_{\bullet\bullet 1} = 408$		
$\overline{X}_{\bullet j(1)}$	19	7	17	17	16	26	$\overline{X}_{\bullet\bullet 1} = 17$		
$\beta_{j(1)}$	−2.75	−5.25	0.00	0.00	−0.25	8.25	$\hat{\gamma}_1$		

	B_1	B_2	B_3	B_4	B_5	B_6	$t_{i\bullet(2)}$	$t^*_{i(2)}$	$\hat{\alpha}_{i(2)}$	$\hat{\alpha}_i$
								C_2		
A_1	22 16		26 13		29 20		126	102	6.00	5.000
A_1	21 13			23 13		36 29	135	117	4.50	5.000
A_3		7 8	16 13			20 17	81	100	−4.75	−4.375
A_4		4 11		13 11	8 7		54	77	−5.75	−5.625
$t_{\bullet j(2)}$	72	30	68	60	64	102	$t_{\bullet\bullet 2} = 396$	$T = 804$		
$\overline{X}_{\bullet j(2)}$	18.0	7.5	17.0	15.0	16.0	25.5	$\overline{X}_{\bullet\bullet 2} = 16.5$	$\overline{X}_{\bullet\bullet\bullet} = 16.75$		
$\hat{\beta}_{j(2)}$	−3.75	−3.75	−.125	−.875	−.625	9.125	$\hat{\gamma}_2 = -.25$			
$\hat{\beta}_j$	−3.25	−4.50	−.06	−.44	−.44	8.69				

	RS	SS	df	MS	F	α
m		13467.00	1			
a	14937.17	806.25	3	268.75	16.43	.01
b	14871.00	740.08	5	148.02	9.05	.01
c	13470.00	3.00	1	3.00	0.18	
rem		621.75	38	16.36		
t	16302.00					

where $\hat{\alpha}_{i(k)}$ and $\hat{\beta}_{j(k)}$ are the estimates obtained from the balanced incomplete blocks design under the kth level of Factor C. The value SS_a is then found as $ngHK \Sigma_i \hat{\alpha}_i^2$. To find SS_b, we have to slightly modify the formulas for RS_a and RS_b (given in Table 9-7 and Eqs. 9-5 and 9-6):

$$RS_a = \Sigma_i t_{i..}^2/(NgK)$$

$$RS_b = \Sigma_j t_{.j.}^2/(nhK)$$

(In the first equation $t_{i..}$ is the sum over all cells in the entire design that are paired with level A_i; in the second equation $t_{.j.}$ is similarly defined with respect to B_j.)

Then SS_b is found by Equation 9-7 as before:

$$SS_b = RS_b + SS_a - RS_a.$$

The C main effect is found much as it would be in a three-way design. Each $\hat{\gamma}_k$ is found by averaging the values in all cells paired with level C_k and subtracting $\mu = \overline{X}_{....}$ from the average:

$$\hat{\gamma}_k = t_{..k}/(nhJK) - \overline{X}_{....}.$$

SS_c, with $(K - 1)$ degrees of freedom, is found by

$$SS_c = nhJ \Sigma_k \hat{\gamma}_k^2 = [\Sigma_k t_{..k}^2/(nhJ)] - (T^2/N) \qquad \textbf{(9-42)}$$
$$= RS_c - SS_m.$$

Finally, as before, the denominator sum of squares is found by summing the squares of all observations and subtracting the various effects from this total. The formula is Equation 9-18, with SS_c subtracted from the result:

$$SS_{rem} = RS_t - RS_a - SS_b - SS_c$$

The degress of freedom for SS_{rem} are $(N - I - J + 2)$, as can be seen by subtracting the degrees of freedom for the various effects from N.

Warning: It is important that exactly the same design be repeated under each level of C; if levels A_i and B_j are paired (i.e., if AB_{ij} contains data) under one level of C, then they must be paired under all levels of C. The pairings must be determined randomly, like in the simple balanced incomplete blocks design, but once determined the same pairings must be repeated under all levels of C.

HIGHER-WAY DESIGNS

The approach used in the repeated balanced incomplete blocks design can be applied to any of the designs discussed in this chapter. Either the Youden square or the latin square can be expanded to a four-way design by repeating the basic three-way design over the levels of a fourth factor. For the latin square, the analysis of such a repeated design is especially simple. To find the sum of squares for say the A main effect, find RS_a by first finding the total of the observations ($t_{i...}$) for all cells under each level of A_i. Then square each sum, $t_{i...}$, add the squares, and divide the final total by the number of observations over which you summed to get each $t_{i....}$. In other words, RS_a is found by basically the same formulas as those in Table 9-12, except that the sum is taken over the levels of Factor D as well as the levels of B and C. To reflect the extra summation, the divisor is changed from (nJ) to (nJL), where L is the number of levels of the fourth factor. We then find SS_a, as usual, by $SS_a = RS_a - SS_m = RS_a - T^2/N$. The sums of squares, SS_b, SS_c, and SS_d are found the same way. This is, of course, the same method used to find SS_c in the repeated balanced incomplete blocks design. The SS for the factor over which the design is repeated is always found in this way.

In the repeated Youden square design Factors A and B are treated just as in the repeated balanced incomplete blocks design. Factors C and D are treated in the manner described in the preceding paragraph.

The balanced incomplete blocks design can also be extended to a four-factor design. Suppose, for example, we want to add two more factors, C with three levels and D with two levels, to a basic A × B balanced incomplete blocks design. The three levels of C and the two levels of D combine to form a six-celled table; thus, six repetitions of the balanced incomplete blocks design (one for each combination of levels of C and D) are needed. To analyze the A and B main effects in such a design, we treat the six combinations of levels of Factors C and D as though they were six levels of a single factor, and we follow the procedure outlined on pp. 210–214. The C and D main effects are analyzed by finding RS for each and then subtracting SS_m as described above.

When the design is repeated over two factors, calculations for the C and D main effects are usually simplified by first listing the total of all observations for each CD_{kl} combination in the cells of a C by D table just as though they were the cell totals in an ordinary two-way analysis of variance. Sums of squares for the C and D main effects are then found just as they would be for such a two-way table (as described in Chapter 6) if there were no A or B factors. In addition, a sum of squares for the CD interaction can also be calculated in the usual way, from the C × D table. This, then, is an exception to the usual rule that no sums of squares for interactions can be calculated in specialized incomplete designs.

This same approach can, of course, be used to extend the balanced incomplete blocks design to five or more factors, by repeating it over three or more additional factors. In addition, it can be used to extend the Youden square and the latin square to designs of five or more factors. In every case the same denominator mean square is used for testing all of the effects. The denominator sum of squares is found by first finding RS_t, the sum of the squared observations. From this is subtracted each SS calculated, including SS_m. For the balanced incomplete blocks and Youden square design, however, it is necessary to remember that SS_a and SS_b are not independent and cannot simply be subtracted; in these cases the subtraction is handled as described in this chapter. Similarly, the degrees of freedom for the denominator sum of squares are found by subtracting the degrees of freedom for each effect from N. The denominator mean square is then the sum of squares divided by the degrees of freedom.

The designs described in this chapter are only a few of the possible specialized designs available. To describe all of the available designs would require an entire volume in itself. In fact, the larger portion of one entire textbook (Cochran and Cox, 1957) is devoted to such specialized designs, and a number of new designs have been devised since that text was published. Most of these designs are so specialized and require such strict assumptions (regarding interactions, etc.) that they are very rarely used. The reader who is interested in such designs or who has very specialized needs should refer either to Cochran and Cox or to more recent statistical literature such as the *Journal of the American Statistical Association*.

Exercises

1. Surgical lesions were made in ten areas of the brain in 30 cats (the ten levels of Factor A). The cats were then tested for visual (B_1), auditory (B_2), tactual (B_3), gustatory (B_4), and olfactory (B_5) discrimination. The data are total correct discrimination in 20 learning trials.

A_1	A_2	A_3	A_4	A_5	A_6	A_7	A_8	A_9	A_{10}
B_1 9	B_1 7	B_1 5	B_2 8	B_1 1	B_1 10	B_1 6	B_2 2	B_2 13	B_3 2
B_2 10	B_2 2	B_2 4	B_3 10	B_3 6	B_3 1	B_4 4	B_3 8	B_4 4	B_4 2
B_3 8	B_4 2	B_5 1	B_4 11	B_5 0	B_4 2	B_5 0	B_5 0	B_5 10	B_5 0

(a) What is λ?

(b) What is the relative efficiency of this design?

(c) Analyze and interpret the data.

(d) What assumptions are needed for this analysis? Comment on the probable validity of the assumptions for this particular study.

2. In a study on judgments of TV sets, each of four different screen sizes of seven different brands of sets were tested. There were seven pairs of judges in all, but an incomplete design was used to save time. In the following data matrix Factor A is the four screen sizes, Factor B is the seven brands, and Factor C is the seven pairs of judges (subjects). The scores are ratings of picture quality on a ten-point scale.

C

		1	2	3	4	5	6	7
		B_7	B_2	B_6	B_5	B_4	B_1	B_3
	1	3	5	4	3	3	5	5
		6	1	4	1	3	2	5
		B_1	B_3	B_7	B_6	B_5	B_2	B_4
	2	5	2	5	3	3	3	5
		6	1	3	4	4	5	4
A		B_5	B_7	B_4	B_3	B_2	B_6	B_1
	3	5	5	5	3	2	4	4
		6	3	5	2	3	6	6
		B_2	B_4	B_1	B_7	B_6	B_3	B_5
	4	4	3	5	5	2	5	7
		5	4	7	5	5	7	6

(a) Analyze these data, assuming that the two observations from each pair of subjects are independent.

(b) Interpret these results. What can be said, on the basis of your analysis, about differences between sets and screen sizes?

3. Reanalyze and reinterpret the data in Problem 2. Assume the pairs of subjects are husbands and wives, the upper value being that of the wife and the lower score that of the husband. Treat the husband-wife dichotomy as the two levels of a fourth factor, D.

4. Reanalyze the data in Problem 2, assuming that the two scores in each cell are independent. However, in this analysis ignore Factor A, treating the design as a 7×7 balanced incomplete blocks design. What is the value of λ for this design? What is the relative efficiency?

5. Analyze the data in Problem 2 once more. Ignore Factor A, as in Problem 4, but treat the two scores within each cell as coming from husband-wife pairs, as in Problem 3.

6. The first latin square below is a study of the effects of different kinds of music on college students' studies. Factor A represents five different subject matters that the students were asked to study, Factor B represents five different kinds of music played during study, and Factor C is the five students. The observations are scores on standardized tests. The second latin square is a replication of the first study using five different subjects.

	B$_1$	B$_2$	B$_3$	B$_4$	B$_5$
C$_1$	A$_3$ 83	A$_1$ 77	A$_4$ 80	A$_5$ 83	A$_2$ 85
C$_2$	A$_1$ 80	A$_5$ 85	A$_2$ 85	A$_4$ 81	A$_3$ 79
C$_3$	A$_2$ 82	A$_3$ 97	A$_5$ 76	A$_1$ 84	A$_4$ 76
C$_4$	A$_4$ 81	A$_2$ 93	A$_1$ 81	A$_3$ 91	A$_5$ 83
C$_5$	A$_5$ 74	A$_4$ 87	A$_3$ 89	A$_2$ 88	A$_1$ 72

	B$_1$	B$_2$	B$_3$	B$_4$	B$_5$
C$_1$	A$_2$ 75	A$_4$ 77	A$_3$ 78	A$_1$ 74	A$_5$ 75
C$_2$	A$_3$ 90	A$_2$ 91	A$_1$ 82	A$_5$ 82	A$_4$ 79
C$_3$	A$_1$ 72	A$_5$ 80	A$_4$ 75	A$_3$ 86	A$_2$ 79
C$_4$	A$_4$ 81	A$_3$ 93	A$_5$ 76	A$_2$ 88	A$_1$ 73
C$_5$	A$_5$ 77	A$_1$ 80	A$_2$ 79	A$_4$ 79	A$_3$ 82

(a) Analyze each set of data separately.

(b) Combine the two sets of data into a single generalized latin square with ten levels of C and analyze the result.

(c) These data, while artificial, were chosen so that the second set of data do in fact represent a replication of the first data. Comment, accordingly, on (a) the value of replicating small studies, and (b) the value of larger rather than smaller studies when larger studies are feasible.

7. An experimenter is interested in problem-solving activities of groups. His method consists of bringing a number of children of about the same age together in a room, giving them a problem, and measuring the length of time required to solve it. He wants to use four different group sizes (3, 5, 7 and 9 subjects), four different age levels (grades 1, 2, 3, and 4, each grade representing an age level), and three different problems. Because learning effects would produce confounding, he cannot give more than one problem to any group. A complete $4 \times 4 \times 3$ design would require 48 different groups, and a total of 288 subjects, or 72 at each grade level. The school where he is doing the study can only provide 60 subjects at each grade level, or 240 subjects in all. If he divides these equally among the four group sizes, he can get only 40 groups, or ten at each grade level.

(a) Briefly describe two basically different designs that the experimenter might use that would enable him to perform the study using all four group sizes, all four grades, and all three problems, without using more than 40 groups in all. (Note: you may use fewer than 40 groups.)

(b) List the assumptions the experimenter must make to use each of the designs you proposed and discuss their importance for this particular problem.

(c) Tell which of the two designs you would recommend most highly, and why.

8. An educational psychologist wants to study differences in performance of students on four supposedly equivalent tests. He has been given the use of four different sixth-grade classes for about 50 minutes each. He wants to give each student each test but, unfortunately, each test takes 15 minutes so that each student can be given only three tests. What kind of design could he use? What assumptions would he have to make in order to be able to use the design you have chosen?

9. An experimenter wants to test five different drugs for their effects on visual acuity, using subjects with a certain rare eye disease. He can find only twelve subjects who have this disease. He can test each subject under all five drugs (one at a time, of course), but, while he expects no interactions among the effects of the drugs, he does expect a learning effect. Consequently, the sequence in which the drugs are given is important. What experimental design could he use?

10

One-way Designs with Quantitative Factors

In the designs discussed so far only the observations themselves have had meaningful numerical values. Although the levels of the factors have been identified by numerical subscripts (A_1, A_2, etc.), the numbers in these cases served only as arbitrary labels of the factor levels. In some designs, however, meaningful numerical values can be assigned to the factor levels themselves. An example of this might be an extinction experiment in which the data are the times required for extinction of a learned response after 10, 20, 30, 40, 50, and 60 learning trials. The six totals of learning trials are the six levels of the factor being studied; the data are the numbers of trials to extinction. The labels on the factor levels in this case are meaningful numerical values—20 trials are twice as many as 10, 60 are three times as many as 20, and so on. If the cell means from such an experiment are plotted in a graph, they might look like those in Figure 10-1 (taken from the data in Table 10-1); in this graph the numerical values of the factor levels dictate both their order and their spacing along the x axis. By contrast, for the data plotted in Fig. 4-1, both the ordering and the spacing of the factor levels on the x axis were arbitrary.

Other examples of experiments with numerically valued factors would be studies of the effects of a person's age or the amount of some drug on the performance of a task. The *data* in both cases would be the values of some index of performance on that task. The *factor levels* in the former case would be the ages of the people being tested, e.g., people might be tested at ages 20, 30, 40, and 50 years, and in the latter case would be the amounts of the drug they had been given. In either experiment the levels of the factor would be readily characterized by specific numerical values.

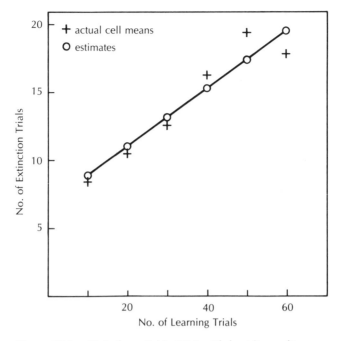

Figure 10-1. Data from Table 10-1 with best linear fit.

TREND ANALYSIS OF ONE-WAY FIXED EFFECTS

In any such design we can, of course, ignore the numerical values of the factors and perform a conventional one-way analysis of variance — perhaps testing for some specific differences with planned comparisons. However, when the data from such an experiment are graphed as in Figure 10-1, they suggest that the cell means might be related to the numerical values of the factors by some specific continuous function. Trend analysis is one method that has been developed to provide evidence on the shape of the function relating numerical factor levels to cell means.

The Model Equation

Trend analysis uses a polynomial function to describe the relationship between the cell means and the numerical values of the factor levels. If we let V_i be the numerical value of the ith factor level (e.g., in Table 10-1, $V_1 = 10$, $V_2 = 20$, etc.), and we let $\bar{V}_.$ be the average of the V_i (in Table 10-1, $\bar{V}_. = (10 + 20 + 30 + 40 + 50 + 60)/6 = 35$), then the polynomial function used can be expressed as

$$\mu_i = \mu + a_1(V_i - \bar{V}_.) + a_2(V_i - \bar{V}_.)^2 + a_3(V_i - \bar{V}_.)^3 + \cdots \qquad \textbf{(10-1)}$$

Recall that by the linear model in Chapter 3 $\mu_i = \mu + \alpha_i$. The model in Equation 10-1 is basically the same model, but we assume in Equation 10-1 that the α_i are a polynomial function of the V_i:

Table 10-1. Hypothetical extinction data illustrating trend analysis.

		A_1	A_2	A_3	A_4	A_5	A_6	
	Number of learning trials:	10	20	30	40	50	60	
		13	12	13	14	18	17	
		5	3	10	13	18	19	
		9	9	17	21	18	18	
Number of		7	10	11	18	14	12	
trials to		12	9	11	14	26	21	
extinction		3	16	11	12	19	20	
		10	10	13	17	18	15	
		10	15	14	17	21	19	
		9	12	13	15	21	20	
		10	11	14	22	22	17	
t_i		88	107	127	163	195	178	858
$\bar{X}_{i\bullet}$		8.8	10.7	12.7	16.3	19.5	17.8	14.3

	RS	SS	df	MS
m		12269.4	1	
bet	13160.0	890.6	5	178.1
w		496.0	54	9.185
t	13656.0	1386.6	59	

$$\alpha_i = a_1(V_i - \bar{V}_{\bullet}) + a_2(V_i - \bar{V}_{\bullet})^2 + a_3(V_i - \bar{V}_{\bullet})^3 + \cdots \qquad (10\text{-}2)$$

Inserting Equation 10-2 into Equation 3-13 gives us the fundamental model for trend analysis:

$$X_{ij} = \mu + a_1(V_i - \bar{V}_{\bullet}) + a_2(V_i - \bar{V}_{\bullet})^2 + \cdots + \epsilon_{ij}.$$

Trend analysis is a technique for testing, for each a_k in the equation, the null hypothesis that $a_k = 0$.

Warning. A polynomial function is one function that might describe the obtained data; however, it might not be the best function for this purpose. In a particular experiment, for example, a logarithmic function, or some other function, might also provide a good fit to the data. As we will show, any set of points can be fit perfectly by a polynomial function containing enough terms. In particular, the data from an experiment with I groups can always be fitted perfectly by a polynomial having $(I - 1)$ values of a_k. Fitting a function by $(I - 1)$ different parameters, however, can be an awkward procedure at best. Sometimes some function other than the polynomial will provide a more "natural" fit, requiring fewer parameters to achieve a good fit to the data. By automatically applying trend analysis to his data, the experimenter may miss this more natural fit. Nevertheless, finding a more natural fit is often difficult, requiring a greater familiarity with various kinds of functions than many experimenters have. Trend analysis has the virtue of providing a relatively straightforward method of fitting the data to a relatively easily understood function. If, however, it is found that a very large number of terms have to be incorporated in the function, one might be advised to look for a different function that will provide a better fit. Fitting it would, of course, require more sophisticated techniques than those described here.

Linear Trend

Trend analysis is most easily illustrated by limiting the model in Equation 10-3 to only the first term in the sum. The model then is

$$X_{ij} = \mu + a_1(V_i - \bar{V}_.) + \epsilon_{ij}. \tag{10-3}$$

This model is applicable, of course, only if we have good *a priori* reasons to assume that the μ_i are related to the V_i by a linear function, i.e., a straight line. The points in Figure 10-1, for example, fit a straight line only rather crudely, but the crudeness of the fit could be due to errors in the data. If we had good *a priori* reasons for believing that the population means were related to the V_i by a straight line, Equation 10-3 would be justified.

Our problem, given this model, is to estimate μ and a_1 and to test the null hypothesis

$$H_0(a_1): a_1 = 0. \tag{10-4}$$

Since the test depends directly on the estimates, we will find them first.

Parameter Estimates, Equal Samples

The best estimates of μ and a_1, for the purposes of trend analysis, are defined (as usual) to be those that minimize the variance of the estimated errors. The best estimates, for equal samples, can be shown to be

$$\hat{\mu} = \bar{X}_{..} \tag{10.5}$$

$$\hat{a}_i = \Sigma_i \left[\frac{V_i - \bar{V}_.}{\Sigma_i (V_i - \bar{V}_.)^2} \right] \bar{X}_{i.}$$

For the data in Table 10-1, $\hat{\mu} = \bar{X}_{..} = 14.3$, and

$$\bar{V}_. = 35, \Sigma_i (V_i - \bar{V}_.)^2 = (10 - 35)^2$$
$$+ (20 - 35)^2 + \cdots + (60 - 35)^2 = 1750,$$

so that

$$\hat{a}_1 = \left[\frac{10 - 35}{1750} \right] 8.8 + \left[\frac{20 - 35}{1750} \right] 10.7 + \cdots + \left[\frac{60 - 35}{1750} \right] 17.8. \tag{10-6}$$
$$= .214.$$

The line representing the function $\hat{X}_{ij} = 14.3 + .214(V_i - \bar{V}_.)$ is shown in Figure 10-1.

Significance Test, Equal ns

Since a_1 is the slope of the best fitting straight line, Equation 10-4 states, as the null hypothesis, that the best fitting straight line has slope zero. In general, we will reject the null hypothesis if $\hat{a}_1$ is very different from zero, and we will fail to reject if $\hat{a}_1$ is close to zero. The method of testing the null hypothesis becomes obvious if we note that Equation 10-5 is a linear contrast in the $\overline{X}_{i.}$ (see pp. 57–58). If we let

$$c_{i1} = \frac{(V_i - \overline{V}_.)}{\Sigma_i (V_i - \overline{V}_.)^2},$$

(10-7)

then

$$\hat{a}_1 = \Sigma_1 \, c_{i1} \overline{X}_{i.},$$

the equation of a linear contrast. The null hypothesis of no linear trend (Eq. 10-4) is thus tested as a simple contrast for which the c_{i1} values are defined in Equation 10-7. For the example in Table 10-1, the c_{i1} are the terms in brackets in Equation 10-6.

Since the theory of planned comparisons was developed in Chapter 4, there is no need to repeat that theory here. The analysis is simplified, however, if we recall that the test of a planned comparison is not changed when we multiply each c_{ik} by a constant. Usually a multiplier can be found that will make the c_{ik} integers. For this experiment, by multiplying every c_{i1} in Equation 10-7 by 350, we can transform them to the more easily handled integers shown in Table 10-2. Table 10-2 also illustrates the test for a linear trend on the data in Table 10-1.

Estimates and Tests, Unequal ns

The theory of estimation and testing for linear trends is the same for unequal as for equal ns; the specific equations, however, take into account the different numbers of observations in the different cells. For unequal ns, for example, $\overline{V}_.$ is a weighted average of the V_i:

$$\overline{V}_. = (1/N) \, \Sigma_i \, n_i V_i.$$

Table 10-2. Test for linear trend in extinction data, Table 10-1.

	A_1	A_2	A_3	A_4	A_5	A_6
Number of learning trials:	10	20	30	40	50	60
$\overline{X}_{i.}$	8.8	10.7	12.7	16.3	19.5	17.8
c_{i1}	-5	-3	-1	1	3	5

$C_1 = (-5)(8.8) + (-3)(10.7) + \cdots + (5)(17.8) = 75.0$
$(C_1')^2 = (10)(75)^2/(25 + 9 + \cdots + 25) = 803.6$
$F_{(1, 54)} = 803.6/9.185 = 87.5, \ \alpha < .01$

As before, $\hat{\mu} = \overline{X}_{..}$, but

$$\hat{a}_1 = \Sigma_i \left[\frac{n_i(V_i - \overline{V}_.)}{\Sigma_i\, n_i(V_i - \overline{V}_.)^2} \right] \overline{X}_{i.}$$

The test of the null hypothesis, Equation 10-4, is a planned comparison with unequal ns and with

$$c_{i1} = \frac{n_i(V_i - \overline{V}_.)}{\Sigma_i\, n_i(V_i - \overline{V}_.)^2}.$$

As with equal ns, however, it is often possible (by multiplying each c_{i1} by a suitable constant) to transform them into more easily handled integer values. Table 10-3 illustrates a test for linear trend on a variation, with some data missing, of the design in Table 10-1. In this case, for the linear trend test, we choose to let

$$c_{i1} = n_i(V_i - \overline{V}_.).$$

Higher-Order Trends

The test for a linear trend merely tells whether there is an overall tendency for the μ_i to increase or decrease with increases in the V_i. Put another way, *if* the function is a straight line, the test for linear trend tests whether that straight line has a slope different from zero. The test for a linear trend *does not* test the appropriateness of using a straight line to fit the data in the first place. We can test whether the relationship is truly linear by using the theory relating to MS_{rem} in Chapter 4. If the relationship is in fact linear, then the linear model of Equation 10-3 should describe the data almost perfectly, and all of the significant differences in the data should be accounted for by the test for a linear trend. A general test of the null hypothesis that there are no differences other than those accounted for by the linear trend can be made by calculating (see pp. 68–70)

$$SS_{\text{rem}} = SS_{\text{bet}} - (C_1')^2 = 890.6 - 803.6 = 87.0,$$
$$MS_{\text{rem}} = SS_{\text{rem}}/(I - 2) = 87.0/4 = 21.7,$$
$$F_{(4,54)} = MS_{\text{rem}}/MS_{\text{w}} = 21.75/9.185 = 2.37,$$
$$\alpha = .07$$

A significance level of .07 is suggestive but not compelling. We saw in Chapter 4, however, that even though an overall test like this is not highly significant, some specific additional contrasts may be significant. Just as the linear trend was tested by a contrast, certain specific nonlinear trends can also be tested by contrasts. In particular, referring again to Equation 10-1, for each a_k in that equation we can test the null hypothesis

$$H_0(a_k): a_k = 0.$$

Since a_2 is associated with the squares of the V_i, a test of $H_0(a_2)$ is called a test for *quadratic trend*. Similarly, a test of $H_0(a_3)$ is a test for cubic trend, a test of $H_0(a_4)$ a test of quartic trend, and so on.

Table 10-3. Linear trend analysis, unequal ns.

		A_1	A_2	A_3	A_4	A_5	A_6	
	Number of learning trials:	10	20	30	40	50	60	
		13	12	13	14	18	17	
		5	3	10	13	18	19	
		9	9	17	21	18	18	
		7	10	11	18	14	12	
Number of trials		12	9	11	14	26	21	
to extinction		3	16	11	12		20	
		10		13	17			
		10		14				
		9		13				
				14				
n_i		9	6	10	7	5	6	43
t_i		78	59	127	109	94	107	574
$\overline{X}_i.$		8.67	9.83	12.70	15.57	18.80	17.83	13.35
c_{i1}		-203.0	-75.3	-25.6	52.1	87.2	164.7	$C_1 = 2562$

$$(C_1')^2 = \frac{(2562)^2}{(-203)^2/9 + (-75.3)^2/6 + \cdots + (164.7)^2/6} = 546.1$$

	RS	SS	df	MS
m		7662.2	1	
bet	8241.7	579.5		115.9
w		400.3	37	10.82
t	8642.0	979.8	42	

$F_{(1,37)} = 546.1/10.82 = 50.5$, $\alpha < .01$
$SS_{\text{rem}} = 579.5 - 546.1 = 33.4$, $MS_{\text{rem}} = 33.4/4 = 8.35$
$F_{\text{rem}} = 8.35/10.82 < 1$

The test of quadratic trend (a_2) will tend to be significant if there is a large curvature in a single direction. Figure 10-2 shows three examples of quadratic curvature. Notice that for each curve all of the curvature is in the same direction—none of the curves have *inflection points* (places where the direction of curvature changes).

If there is a cubic trend, there will usually be some point (called an inflection point) at which the direction of curvature changes. Figure 10-3 shows some examples of cubic curvature. Of course, linear, quadratic, and cubic effects may all be present in the same data. In Functions B and C of Figure 10-2, for example, since the overall slope of each function is

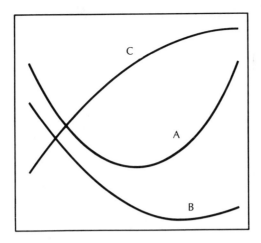

Figure 10-2. Examples of quadratic trends. Curve A is a pure quadratic trend; curves B and C have linear components.

different from zero, a linear trend is present in addition to the quadratic trend. Function A, however, has no linear trend. In Figure 10-3, even though Functions C and D have inflection points, most of the curvature is in only one direction; this indicates the presence of a quadratic effect. By contrast, Functions A and B have no quadratic effect.

Similar considerations apply to higher-order trends. If there is a quartic (fourth power) trend, there will tend to be two inflection points, i.e., the curve will tend to change direction twice; if there is a fifth-power trend, there will usually be three inflection points, and so on. Figure 10-4 illustrates some of these trends.

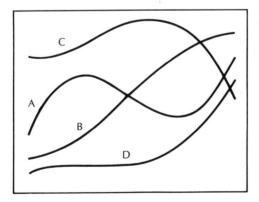

Figure 10-3. Examples of cubic trends. Curve A is a pure cubic trend; curve B has a linear component; curve C has a quadratic component; curve D has both linear and quadratic components.

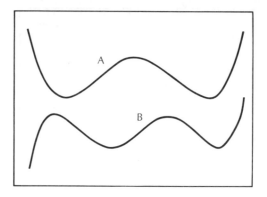

Figure 10-4. Examples of fourth-power (Function A) and fifth power (Function B) trends.

Trend Tests, Equal *ns*, Equal Intervals

Each trend is tested by a planned comparison. The c_{ij} for higher-order trends, however, are not as easily derived as those for linear trends. Their derivation requires finding the solution to a complex set of multiple linear equations. For one special case the values have been derived and tabled; this is the common case in which the n_i are all equal and the intervals between the adjacent V_i are also equal. In the example in Table 10-1 each cell contains the same number of observations and the V_i are all exactly 10 units apart, so they are at equal intervals. In the example of Table 10-3 these requirements are not met, since the n_i are not all equal. Similarly, suppose the numbers of learning trials had been 2, 4, 8, 16, 32, and 64 instead of 10, 20, 30, etc. Then, even though each cell contained the same number of observations, the method would not apply because the intervals between the V_i are not equal. For the method described here to apply, both the n_i and the intervals between the factor values V_i must be equal.

Warning. The intervals between V_i depend on the scale on which the V_i are measured. In the example just given, with the numbers of trials equal to 2, 4, 8, 16, and 32, the intervals are not equal. If, however, we chose to let the V_i be the logarithms of the numbers of trials, the V_i would be $V_1 = \log 2 = .301$, $V_2 = \log 4 = .602$, $V_3 = \log 8 = .903$, etc. The intervals between the V_i would then be equal (each interval equals .301), and the method described here could be used. However, the interpretation of the results would then be very different. Such transformations on the factor values have effects similar to those described for transformations on the data (pp. 105–106); effects that do not exist for the untransformed factor values may exist for the transformed values, and vice versa. A test on the logarithms of the factor values may show a significant quadratic trend, for example, while a test on the untransformed values may not; the reverse may also be true. This is not intended to be an argument against transforming the factor values; the choice of factor values is always somewhat arbitrary, and the experimenter may change a difficult task into a relatively simple one by a transformation like the logarithmic transformation used above. He might even get a better fit with fewer terms to the transformed factor values than he would have gotten to the untransformed values. However, a linear trend, for example, then implies that the data are a linear function of the transformed scores, not of the original scores.

Calculating Formulas

Table A-6 in the appendix gives c_{ik} values for trend tests. If there are I groups, only $I - 1$ orthogonal contrasts can be tested, so the highest power at which a trend can be tested is $(I - 1)$. Each of these $(I - 1)$ trends is tested as a planned comparison using the c_{ik} values in Table A-6. Table 10-4 shows the trend analysis of the data in Table 10-1.

It is clear in this case that most of the systematic variance is accounted for by the linear trend. The cubic trend is also significant, although it does not contribute an especially high proportion of the variance. The quadratic trend approaches the 10 percent level of significance but accounts for only a very small proportion of the variance; it can probably be ignored for practical purposes. Both the fourth-power and the fifth-power components are nonsignificant.

Estimation

All of the theory in Chapter 4, including calculation of power, estimation of ω^2, and so on, applies to trend analysis. Estimation of the cell means, however, is usually done by a special formula. The formula gives estimates that can be connected by a smooth continuous function. For the example in Table 10-1, for instance, we concluded from the tests that the data could be almost completely accounted for by a linear and a cubic trend. That is, they can be accounted for by a function of the form

$$X_{ij} = \mu + a_1(V_i - \overline{V}_.) + a_3(V_i - \overline{V}_.)^3 + \epsilon_{ij}$$

so that

$$\mu = \mu_i + a_1(V_i - \overline{V}_.) + a_3(V_i - \overline{V}_.)^3. \tag{10-8}$$

If we could estimate a_1 and a_3, we could estimate the μ_i from this function. More generally, whenever we conclude that the μ_i are the sum of certain trends, we can estimate the μ_i if we can estimate the coefficients (a_k) of the trends. Unfortunately, the a_k are difficult to estimate directly. The μ_i, however, can be estimated easily if we change the form of Equation 10-8 to

$$\mu_i = \mu + a_1^* c_{i1} + a_3^* c_{i3}. \tag{10-9}$$

This change is possible because the c_{i1} values are directly related to the linear trend and the c_{i3} values are directly related to the cubic trend. Again, more generally, an equation of the form

Table 10-4. Trend analysis on data from Table 10-1.

		A_1	A_2	A_3	A_4	A_5	A_6	C_k
Number of learning trials:		10	20	30	40	50	60	
$\overline{X}_{i.}$		8.8	10.7	12.7	16.3	19.5	17.8	
	1	−5	−3	−1	1	3	5	75.0
	2	5	−1	−4	−4	−1	5	−13.2
c_{ik}	3	−5	7	4	−4	−7	5	−31.0
	4	1	−3	2	2	−3	1	−6.0
	5	−1	5	−10	10	−5	1	1.0

		C_k	$(C_k')^2$	F	α	$\hat{\omega}^2$	$\hat{a}*$
	1	75.0	803.6	87.5	.01	.57	1.071
	2	−13.2	20.7	2.25		.01	
Tests	3	−31.0	53.4	5.81	.02	.03	−.172
	4	−6.0	12.9	1.40		.00	
	5	1.0	0.0	0.00		.00	

	Estimates					
	A_1	A_2	A_3	A_4	A_5	A_6
Number of learning trials:	10	20	30	40	50	60
$\overline{X}_{i.}$	8.8	10.7	12.7	16.3	19.5	17.8
$\hat{\mu}_i$	9.8	9.9	12.5	16.1	18.7	18.8

$$\mu_i = \mu + a_1(V_i - \overline{V}) + a_2(V_i - \overline{V})^2 + a_3(V_i - \overline{V})^3 + \cdots \quad \text{(10-10)}$$

can always be changed to an equivalent equation of the form

$$\mu_i = \mu + a_1^* c_{i1} + a_2^* c_{i2} + a_3^* c_{i3} + \cdots \quad \text{(10-11)}$$

When we say that Equations 10-10 and 10-11 are equivalent, we mean that if we insert appropriate values for the a_k^* into Equation 10-11, we will obtain the same values of μ_i that we would have obtained from the a_k in Equation 10-10. Similarly, if we obtain best unbiased estimates of the a_k^* and use them in Equation 10-11 to estimate the μ_i, we will obtain the same μ_i estimates that we would have obtained from inserting estimates of the a_k into Equation 10-10. Thus, while we cannot estimate the a_k directly, we can estimate the μ_i.

In practice, of course, only those components that contribute meaningfully to the overall trend, i.e., those that are significant and account for a reasonably large proportion of the variance, are included. In our example only the linear and cubic trends are included. The estimate of a_k^* is a relatively simple function of the c_{ik}. The formula is

$$\hat{a}_k^* = C_k / (\Sigma\, c_{ik}^2), \quad \text{(10-12)}$$

where C_k and c_{ik} are the values used in testing for the kth trend component.

To illustrate, for the data in Table 10-1, we must estimate a_1^* and a_3^* (cf. Eq. 10-9). The estimates use the c_{ik} and C_k values in Table 10-4:

$$\hat{a}_1^* = 75/[(-5)^2 + (-3)^2 + \cdots + (5)^2] = 75/70 = 1.071$$
$$\hat{a}_3^* = -31/[(-5)^2 + (7)^2 + \cdots + (5)^2] = -31/180 = -.172.$$

Since the estimate of μ (from Table 10-1) is $\overline{X}_{..} = 14.3$, the estimation formula is

$$\hat{\mu}_i = 14.3 + 1.071c_{i1} - .172c_{i3}.$$

From this equation we estimate μ_1 by inserting $c_{11} = -5$ and $c_{13} = -5$ to obtain

$$\hat{\mu}_1 = 14.3 + 1.071\ (-5) - .172\ (-5) = 9.8.$$

Similarly, we estimate μ_2 by inserting $c_{21} = -3$ and $c_{23} = 7$ into the same equation to obtain

$$\hat{\mu}_2 = 14.3 + 1.071(-3) - .172(7) = 9.9.$$

The remaining four $\hat{\mu}_i$, found the same way, are shown in Table 10-4. These predicted values are plotted, along with the obtained means, in Figure 10-5. It can be seen that the cubic function provides a somewhat better fit than did the linear function in Figure 10-1; it also gives a smoother function than that provided by the original cell means.

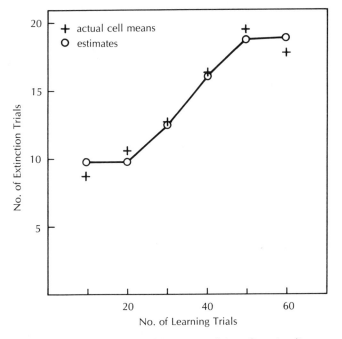

Figure 10-5. Data from Table 10-1 with best fit, using linear and cubic trend components.

Note on Interpretation of Polynomial Functions

The use of straight lines to represent the function in Figure 10-5 may seem strange, since the purpose of the estimation equation is to generate a smooth curve. Nevertheless, my use of straight lines to connect the predicted values of the μ_i helps to make an important point.

The use of a polynomial function to describe the data implies that that function would also describe the data at values of V_i other than those actually sampled. For example, from Figure 10-6, in which is drawn the smooth curve actually represented by the obtained cubic function, we would predict that given 45 learning trials the expected number of extinction trials would be 17.6. However, there is nothing in either the data or the statistical analysis to justify such a prediction on *statistical* grounds. Intuitively, we would expect such a prediction to be approximately correct, but there is no logical or statistical basis for using the obtained function to predict to conditions other than those actually studied in the experiment.

In fact, the danger of extending the conclusions to conditions that were not studied is clearly shown in Figure 10-6, which shows that the predicted number of trials to extinction reaches a peak after about 55 learning trials, after which it declines rapidly. Most experts on learning, I am sure, would hesitate to take very seriously a function that predicts that after a certain number of learning trials additional trials will decrease the time to extinction.

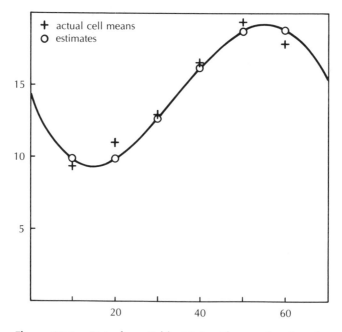

Figure 10-6. Data from Table 10-1 with complete function of best linear and cubic fit.

The situation at the lower end of the scale is even worse. If we take the function seriously here, we will be forced to conclude that it takes more than fourteen trials to extinguish a response that has never been learned!

Finally, consider Figure 10-7. The function in this figure fits the predicted points just as well as the function in Figure 10-6. It would lead, however, to considerably different predictions for points other than those that were actually studied. In short, there are usually good intuitive and theoretical reasons for assuming that a relatively simple function would describe potential data from points that were not actually studied. Nevertheless, such reasons are extrastatistical; there are no strictly logical or statistical reasons for preferring one such curve over another. In practice, a smooth curve like that in Figure 10-6 will most often appear most reasonable. The exact shape of that curve should not be taken too seriously, however. In many cases a "curve" consisting simply of line segments like those in Figure 10-5 will fit the function about as well as the smooth curve. Other times such a smooth curve connecting the points might not be appropriate at all.

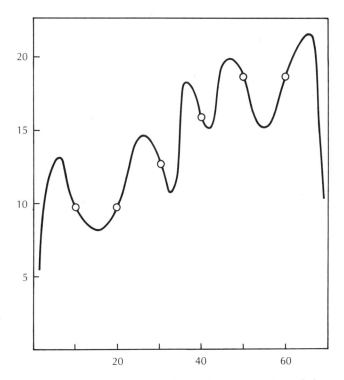

Figure 10-7. Possible curve fitting the same estimated data points as those in Figure 10-6.

ONE OBSERVATION PER CELL

In order to calculate MS_w it is necessary to have more than one observation in each cell of a one-way design. Consequently, in an ordinary one-way analysis of variance no test could be made with only one observation per cell. When the factor levels have numerical values, however, there is a trick that allows us (provided our assumptions are reasonable) to analyze data for which there is only one observation per cell.

The use of a special method to deal with only one observation per cell is best demonstrated by an example. Figure 10-8 shows the numbers of applications for admission to the University of Knutsta (affectionately known as "Knutsta U") during the years 1960 to 1968. The officials noticed what appeared to be a sharp rise in applications during the early sixties and wondered if it might be due to the "baby boom" of the late forties. They have two questions to ask of the data. (1) Was there a significantly greater rise in applications during the years after 1962 or 1963 than previously? (2) Was there a significantly smaller rise in applications during the years after 1966 than between 1963 and 1965? If the answer to the first question is yes, the increase can probably be reasonably attributed to the baby boom; if the answer to the second question is yes, the officials can expect that the baby boom has about ended and that a relatively steady but smaller growth rate in applications will occur in the future.

One way to answer these questions would be to use a trend analysis. We would expect a significant linear trend, of course, since applications are increasing steadily throughout the period. The answer to the second question would be yes if there were a significant nonlinear trend — probably

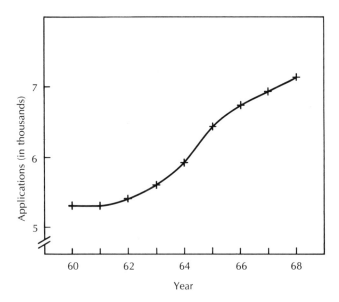

Figure 10-8. Hypothetical data, applications for admission to Knutsta University, 1960–1968.

quadratic or cubic, from the looks of the data. If the first answer is yes, the second question can be answered by testing specifically for a cubic trend. If the baby boom is in fact ending, the initial increase followed by a tendency to taper off would show up as a significant cubic trend (compare Fig. 10-8 with curve B of Fig. 10-3). If it is not tapering off, there should be a significant quadratic trend but no cubic trend.

Trend Tests

One simple and straightforward way to fit these data into the framework of the analysis of variance would be to consider the numbers of applications as the observations in a one-way analysis of variance for which the years are the levels. This arrangement of the data in an ordinary one-way design is shown in Table 10-5. With such a design, however, there is only one observation per cell. Thus, special techniques will be needed to find a denominator for our F tests on trends.

Even with only one observation per cell, $(C_k')^2$ values for tests on trends can be calculated. Table 10-5 shows the values for the first three trend components. Notice that for this case nearly all of the variance (in fact, 99.7 percent) is accounted for by the first three components. Furthermore, there are no a priori reasons to expect large trends higher than the third order. We thus have reason to believe—and the data support our belief— that the first three trend components account for all, or nearly all, of the true variation in the data. Any remaining variance must then be due largely to pure chance—that is, it must be error variance. The remainder, after subtracting the sum of the $(C_k')^2$ from SS_{bet}, is 15,592 (see Table 10-5), and the above argument makes it reasonable to believe that this variance represents error. We can thus treat it as a sum of squares for error in

Table 10-5. Trend analysis of data in Figure 10-8.

		A_1	A_2	A_3	A_4	A_5	A_6	A_7	A_8	A_9	C_k
	Year:	1960	1961	1962	1963	1964	1965	1966	1967	1968	
	Applications:	5300	5300	5400	5600	5900	6400	6700	6900	7100	
	Linear	−4	−3	−2	−1	0	1	2	3	4	15400
c_{ik}	Quadratic	28	7	−8	−17	−20	−17	−8	7	28	13800
	Cubic	−14	7	13	9	0	−9	−13	−7	14	−10100

$SS_{tot} = SS_{bet} = 4,140,000$, df $= 8$

	$(C_k')^2$	F	α	$\hat{\omega}^2$
Linear	3,952,667	1268	.01	.952
Quadratic	68,701	22.0	.01	.017
Cubic	103,040	33.0	.01	.025

$df_{rem} = 5$, $MS_{rem} = 15592/5 = 3118$

developing an F ratio to test for the three trends in which we are interested. Since there were originally eight degrees of freedom, and the $(C'_k)^2$ for three orthogonal tests have been subtracted out, SS_{rem} has five degrees of freedom. The appropriate denominator for the F ratio thus turns out to be $MS_{rem} = 15592/5 = 3118$. Table 10-5 shows the results of the F tests, with MS_{rem} in the denominator. Although the linear trend is dominant, as we expected, both the quadratic and the cubic trend are significant. We can conclude that applications were affected by the baby boom and that the increase in applications began to taper off near the end of the period studied. However, the positive quadratic trend indicates that the tapering off was not complete, that is, that the growth rate was still higher at the end of the period than it was at the beginning.

In devising a denominator for these tests we merely made use of a principle often repeated in this text: a variance that is not due to systematic effects is error variance and can be treated as such in forming F ratios. The same principle could have been used on the data of Table 10-1 and 10-3. If we had good a priori reasons for believing that there were no true fourth or fifth degree trend components, $(C'_4)^2$ and $(C'_5)^2$ could have been pooled with SS_w to increase the degrees of freedom in the denominator of each F ratio. (The same would hold for the quadratic effect if we had a priori reasons for expecting that there was no true quadratic effect. It is extremely unlikely that we would have such reasons and still expect a cubic effect.) However, the increase in degrees of freedom would only be two (from 54 to 56), and MS_{pooled} would be 9.088, down only very slightly from $MS_w = 9.185$. The very small gain in power in this case must be balanced against all of the risks and cautions discussed in Chapter 5 (pp. 106–108).

This method is therefore useful primarily when there is no other denominator available, when the only alternative to the risks discussed in Chapter 5 (pp. 106–108) is to make no tests at all. In any case we can be reasonably certain whenever we use this approach that the error will tend to make rejection of the null hypothesis less, rather than more, probable.

The chief problem in applying this approach to real data lies in determining which trend components are to be tested and which will comprise the error. In the example above solving this problem was not difficult. The nature of the questions asked required the testing of linear, quadratic, and cubic trends, and the nature of the data made it reasonable to assume that higher-order trends would be small. In other cases, however, such assumptions are not easily made before gathering the data. When they can be made (and the data bear them out), it is usually best to proceed as though the assumptions were true. For cases in which the assumptions cannot be made a special procedure has been developed to give approximate significance levels. Although the procedure is far from satisfactory in many of its aspects, it is sometimes useful when no other procedure is available.

With this procedure, one tests each trend as if no higher-order trends existed. Thus, when testing the linear trend, all of the variance remaining after subtracting the linear trend from SS_{bet} is treated as error. For the example in Table 10-5 the error would be $SS_{rem} = SS_{bet} - (C_1')^2 = 4,140,000$ $- 3,952,667 = 187,333$. In this case SS_{rem} would have seven degrees of freedom, so that $MS_{rem} = 187333/7 = 26,762$, and the linear effect is tested by $F_{(1.7)} = 3952667/26762 = 148$, ($\alpha < .01$).

When testing the quadratic trend, one subtracts the variances for the linear and quadratic trends from SS_{bet} and treats the remainder as error:

$$SS_{rem} = SS_{bet} - (C_1')^2 - (C_2')^2 = 4,140,000$$
$$- 3,952,667 - 68,701 = 118,632$$
$$MS_{bet} = 118,632/6 = 19,772$$
$$F_{(1.6)} = 68,701/19,722 = 3.47$$
$$\alpha > .10.$$

Notice that by this method the quadratic effect is not significant, although in the analysis in Table 10-5 the quadratic effect is significant beyond the .01 level. The analysis usually proceeds until two or three nonsignificant trends in a row have been obtained. A complete analysis of the same data in Table 10-5 is shown in Table 10-6. The analysis in this case was discontinued after finding that the fourth, fifth, and sixth power trends were nonsignificant.

A comparison of Table 10-6 with Table 10-5 illustrates the relative loss of power resulting from an inability to make valid assumptions about the trends. In Table 10-5 all three trends are significant beyond the .01 level. In Table 10-6 the quadratic trend is not even significant at the .10 level. The F ratio for the linear trend is also much smaller in Table 10-6 than in

Table 10-6.   Trend analysis of data in Figure 10-8 (cubic function *not* assumed).

		A_1	A_2	A_3	A_4	A_5	A_6	A_7	A_8	A_9	C_k
	Year:	1960	1961	1962	1963	1964	1965	1966	1967	1968	
	Applications:	5300	5300	5400	5600	5900	6400	6700	6900	7100	
	Linear	-4	-3	-2	-1	0	1	2	3	4	15400
	Quadratic	28	7	-8	-17	-20	-17	-8	7	28	13800
	Cubic	-14	7	13	9	0	-9	-13	-7	14	-10100
c_{ik}	4th degree	14	-21	-11	9	18	9	-11	-21	14	-1500
	5th degree	-4	11	-4	-9	0	9	4	-11	4	2000
	6th degree	4	-17	22	1	-20	1	22	-17	4	2400

$SS_{tot} = SS_{bet} = 4,140,000$, df $= 8$

	$(C_k')^2$	SS_{rem}	df$_{rem}$	MS_{rem}	F	α
Linear	3,952,667	187,333	7	26,762	148	.01
Quadratic	68,701	118,632	6	19,772	3.47	
Cubic	103,040	15,592	5	3,118	33.00	.01
4th degree	1,124	14,468	4	3,617	0.31	
5th degree	8,547	5,921	3	1,974	4.33	
6th degree	2,909	3,012	2	1,506	1.93	

Table 10-7. Calculation of estimated applications, Table 10-5.

$\hat{\mu} = 54600/9 = 6067$

$$\hat{a}_1^* = \frac{15400}{(-4)^2 + (-3)^2 + \cdots + (4)^2} = 256.7$$

$$\hat{a}_2^* = \frac{13800}{(28)^2 + (7)^2 + \cdots + (28)^2} = 4.98$$

$$\hat{a}_3^* = \frac{-10100}{(-14)^2 + (7)^2 + \cdots + (14)^2} = 10.20$$

	A_1	A_2	A_3	A_4	A_5	A_6	A_7	A_8	A_9
Year:	1960	1961	1962	1963	1964	1965	1966	1967	1968
Applications	5300	5300	5400	5600	5900	6400	6700	6900	7100
Estimated applications	5320	5260	5380	5630	5970	6330	6670	6940	7090

Table 10-5, although it is highly significant in both tables. The loss of power derives from the fact that the "error" estimates used for testing the linear and quadratic trends are inflated by inclusion of the cubic effect. Thus they are not pure error at all, but error plus a systematic effect. It would be possible, of course, to redo the analysis, once one had found the highest significant trend (using the final error term to retest all lower trends). It would be possible, but it would not be legitimate, because such a procedure involves choosing one's error term on the basis of the data.

Estimating Trends

The formulas for estimating the trends are the ones for ordinary trend analysis (pp. 230–231). For the example in Table 10-5, we are fitting the data to the equation

$$\hat{X}_i = \hat{\mu} + \hat{a}_1^* c_{i1} + \hat{a}_2^* c_{i2} + \hat{a}_3^* c_{i3}.$$

Table 10-7 illustrates the calculation of the $\hat{a}_k^*$, using Equation 10-12; it also compares the actual and estimated X_i. In this case the actual and estimated values are very close.

RANDOM EFFECTS MODEL – CORRELATION

Analyses for the random-effects model are similar to those for the fixed-effects model. The random-effects model has a serious limitation, however, when applying the methods described above. The analysis of nonlinear trends requires that the values associated with the factor levels, i.e., the V_i, be equally spaced. However, when the factor levels are chosen randomly, the V_i are also chosen randomly, and in general they will not be equally spaced. This fact usually limits us to the test of a linear trend—if we are to avoid much more complicated methods of analysis.

At this point, you may recall the statement in Chapter 5 that planned comparisons were not appropriate for the random-effects model. Tests for trend (and especially tests for linear trend) are an exception to this rule. The reason trend tests are legitimate is that they enable us to infer the function relating the data to the factor levels, and therefore to draw conclusions about factor levels that were not sampled. The validity of our conclusions depends of course on the validity of the model used in the data analysis. The model we are considering, for example, assumes a linear relationship between the observations (X_i) and the factor levels (V_i). If the assumption is incorrect for the data we are studying, our inferences about unsampled factor levels will also be incorrect. Given the correctness of the model, however, the results of the planned comparison for trend can legitimately be extended to inferences about factor levels other than those actually sampled in the experiment. To illustrate, statistics teachers commonly assert that there is a strong relationship between a child's shoe size and his intelligence. The example is usually given to illustrate the dangers of drawing unwarranted conclusions from statistical relationships (there is a relationship only insofar as both variables tend to increase with age). Suppose an experimenter, in an attempt to test this claim, measured the shoe sizes and intelligence of twenty-six school children from the first grade through high school. Shoe size was measured to the nearest inch and intelligence was measured by a "quickie" ten-item test (the maximum possible score was ten, the minimum possible was zero). The data obtained might look like those in Table 10-8; Figure 10-9 shows a scatterplot of the same data.

Linear Trend

For the above experiment, let each subject be a "group," or level of the factor, and let the shoe sizes be the values (V_i) assigned to those levels. Our data, the X_i, are the intelligence test scores. (Alternatively, we could have let the test scores be the V_i and the shoe sizes be the X_i; as we shall see, however, in the linear model we get exactly the same results either way.) We thus have a one-way random design with twenty-six "groups," or cells, and one observation per cell.

The method developed for testing trends does not depend in any way on the assumption of a fixed-effects design. The test of a linear trend is basically the same for both the fixed-effects and the random effects design. However, we are going to develop it further here because the development in itself is instructive and because power calculations are different for the random-effects model.

As we saw earlier (pp. 223–225), the test for a linear relationship is a planned comparison with

$$c_{i1} = \frac{(V_i - \bar{V}_.)}{\Sigma_i \ (V_i - \bar{V}_.)^2}.$$

Table 10-8. Hypothetical data, shoe sizes vs. intelligence test-scores for 26 randomly selected school children.

Shoe size	Test score	Shoe size	Test score	Shoe size	Test score
6	2	9	3	8	7
6	3	8	4	10	7
9	6	8	3	9	6
9	4	11	7	9	7
7	4	9	8	8	6
7	4	9	5	7	5
9	5	9	5	8	3
8	4	9	4	9	7
8	4	9	4		

	V	X	VX
Sum	218	127	
Mean	8.385	4.885	
Sum of squares or cross products	1860	685	1092
s^2 or $C(V, X)$	1.237	2.487	1.044
s or r_{VX}	1.112	1.577	.596

$\hat{\omega}^2 = r^2 = .355$, $F_{(1, 24)} = 13.2$, $\alpha < .01$
$\hat{X}_i = 4.885 + .845(V_i - \bar{V}_.)$
$\quad = 4.885 + .845(V_i - 8.385) = .845V_i - 2.200$

The test is equally valid if we first multiply every c_{i1} by the same constant; since $\Sigma_i (V_i - \bar{V}_.)^2$ is such a constant, the test is equally valid if it is made using $c_{i1} = (V_i - \bar{V}_.)$. Then, with $n - 1$,

$$C_1 = \Sigma_i (V_i - \bar{V}_.) X_i = \Sigma_i V_i X_i - N\bar{V}_.\bar{X}_. \tag{10-13}$$

The ratio C_1/N is of very general importance, as we will see. It is *the covariance of V and X*, symbolized as $C(V,X)$:

$$C(V,X) = (\Sigma_i V_i X_i / N) - \bar{V}_.\bar{X}_. \tag{10-14}$$

From the symmetry discussed above, and evident in Equations 10-13 and 10-14, $C(V,X) = C(X,V)$. Table 10-8 shows these calculations for the data on shoe size and intelligence.

The formula for covariance is very similar to that for variance. In fact, using the raw score formula, the variance of the observed sample of X_i is

$$s_X^2 = (\Sigma_i X_i^2 / N) - \bar{X}_.^2 = (\Sigma_i X_i X_i / N) - \bar{X}_.\bar{X}_. = C(X,X).$$

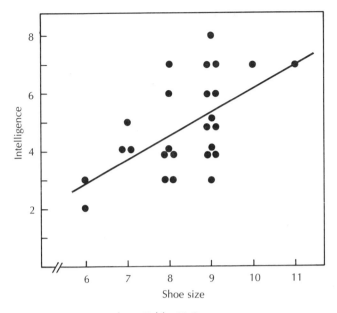

Figure 10-9. Data from Table 10-8.

We can see that the variance of X is simply the covariance of X with itself. From Equations 10-13 and 10-14 we can easily derive $(C_1')^2$:

$$(C_1')^2 = \frac{C_1^2}{\Sigma_i (V_i - \overline{V}_.)^2} = \frac{N[C(V,X)]^2}{s_V^2}$$

where

$$s_V^2 = \left(\frac{1}{N}\right) \Sigma_i (V_i - \overline{V}_.)^2$$

is the variance of the observed V_i values. (Remember that because the groups were selected randomly the V_i are themselves values of a random variable.)

Since there is only one observation per cell, we must find a denominator by the method described on pp. 234–238. Note, however, that the application of the method to this problem requires the very strong assumption that a *strictly linear* relationship exists between V and X. Any tendency for the relationship to be nonlinear will reduce the power of the test considerably because all regular, nonlinear effects are treated as error.

To find the denominator, we calculate SS_{rem}, noting that with only one observation per cell, $SS_t = Ns_X^2$:

$$SS_{\text{rem}} = SS_t - (C_1')^2 = Ns_X^2 - N[C(V,X)]^2/s_V^2$$
$$= \left(\frac{N}{s_V^2}\right) \{s_X^2 s_V^2 - [C(X,V)]^2\}.$$

The degrees of freedom for SS_{rem} are $(N-1)-1=N-2$, so that

$$MS_{\text{rem}} = SS_{\text{rem}}/(N-2),$$

and the F ratio for testing the null hypothesis is

$$F_{(1,N-2)} = (C_1')^2/MS_{\text{rem}},$$

which, with some algebraic manipulation, becomes

$$F_{(1,N-2)} = \frac{(N-2)[C(V,X)]^2}{s_V^2 s_X^2 - [C(V,X)]^2}.$$

This formula can be simplified further, if we define

$$r_{VX} = \frac{C(V,X)}{s_V s_X}.$$

Then

$$F_{(1,N-2)} = \frac{(N-2)r_{VX}^2}{1 - r_{VX}^2}.$$

Table 10-8 shows these calculations for shoe sizes and intelligence.

Linear Correlation

The value r_{VX} is highly interesting and useful in itself. It is called the *Pearson product-moment correlation coefficient,* and it varies from plus one (when X is perfectly predictable from V), through zero (when there is no predictable relationship), to minus one (when X is perfectly predictable but inversely related to V). Values different from zero, plus one, and minus one represent varying degrees of predictability of X from V.

The degree of predictability for a given r_{VX} can be measured in terms of ω^2, which in this case is the proportion of variance of X accounted for by its relationship to V. It can be shown that if ρ_{VX} is the correlation between V and X in the whole population, then $\rho^2 = \omega^2$, the proportion of variance accounted for. From this and Equation 4-23 (remembering that with only one observation per cell, $MS_w = 0$), it follows that $\hat{\omega}^2 = r_{VX}^2$ is an approximately unbiased estimate of ω^2. Thus, the larger the absolute value of r_{VX}, the larger the proportion of variance accounted for.

Figure 10-10 shows several scatterplots (like that in Figure 10-9), illustrating various values of r_{VX}. Generally, such a scatterplot will take on the form of an ellipse. If the data are plotted so that one unit equals one standard deviation on each axis, then the ratio of the width to the length of the ellipse will be about

$$\frac{w}{l} \approx \sqrt{\frac{1 - |r|}{1 + |r|}}.$$

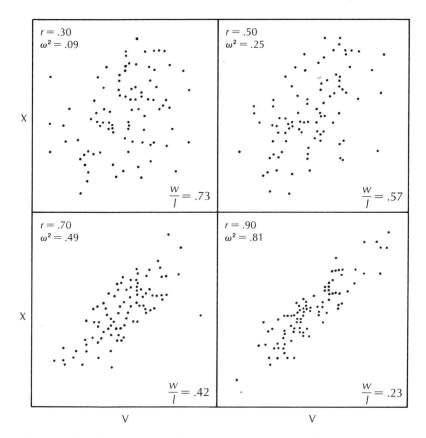

Figure 10-10. Sample scatterplots for selected values of the correlation coefficient, r_{vx}.

With $r = .596$, the ratio of the width to the length of the ellipse in Figure 10-9 should be about

$$\frac{w}{l} = \sqrt{\frac{1 - .596}{1 + .596}} \approx .50.$$

The ellipse in Figure 10-11 has this ratio of width to length and just about covers the points.

Estimation

The best estimates of μ and a_1 are the same as those for the fixed-effects model (p. 223). For the data in Table 10-8, $\hat{\mu} = \overline{X}. = 4.885$. The formula for $\hat{a}_1$, however, can be expressed most easily in terms of either $C(V,X)$ or r_{VX}:

$$\hat{a}_1 = \frac{C(V,X)}{s_V^2} = (s_X/s_V)r_{VX}. \tag{10-15}$$

For the data in Table 10-8,

$$\hat{a}_1 = (1.577/1.112).596 = .845.$$

The prediction function, $\hat{X}_i = 4.885 + .845(V_i - \overline{V}_.)$, called the *regression line,* is plotted against the scatterplot of obtained data in Figure 10-9.

Notice that the estimation formula (Eq. 10-15) is not symmetric in V and X. If we were predicting V_i from X_i, our estimates of μ and a_1 would be

$$\hat{\mu}^* = \overline{V}_. = 8.385$$
$$\hat{a}_1^* = (s_V/s_X)r_{VX} = (1.112/1.577).596 = .420.$$

In other words, for a $V_i = 10$, our best estimate of X_i is 6.25; yet for an actual $X_i = 6.25$, our best estimate of V_i is not 10, but 8.96. Formally, this · is called *regression toward the mean;* informally, it is a kind of "hedging of bets" by predicting a value a little closer to the mean than one would expect if the correlation were perfect. The smaller the absolute value of r_{VX}, the greater the regression toward the mean. For r_{VX} equal to one or minus one, there is no regression toward the mean at all. For r_{VX} equal to zero, the regression is complete; since $a_i = 0$, the best estimate of X_i is then $\overline{X}_.$, no matter what the value of V_i.

A visual explanation of regression toward the mean can be found in Figure 10-11, where Figure 10-9 is reproduced with an added ellipse that covers the "cloud" of points. The ellipse represents the region in which we could normally expect about 95 percent of the points to fall. Consider, now, the heavy vertical line just above $V_i = 10$. If we know that a person's shoe size (V_i) is 10 and we wish to guess his intelligence score (X_i), our best guess would be the mean of the distribution of the X_i along that line. The mean, however, will generally lie about half-way between the two points where the line intersects the ellipse, that is, just about where

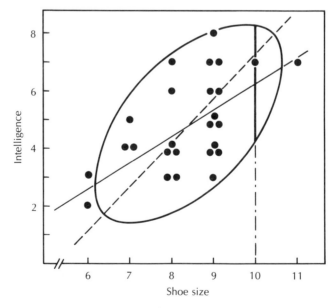

Figure 10-11. Illustration of regression toward the mean.

it intersects the regression line. If there were no regression toward the mean, the best prediction would be the value of X_i at the intersection of the vertical line and the diagonal dotted line.

Other Tests, Power, Confidence Intervals

We saw in Chapter 6 that power calculations are not the same for the random-effects model as for the fixed-effects model. Exact power calculations for tests of correlation are extremely complex. Similarly, in Chapter 6, tests of the null hypothesis H_0: $\omega^2 = \omega^{*2}$ were found to be rather simple. Analogous tests for correlation would be of the form:

$$H_0: \rho_{VX} = \rho^*_{VX}. \tag{10-16}$$

Like power calculations, exact tests of null hypotheses like Equation 10-16 are extremely complex. However, a transformation on r, called *Fisher's Z transformation*, makes approximate tests, power calculations, and confidence intervals possible. The transformation is

$$Z = (1/2) \log_e\left(\frac{1+r}{1-r}\right).$$

Under the rather restrictive assumptions that V and X are both normally distributed and no curvilinear relationship exists between them, Z is approximately normally distributed with mean and variance

$$E(Z) = \xi = (1/2) \log_e\left(\frac{1+\rho}{1-\rho}\right), \tag{10-17}$$

$$V(Z) = 1/(N-3).$$

A slightly better approximation is

$$E(Z) = \xi + \frac{\rho}{2(N-1)}, \tag{10-18}$$

with the variance the same as above. With this approximation, all of the theory of the normal distribution is at our disposal for making hypothesis tests, estimating power, finding confidence intervals, and so on. Fisher's Z transformation is in Table 10 of the appendix. We will illustrate its use by finding the approximate 95 percent confidence interval for r_{VX}, using the data on shoe sizes and intelligence. We will find the confidence interval first for the more crude approximation in Equation 10-17, and then for the better one in Equation 10-18. To find the 95 percent confidence interval for the first approximation, we first convert $r_{VX} = .596$ to $Z = .687$, using the table. We then have to find the limits on ξ such that

$$|Z - \xi| \sqrt{N-1} = |.687 - \xi| \sqrt{23} \leq 1.960.$$

Found this way, the limits on ξ are .278 and 1.096. Finally, we look these up in the body of the table of the inverse Z transformation to convert them back to values of ρ_{VX}; the 95 percent confidence interval is then $.27 < \rho_{VX}$

< .80. For the second approximation, we have to find the limits on ρ_{VX} such that

$$\left| .687 - \xi - \frac{\rho_{VX}}{2(N-1)} \right| \sqrt{23} \leq 1.960.$$

Since ξ is a complex function of ρ_{VX}, this equation cannot be solved directly. It is fairly easy to solve it by trial and error, choosing values of ρ_{VX}, finding the corresponding values of ξ from the table, and inserting them in the equation. The limits found from the first approximation are a convenient starting point. The final result is

$$.266 < \rho_{VX} < .794.$$

In this case, and in most cases in which $N \geq 25$, the difference between the results of the two approximations is too small to be of consequence.

Exercises

1. Do trend analyses on: (a) Problem 1, Chapter 3; (b) Problem 2, Chapter 3; (c) Problem 4, Chapter 3. Assume that the V_i are equally spaced.

2. Test for (a) a linear trend; and (b) a nonlinear trend on the data in Problem 1, Chapter 3, assuming the following V_i values:

	A_1	A_2	A_3	A_4	A_5	A_6
V_i	0	1	2	4	8	10

3. Test for linear and nonlinear trends on the data in Problem 3, Chapter 3, assuming that the V_i are equally spaced.

4. A theorist believes that for the data of a certain one-way experiment with five equally spaced levels (i.e., the single factor is quantitative) the appropriate model is

$$X_{ij} = a_1(V_i - \bar{V}) + \epsilon_{ij}.$$

That is, he believes that the only effect is linear, and $\mu = 0$. You suspect that the data are nonlinear and the true mean is not zero, so you repeat the experiment. Financial considerations restrict you to only one observation at each of the five levels.

(a) Describe what you consider to be the best available way to test his theory, using the obtained data. Tell what assumptions your test(s) requires, and the effects of violation of the assumptions.

(b) The data are as follows:

	A_1	A_2	A_3	A_4	A_5
a_1	−10	−5	0	5	10
X_i	−3	0	−1	2	5

Perform the analysis (analyses) that you described in part a and draw whatever conclusions you can. Justify your conclusions (or lack of them).

(c) Suppose you had two observations in each cell. How would you then answer part a?

5. Several pairs of fathers and sons were obtained, in which both father and son were working full-time. The annual income levels of each father and son were ascertained; the data are as follows (in thousands of dollars):

Father	Son	Father	Son	Father	Son
29	77	7	26	66	44
7	79	59	63	2	37
17	85	39	70	43	25
64	61	13	64	12	12
27	35	13	7	36	72
15	38	27	3	64	85
22	80	29	1	21	51
13	12	68	98	50	33

(a) Test the null hypothesis that the average income of all the sons is the same as the average income of all the fathers.

(b) Test the null hypothesis that there is no linear relationship between the incomes of fathers and sons.

(c) Find the best linear function relating all the sons' incomes to the fathers' incomes; then find the best linear function relating all the fathers' incomes to the sons' incomes, and compare the two.

(d) Estimate the proportion of variance accounted for by the linear relationship between the incomes of fathers and sons.

(e) Test the null hypothesis that there is a nonlinear relationship between the incomes of fathers and sons.

(f) Draw the scatterplot of fathers' vs. sons' incomes; comment on (1) the validity of the assumptions made in the above tests, and (2) the appropriateness of the tests themselves for revealing the important information in the data.

11

Trend Analyses in Multifactor Designs

Frequently, a multifactor design has one or more numerically valued factors. When this occurs, trend analyses can be performed on main effects and interactions involving these factors. Just as in the one-way design, the theory of trend analysis differs for fixed and random numerical factors. For multifactor designs, however, the problem is compounded by the fact that numerically valued random factors exert important influences on tests involving only fixed effects. In this chapter we will discuss designs in which all of the numerical factors are fixed; in the next chapter we will consider numerical random factors.

In the preceding chapters we found that planned comparisons on main effects in multifactor designs followed basically the same formulas as planned comparisons among groups in a one-way design. The same holds true for trend analyses among main effects in multifactor designs.

MAIN EFFECTS

Since any trend analysis is basically a contrast, the theory for contrasts holds for trend analyses on main effects. The c_{ik} values are the same as in Chapter 10, and the formulas are the same as those for contrasts on main effects.

The data in Table 11-1 illustrate a trend analysis in a two-way design. The hypothetical data are scores on standard arithmetic tests in three different schools in a medium-sized town. The tests were given to students whose measured IQs fell into the ranges 80–90, 90–100, 100–110, and 110–120. The IQ ranges are the levels of Factor A, and the schools are the levels of Factor B. Ten pupils in each IQ range were tested in each school,

Table 11-1. Hypothetical mean arithmetic scores for students in four IQ ranges (Factor A) at three schools (Factor B), $n = 10$.

	A_1	A_2	A_3	A_4	Mean
B_1	31	43	48	58	45.00
B_2	33	50	62	66	52.75
B_3	37	59	66	69	57.75
Mean	33.67	50.67	58.67	64.33	51.83

	SS	df	MS	F	α	$\hat{\omega}^2$
m	322,403.3	1				
a	16,030.0	3	5343	133.6	< .001	.66
b	3,301.7	2	1651	41.27	< .001	.13
ab	605.0	6	100.8	2.52	.05	.00
w	4,320.0	108	40.0		—	
t	24,256.7	119				

a total of 120 pupils. The values in Table 11-1 are the average scores for the pupils in each cell. The results of a standard analysis of variance are shown below the table. Although all three effects are significant, the experimenter is not interested in the overall tests on the A main effect and the interaction. Instead, he is interested first in knowing what trends are significant in the A main effect, and, second, in knowing whether the trends differ among schools. The first question involves the A main effect; the second involves the AB interaction, as we shall see.

The trend analysis on the A main effect is simply a set of planned comparisons, using the coefficients for the linear, quadratic, and cubic trends given in the appendix. The tests can be made directly on the $\overline{X}_{i..}$ values, as in Chapter 5, remembering that since each $\overline{X}_{i..}$ is the mean of nJ observations (n observations at each of the J levels of B),

$$(C'_k)^2 = nJ(\Sigma_i c_{ik} \overline{X}_{i..})^2 / (\Sigma_i c_{ik}^2).$$

Table 11-2 gives these calculations for the linear, quadratic, and cubic

Table 11-2. Trend analysis on Factor A, data from Table 11-1.

		A_1 80–90	A_2 90–100	A_3 100–110	A_4 110–120	Σc^2	C	$(C')^2$	F	α	$\hat{\omega}^2$
Mean		33.67	50.67	58.67	64.33						
c_{ik}	Linear	−3	−1	1	3	20	100	15000	375	< .001	.62
	Quad	1	−1	−1	1	4	−11.33	963.3	24.08	< .001	.04
	Cubic	−1	3	−3	1	20	6.67	66.7	1.67	—	.00

trends on A. The linear and quadratic trends are significant, but the cubic trend is not.

Estimates of the true variance of each trend component are found in the usual way by Equation 4-22, and estimates of proportion of variance accounted for, listed in Table 11-2, are found by Equation 4-23.

Note that the calculations shown are those for a design in which Factor B is fixed. If Factor B were random, the appropriate denominator for the trend tests would be MS_{ab} instead of MS_w. This is because the trends are part of the A main effect, and the appropriate denominator for the A main effect would be MS_{ab} in a mixed-effects design. The F ratio for the linear trend would then be $15000/100.8 = 148.7$, with 6 df in the denominator. The proportion of variance accounted for would be .57. Values for the other two trends would be calculated the same way.

TWO-WAY INTERACTIONS WITH ONE NUMERICAL FACTOR

Significance Tests

The second question in the introduction to this chapter asks whether scores in all three schools can be described by a single function. To answer that question, we first separately calculate $(C'_k)^2$ values for each trend at each level of B. To make the calculations, we assume that each level of B constitutes a complete experiment in itself (much as we did when testing simple effects in Chapter 5). Then, since each $\overline{X}_{ij}$ is the average of only n rather than nJ observations, the multiplier for the separate levels is n rather than nJ. These calculations, for each trend and each level of B, are shown in Table 11-3.

Table 11-3. Dividing interaction into trend components, data from Table 11-1.

	Linear		Quadratic		Cubic	
	C_1	$(C'_1)^2$	C_2	$(C'_2)^2$	C_3	$(C'_3)^2$
B_1	86	3698.0	-2	10.0	12	72.0
B_2	111	6160.5	-13	422.5	-3	4.5
B_3	103	5304.5	-19	902.5	11	60.5
Sum		15163.0		1335.0		137.0
A main effect[a]		15000.0		963.3		66.7
SS		163.0		371.7		70.3
MS		81.5		185.8		35.2
F		2.04		4.65		0.88
α		—		.02		—
$\hat{\omega}^2$		.00		.01		.00

[a]From Table 11-2.

Since each $(C'_k)^2$, calculated this way, is mathematically equivalent to that for a contrast in a simple one-way analysis of variance, each $(C'_k)^2$ is proportional to chi-square with one degree of freedom; hence, their sum will be distributed as chi-square.

To test the null hypothesis that the value of the *linear* trend is the same for all three schools, we first sum the $(C'_1)^2$ (the values for the linear trend components) for the three separate levels of B. The resulting sum is proportional to chi-square with three degrees of freedom. We then subtract the $(C'_1)^2$ for the A main effect to obtain a statistic proportional to chi-square with two degrees of freedom. The actual calculations, in this case, are

$$SS_{\text{a linear (b)}} = 3698.0 + 6160.5 + 5304.5 - 15000 = 163.0$$

We divide this by its degrees of freedom to obtain

$$MS_{\text{a linear (b)}} = 163/2 = 81.5.$$

Finally, we divide this by MS_w to obtain the F ratio

$$F_{(2,108)} = 81.5/40 = 2.04.$$

Table 11-3 shows these calculations, including calculations for the quadratic and cubic effects.

The general method for testing trend interactions should now be clear. First, the $(C'_k)^2$ for the trend in question are calculated separately for each level of Factor B (assuming that Factor B is the nonnumerical factor). These are summed, and the $(C'_k)^2$ for the same trend on the A main effect is subtracted from their total. The result is treated as a sum of squares with $(J-1)$ degrees of freedom. The appropriate denominator is MS_w whether B is fixed or random, since MS_w is always the denominator for the AB interaction in a two-way design.

The relationship between these tests and the test of the AB interaction can be seen from Table 11-3. Note, first, that for I levels of A there will be $(I-1)$ trend components to test, and each test has $(J-1)$ degrees of freedom. The total degrees of freedom for all of the tests is thus $(I-1)(J-1)$, which is exactly equal to the degrees of freedom of the AB interaction. Furthermore, if we add the sums of squares in Table 11-3, we get $163.0 + 371.7 + 70.3 = 605.0$, which is equal to SS_{ab}. We have just described, in other words, a method for dividing the AB interaction into $(I-1)$ orthogonal tests, each with $(J-1)$ degrees of freedom.

Estimates of the variance of each effect are obtained directly from the mean squares in Table 11-3 by subtracting MS_w from each mean square in turn, multiplying each difference by the degrees of freedom of the mean square, and dividing that result by N. The estimated variance of the quadratic effect, for example, is $(186-40)(2)/120 = 2.4$. To find the proportion of variance accounted for, we divide this value by $\hat{\sigma}_t^2 = 202$ to get .012. These estimates, like the significance tests, are the same whether Factor B is fixed or random. Table 11-4 summarizes the trend analysis on the data in Table 11-1.

Table 11-4. Summary of trend analysis on data in Table 11-1 (bold numbers are totals).

Effect		SS	df	MS	F	α	$\hat{\omega}^2$
m		332,403.3	1				
a		**16,030.0**	**3**				**.66**
	linear	15,000	1	15,000	375	<.001	.62
	quad.	963.3	1	963.3	24.1	<.001	.04
	cubic	66.7	1	66.7	1.67		.00
b		3,301.7	2	1651	41.3	<.001	.13
ab		**605.0**	**6**				**.00**
	linear	163.0	2	81.5	2.04		.00
	quad.	371.7	2	185.8	4.65	.02	.01
	cubic	70.3	2	35.2	0.88		.00
w		4,320.0	108	40.0			.20
t		24,256.7					

The interpretation of interaction trends is straightforward. If there is a linear trend in the interaction, the size of the linear trend, i.e., the overall slope of the best-fitting straight line, varies across levels of the nonnumerical factor. The presence of a singificant quadratic trend in the data of Figure 11-1 confirms the suspicion that the overall curvature of the best-fitting function varies across levels of the nonnumerical factor (in this case, schools). Basically, this means that a different estimate of A_2 should be used for each level of B. Similar considerations would apply to higher-order trends in the interaction.

Estimation

Since both the B main effect and the quadratic trend component of the AB interaction are significant, we cannot estimate cell means from the trends on the A main effect alone. One solution to this problem might lie in a separate trend analysis, across A, at each level of B. Such an approach would be in the spirit of the simple-effects tests described in Chapter 6. However, it would not be the most parsiminious approach because it would not take into account the fact that the *linear trend* is approximately the same for all schools, i.e., there is no linear trend component in the AB interaction. A more parsimonious approach is to make our estimates from the model equation:

$$\hat{\mu}_{ij} = \hat{\mu} + \hat{a}_i(V_i - \bar{V}) + \hat{a}(b_j)_2(V_i - \bar{V})^2 + \hat{\beta}_j.$$

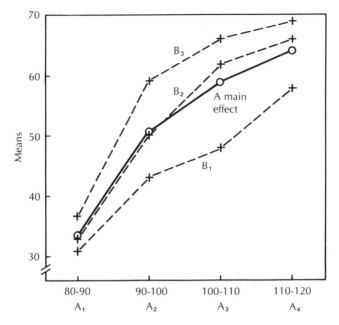

Figure 11-1. Obtained means for A main effect and for each level of B; data from Table 11-1.

In this equation $\hat{a}(b_j)_2$ indicates that because the quadratic component of the interaction is significant a different estimate of a_2 must be used for each level of B. Only one estimate of a_1 is needed, however.

There is no need to include $\hat{a}_2$ in this equation because it is implicitly included in $\hat{a}(b_j)_2$. Whenever a trend component is significant both in the main effect and in the interaction, only the interaction term appears in the equation. To see how the quadratic interaction includes the quadratic component of the A main effect, consider the average of the C values in Table 11-3 for the quadratic interaction term $(-2 -13 -19)/3 = -11.33$, which, we can see from Table 11-2, is the C value for the quadratic component of the A main effect. Since the estimates of the effects will be based on these C values, the A main effect trend will be automatically included in the average of the estimates for the interaction.

The equation can be shortened slightly by noting that

$$\overline{X}_{.j.} = \hat{\mu} + \hat{\beta}_j,$$

so that

$$\hat{\mu}_{ij} = \overline{X}_{.j.} + \hat{a}_1(V_i - \overline{V}_.) + \hat{a}(b_j)_2(V_i - \overline{V}_.)^2.$$

As in Chapter 10, some of these values are difficult to estimate directly. However, estimates can easily be made if we change the equation to

$$\hat{\mu}_{ij} = \overline{X}_{.j.} + \hat{a}_1^* c_{i1} + \hat{a}^*(b_j)_2 c_{i2}.$$

Table 11-5. Estimated effects and cell means for data in Table 11-1, Factor A numerical.

Estimates of effects:

$\hat{\mu}$	$\hat{\beta}_1$	$\hat{\beta}_2$	$\hat{\beta}_3$	a_1
51.83	−6.83	0.92	5.92	5.00

$a^*(b_1)_2$	$a^*(b_2)_2$	$a^*(b_3)_2$
−0.50	−3.25	−4.75

Obtained (upper value) and estimated (lower value) cell means:

	A_1	A_2	A_3	A_4
B_1	31 29.5	43 40.5	48 50.5	58 59.5
B_2	33 34.5	50 51.0	62 61.0	66 64.5
B_3	37 38.0	59 57.5	66 67.5	69 68.0

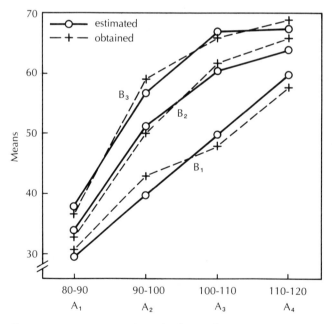

Figure 11-2. Estimated and obtained means, Factor A numerical; data from Table 11-1.

In this equation c_{i1} is the coefficient of $\overline{X}_{i..}$ when testing for a linear trend, and c_{i2} is the coefficient for the quadratic trend.

The value of a_i^* is estimated from C_1 for the A main effect just as in Chapter 10:

$$a_1^* = C_1/\Sigma_i \, c_{i1}^2 = 100/20 = 5.$$

To estimate $a^*(b_j)_2$, we use the same formula but we use a different C_2 for each level of B. The values we use are in the quadratic column in Table 11-3:

$$a^*(b_1)_2 = -2/\Sigma_i \, c_{i2}^2 = -2/4 = -.50$$
$$a^*(b_2)_2 = -13/\Sigma_i \, c_{i2}^2 = -13/4 = -3.25$$
$$a^*(b_3)_2 = -19/\Sigma_i \, c_{i2}^2 = -19/4 = -4.75.$$

The estimate of μ_{11} would then be $\hat{\mu}_{11} = 45.00 + 5(-3) - .50(1) = 29.50$. The estimate of μ_{12} would be $\hat{\mu}_{12} = 52.75 + 5(-3) - 3.25(1) = 34.50$. The other estimates, shown in Table 11-5 and plotted in Figure 11-2, are obtained similarly.

HIGHER-WAY INTERACTIONS WITH ONE NUMERICAL FACTOR

The principles described here are easily extended from two-way to higher-way designs. In a multi-way design tests for trends on main effects of numerical factors are carried out just as planned comparisons are performed on main effects. In addition, any two-way interaction involving one numerical and one nonnumerical factor can be divided into trend components as described above, by doing a separate trend test on each level of the non-numerical factor, summing the $(C')^2$ values, and subtracting the $(C')^2$ for the same trend on the main effect. The number of numerator degrees of freedom for each such test will be one less than the number of levels of the nonnumerical factor.

These calculations are illustrated in Tables 11-6, 11-7, and 11-8, for an imaginary study on the buying habits of American housewives. For the study, the nation was divided into three areas, West, Midwest, and East; 75 stores within each area were selected for testing. Twenty-five stores each were located in low-income, middle-income, and high-income neighborhoods. A new brand of detergent was offered in each of these stores (the same brand in every store). The stores in each group of 25 were divided into subgroups of five each, in which the detergent sold for different prices. The five prices were 29¢, 39¢, 49¢, 59¢, and 69¢.

In this design Factor A, with three levels, is the area of the country; Factor B, with three levels, is the income level; and Factor C, the numerical factor with five levels, is the price. The five stores offering the detergent at the same price provide within-cell variance, so $n = 5$. The data are mean numbers of boxes purchased in each type of store at each price level in a single day. The same data are plotted in Fig. 11-3.

Table 11-6. Hypothetical data, cell means from a study of buying habits, $n = 5$.

<table>
<thead>
<tr><th colspan="2"></th><th colspan="5">ABC</th></tr>
<tr><th colspan="2"></th><th>C_1</th><th>C_2</th><th>C_3</th><th>C_4</th><th>C_5</th></tr>
</thead>
<tbody>
<tr><td></td><td>B_1</td><td>24.0</td><td>28.8</td><td>22.8</td><td>13.2</td><td>4.8</td></tr>
<tr><td>A_1</td><td>B_2</td><td>31.8</td><td>20.4</td><td>15.6</td><td>19.2</td><td>30.6</td></tr>
<tr><td></td><td>B_3</td><td>2.4</td><td>13.2</td><td>32.4</td><td>33.6</td><td>34.8</td></tr>
<tr><td></td><td>B_1</td><td>26.4</td><td>26.4</td><td>16.8</td><td>10.8</td><td>3.6</td></tr>
<tr><td>A_2</td><td>B_2</td><td>7.2</td><td>24.0</td><td>30.0</td><td>27.6</td><td>9.6</td></tr>
<tr><td></td><td>B_3</td><td>10.8</td><td>21.6</td><td>25.2</td><td>32.4</td><td>32.4</td></tr>
<tr><td></td><td>B_1</td><td>37.2</td><td>18.0</td><td>6.0</td><td>2.4</td><td>3.6</td></tr>
<tr><td>A_3</td><td>B_2</td><td>31.2</td><td>12.0</td><td>10.8</td><td>15.6</td><td>28.8</td></tr>
<tr><td></td><td>B_3</td><td>9.6</td><td>9.6</td><td>18.0</td><td>31.2</td><td>52.8</td></tr>
</tbody>
</table>

A

A_1	21.84
A_2	20.32
A_3	19.12

B

B_1	B_2	B_3
16.32	20.96	24.00

AC

	C_1	C_2	C_3	C_4	C_5
A_1	19.40	20.80	23.60	22.00	23.40
A_2	14.80	24.00	24.00	23.60	15.20
A_3	26.00	13.20	11.60	16.40	28.40

C

C_1	C_2	C_3	C_4	C_5
20.07	19.33	19.73	20.67	22.33

AB

	B_1	B_2	B_3
A_1	18.72	23.52	23.28
A_2	16.80	19.68	24.48
A_3	13.44	19.68	24.24

BC

	C_1	C_2	C_3	C_4	C_5
B_1	29.20	24.40	15.20	8.80	4.00
B_2	23.40	18.80	18.80	20.80	23.00
B_3	7.60	14.80	25.20	32.40	40.00

$\bar{X} = 20.43$

An overall analysis of variance on these data is shown in Table 11-7; for the most part, though, it will only serve as the basis for a more detailed trend analysis. Table 11-8 summarizes the trend analyses on the C main effect and on the AC and BC interactions. The significant B main effect is expected—it simply means that, on the whole, rich people buy more detergent than do poor people. The linear BC interaction is not surprising either. From Figure 11-3 it is clear that rich people tend to buy the detergent most readily when it is most expensive (a positive linear trend), perhaps believing that that which is more expensive must be better. Poor people, on the other hand, buy most readily when it is least expensive (a negative linear trend).

The significant quadratic trend in the AC interaction most probably comes from the curves for the middle-income group, which show a pronounced tendency for Midwestern buyers to prefer the moderately priced

Table 11-7. Overall analysis of data in Table 11-6.

	SS	df	MS	F	α	$\hat{\omega}^2$
m	93881.0	1				
a	278.7	2	139.4	0.63		.00
b	2243.8	2	1121.9	5.10	.008	.03
c	247.4	4	61.8	0.28		−.01
ab	344.3	4	86.1	0.39		−.01
ac	4846.1	8	605.8	2.75	.008	.05
bc	16880.2	8	2110.0	9.59	<.001	.22
abc	3840.5	16	240.0	1.09		.00
w	39600.0	180	220.0			
t	68281.0	224				

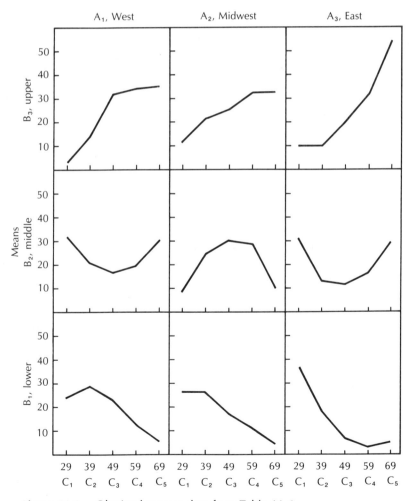

Figure 11-3. Obtained means; data from Table 11-6.

Table 11-8. Trend analysis on C main effect and AC and BC interactions, data from Table 11-6.

		C Main Effect									
	C_1	C_2	C_3	C_4	C_5						
	29	39	49	59	69						
Mean	20.07	19.33	19.73	20.67	22.33	Σc^2	C	$(C')^2$	F	α	$\hat\omega^2$
c_{ij} Linear	−2	−1	0	1	2	10	5.87	154.9	0.70		0
Quadratic	2	−1	−2	−1	2	14	5.33	91.4	0.42		0
Cubic	−1	2	0	−2	1	10	−0.40	0.7	0.00		0
Quartic	1	−4	6	−4	1	70	0.80	0.4	0.00		0

	AC Interaction							
	Linear		Quadratic		Cubic		Quartic	
	C_1	$(C_1')^2$	C_2	$(C_2')^2$	C_3	$(C_3')^2$	C_4	$(C_4')^2$
A_1	9.20	127.0	−4.40	20.7	1.60	3.8	13.20	37.3
A_2	0.40	0.2	−35.60	1357.9	1.20	2.2	−16.40	57.6
A_3	8.00	96.0	56.00	3360.0	−4.00	24.0	5.60	6.7
Sum		223.2		4738.6		30.0		101.6
C Main Effect		154.9		91.4		0.7		0.4
SS		68.3		4647.2		29.3		101.2
df		2.		2.		2.		2.
MS		34.2		2323.6		14.6		50.6
F		0.16		10.56		0.07		0.23
α		—		<.001		—		—
$\hat\omega^2$		−.01		.06		−.01		.00

	BC Interaction							
	Linear		Quadratic		Cubic		Quartic	
	C_1	$(C_1')^2$	C_2	$(C_2')^2$	C_3	$(C_3')^2$	C_4	$(C_4')^2$
B_1	−66.00	6534.0	2.80	8.4	−6.00	54.0	−8.40	15.1
B_2	1.20	2.2	15.6	260.7	−4.40	29.0	0.80	0.1
B_3	82.40	10184.6	−2.4	6.2	−2.80	11.8	10.00	21.4
Sum		16720.8		275.3		94.8		36.6
C Main Effect		154.9		91.4		0.7		0.4
SS		16565.9		183.9		94.1		36.2
df		2.		2.		2.		2.
MS		8283.0		92.0		47.0		18.1
F		37.65		0.42		0.21		0.08
α		<.001		—		—		—
ω^2		.24		.00		−.01		−.01

product, whereas Easterners and Westerners buy most readily when the price is either very high or very low.

Again, in this design both the F test and the estimate of ω^2 depend on whether Factors A and B are random or fixed. The analyses made above assume that they are fixed; if one or both were random, the denominators for the F tests would be those appropriate for that type of design. In addition, the estimate of the variance of each effect would be different. Instead of subtracting MS_w from the mean square for the effect, one would subtract whatever mean square was used in the denominator for testing that effect. If A were random, for example, MS_{ac} would be subtracted from the $(C')^2$ for the trend components of the C main effect, and MS_{abc} would be subtracted from the mean squares for the BC interaction. The procedure after that would be the same as in the fixed effects model; the difference would be multiplied by the degrees of freedom and divided by N to estimate the variance of the effect. Dividing that variance by $\hat{\sigma}_t^2$ would give $\hat{\omega}^2$.

Significance Tests

The procedure for analyzing the two-way interaction into trend components is extended to three-way and higher interactions with only minor modifications. The procedure for the three-way interaction, for the data in Table 11-6, is illustrated in Table 11-9. First, values of C and $(C')^2$ are calculated for each trend component and for every combination of levels of Factors A and B (there are nine such combinations), making 36 values in all (nine for each of the four trend components).

To calculate the linear trend component, $(C_1')^2$ is calculated separately for each of the nine combinations of levels of A and B. These nine values are then summed to give the value 17720.6. From this sum are subtracted the sums of squares for the linear components of the C main effect, the AC interaction, and the BC interaction. The total of these three sums of squares is 16789.1, and the difference, when this total is subtracted from 17720.6, is 931.5. The total number of $(C_1')^2$ that were summed is nine, and the total of the degrees of freedom of the terms subtracted out is five, leaving four degrees of freedom for the linear component of the ABC interaction. This gives a mean square of 232.9 and an F of 1.06, which is not significant.

The general rule for deciding which sums of squares to subtract is to find all lower-order interactions and main effects involving all of the *numerical* factors in the interaction in question. In our example only Factor C is numerical, so we find all lower-order interactions involving C. These, as stated above, are the C main effect and the AC and BC interactions. The sums of squares for the trend components of these effects are subtracted from the totals of the $(C')^2$ for the corresponding trends on the ABC interaction.

The quadratic, cubic, and quartic components are calculated in exactly the same way. The result of each set of calculations is a sum of squares with four degrees of freedom, and since there are four such sums of squares,

Table 11-9. Trend analysis on three-way interaction, data from Table 11-6.

		Linear		Quadratic		Cubic		Quartic	
		C_1	$(C_1')^2$	C_2	$(C_2')^2$	C_3	$(C_3')^2$	C_4	$(C_4')^2$
A_1	B_1	−54.0	1458.0	−30.0	321.4	12.0	72.0	−2.4	0.4
	B_2	−3.6	6.5	54.0	1041.4	1.2	.7	−2.4	0.4
	B_3	85.2	3629.5	−37.2	494.2	−8.4	35.3	44.4	140.8
A_2	B_1	−61.2	1872.7	−10.8	41.7	8.4	35.3	−18.0	23.1
	B_2	8.4	35.3	−78.0	2172.9	−4.8	11.5	−9.6	6.6
	B_3	54.0	1458.0	−18.0	115.7	0.0	0.0	−21.6	33.3
A_3	B_1	−82.8	3427.9	49.2	864.5	−2.4	2.9	−4.8	1.6
	B_2	−1.2	0.7	70.8	1790.2	−9.6	46.1	14.4	14.8
	B_3	108.0	5832.0	48.0	822.9	0.0	0.0	7.2	3.7
Sum			17720.6		7664.9		203.8		224.7
C Main Effect			154.9		91.4		0.7		0.4
AC Interaction			68.3		4647.2		29.3		101.2
BC Interaction			16565.9		183.9		94.1		36.2
Sum			16789.1		4922.5		124.1		137.8
SS			931.5		2742.4		79.7		86.9
df			4		4		4		4
MS			232.9		685.6		19.9		21.7
F			1.06		3.12		0.09		0.10
α			—		.02		—		—
$\hat{\omega}^2$			.00		.03		−.01		−.01

their total degrees of freedom are 16 − the degrees of freedom of the ABC interaction. Moreover the total of the sums of squares is $931.5 + 2742.4 + 79.7 + 86.9 = 3840.5 = SS_{abc}$.

Effect variances and $\hat{\omega}^2$ values are calculated as described on pp. 250–252. Each variance estimate is equal to the corresponding mean square minus MS_w, multiplied by the degrees of freedom, and divided by N. The complete trend analysis on the data in Table 11-6 is summarized in Table 11-10.

Estimation

The significant effects are the B main effect, the quadratic component of the AC interaction, the linear component of the BC interaction, and the quadratic component of the ABC interaction. In the three-way design, however, the estimates for the quadratic components of the ABC interaction will include those for the AC interaction. Thus, only the B main effect, the linear BC interaction, and the quadratic ABC interaction need to be included in the estimation equation.

To see that the quadratic components of the ABC interaction include those for the AC interaction, take the average, over the levels of B, of the C values for the quadratic ABC interaction. The average for level A_1, for example, is $(−30.0 + 54.0 − 37.2)/3 = −4.4$ Table 11-8 shows that −4.4 is the C value of A_1 for the quadratic component of the AC interaction.

Table 11-10. Summary of trend analyses on data in Table 11-6 (bold numbers are totals).

Effect		SS	df	MS	F	α	$\hat{\omega}^2$
m		93881.0	1				
a		278.7	2	139.4	0.63		.00
b		2243.8	2	1121.9	5.10	.008	.03
c		**247.4**	**4**				**(−.01)**
	linear	154.9	1	154.9	.70		.00
	quadratic	91.4	1	91.4	.42		.00
	cubic	0.7	1	0.7	0.00		.00
	quartic	0.4	1	0.4	0.00		.00
ab		344.3	4	86.1	0.39		**(−.01)**
ac		**4846.1**	**8**				**.05**
	linear	68.3	2	34.2	0.16		(−.01)
	quadratic	4647.2	2	2323.6	10.56	<.001	.06
	cubic	29.3	2	14.6	0.07		(−.01)
	quartic	101.2	2	50.6	0.23		.00
bc		**16880.2**	**8**				**.22**
	linear	16565.9	2	8283.0	37.65	<.001	.24
	quadratic	183.9	2	92.0	0.42		.00
	cubic	94.1	2	47.0	0.21		(−.01)
	quartic	36.2	2	18.1	0.08		(−.01)
abc		**3840.5**	**16**				**.00**
	linear	931.5	4	232.9	1.06		.00
	quadratic	2742.4	4	685.6	3.12	.02	.03
	cubic	79.7	4	19.9	0.09		(−.01)
	quartic	86.9	4	21.7	0.10		(−.01)
w		39600.0	180	220.0			.7
t		68281.0	224				

Consequently, the average of the C values for the quadratic ABC interaction take into account those for the AC interaction. In general, the terms for any trend component of an interaction will include, implicitly, the terms for the same trend on all main effects and lower-order interactions involving all of the *numerical* factors in the higher interaction (Factor C in this case). These are the same terms as those that are subtracted in finding the sum of squares of the higher interaction. For the ABC interaction in this design it includes AC and BC interactions and the C main effect. Of these the only significant quadratic term is of course in the AC interaction.

The estimation equation, using the B main effect, the linear BC interaction, and the quadratic ABC interaction, is

$$\hat{\mu}_{ijk} = \hat{\mu} + \hat{\beta}_j + \hat{c}^*(b_j)_1 c_{k1} + \hat{c}^*(ab_{ij})_2 c_{k2}$$
$$= \overline{X}_{.j..} + \hat{c}^*(b_j)_1 c_{k1} + \hat{c}^*(ab_{ij})_2 c_{k2}.$$

Our symbols are now even more complex, and the lower case c has taken on two meanings. In general, when the term is starred, it refers to the C factor; when it is not, it refers to a coefficient for calculating a trend. Thus, c_{k1} is the kth coefficient for testing the linear trend component ($c_{11} = -2$, $c_{21} = -1, \ldots, c_{51} = 2$), and c_{k2} is the kth coefficient for testing the quadratic trend component. The term $c^*(b_1)_j$ refers to the estimated linear effect over Factor C, at level B_j, and $c^*(ab_{ij})_2$ refers to the estimated quadratic trend over Factor C at level AB_{ij}.

Each of the latter estimates is calculated in the usual way, i.e., the corresponding C value (the value from which $(C')^2$ is calculated) is divided by the sum of the squared coefficients from which it was derived. Thus, the $c^*(b_j)_1$ are calculated from the column labelled C_1 inder BC interaction in Table 11-8 as

$$c^*(b_1)_1 = -66.00/10 = -6.6.$$
$$c^*(b_2)_1 = 1.20/10 = .12,$$
$$c^*(b_3)_1 = 82.40/10 = 8.24.$$

(The divisor, 10, is the sum of the c_{k1}^2 for the linear trend on the C main effect; this can be seen at the top of Table 11-8.

The nine values of $c^*(ab_{ij})_2$ are calculated by the same formula from the values in the column labeled C_2 in Table 11-9:

$$c^*(ab_{11})_2 = -30.0/14 = -2.14$$
$$c^*(ab_{12})_2 = 54.0/14 = 3.86$$
$$\cdot \ \cdot \ \cdot \ \cdot \ \cdot \ \cdot$$
$$c^*(ab_{33})_2 = 48.0/14 = 3.43.$$

These values, along with the estimated cell means, are in Table 11-11. The estimated cell means are plotted, along with the actual cell means, in Figure 11-4.

GENERAL PRINCIPLES

The general principle and its application to higher-way interactions should now be clear. First, $(C')^2$ values are computed for the trend in question for each combination of all of the other factors involved. These $(C')^2$ are then totalled, and from this total is subtracted the total of the sums of squares for all lower-order effects involving the numerical factor (and *not* involving any new factors not in the interaction in question) for the same trend component. The result is a sum of squares whose degrees of freedom

Table 11-11. Estimates of effects and cell means for data in Table 11-6, Factor C numerical.

$\hat{\mu}$	$\hat{\beta}_1$	$\hat{\beta}_2$	$\hat{\beta}_3$	$\hat{c}^*(b_1)_1$	$\hat{c}^*(b_2)_1$	$\hat{c}^*(b_3)_1$
20.43	−4.11	0.53	3.57	−6.60	0.12	8.24

$c^*(ab_{ij})_2$

	a_1	a_2	a_3
b_1	−2.14	−0.77	3.51
b_2	3.86	−5.57	5.06
b_3	−2.66	−1.29	3.43

Actual (upper values) and estimated (lower values) cell means:

		C_1	C_2	C_3	C_4	C_5
A_1	B_1	24.0	28.8	22.8	13.2	4.8
		25.2	25.1	20.6	11.9	−1.2
	B_2	31.8	20.4	15.6	19.2	30.6
		28.4	17.0	13.2	17.2	28.9
	B_3	2.4	13.2	32.4	33.6	34.8
		2.2	18.4	29.3	34.9	35.2
A_2	B_1	26.4	26.4	16.8	10.8	3.6
		28.0	23.7	17.9	10.5	1.6
	B_2	7.2	24.0	30.0	27.6	9.6
		9.6	26.4	32.1	26.6	10.1
	B_3	10.8	21.6	25.2	32.4	32.4
		4.9	17.0	26.6	33.5	37.9
A_3	B_1	37.2	18.0	6.0	2.4	3.6
		36.5	19.4	9.3	6.2	10.1
	B_2	31.2	12.0	10.8	15.6	28.8
		30.8	15.8	10.8	16.0	31.3
	B_3	9.6	9.6	18.0	31.2	52.8
		14.4	12.3	17.1	28.8	47.3

are the number of $(C')^2$ that were summed, minus the total degrees of freedom of the sums of squares that were subtracted. A simple check on the calculations derives from the fact that the resulting sums of squares and degrees of freedom must themselves sum to the sum of squares and degrees of freedom of the interaction being partitioned.

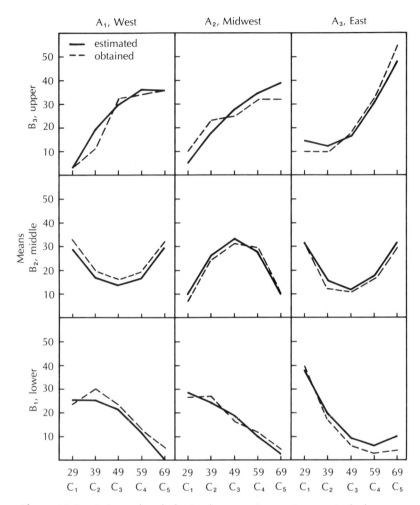

Figure 11-4. Estimated and obtained means, Factor C numerical; data from Table 11-6.

The variances of the effects are estimated by taking the difference between the mean square for the effect and the mean square used in the denominator for testing that effect, multiplying the difference by the degrees of freedom of the effect, and dividing by N. Dividing this estimate by $\hat{\sigma}_t^2$ gives $\hat{\omega}^2$.

The effects are always estimated by dividing the C value for the effect by the sum of the coefficients used in finding that C value. Estimates of a trend component in a higher-order interaction implicitly include those for the same trend in all lower-order interactions and main effects involving the numerical factor (and that do *not* involve any new factors not involved in the higher-order interaction).

TWO NUMERICAL FACTORS

A multifactor design may sometimes contain more than one factor with numerical values. In that case trends on interactions can be more finely partitioned. Suppose, for example, that in the study in Table 11-1 the three schools had been three different, evenly spaced sizes (say, 400, 800, and 1200 pupils), and we had wished to study the effects of size of school as well as student IQ. The trend analysis on the A main effect (IQ level) would then be the same as before, but, in addition, we could perform a trend analysis on school size as well. The procedure is exactly the same as that for the trend analysis on Factor A, and the analysis is shown in Table 11-12.

Two-way Interactions

The two-way interaction can be divided into as many trend components as there are degrees of freedom, with each trend component being tested by a planned comparison. More specifically, each trend in Table 11-3 can be divided into two components, each with one degree of freedom. The linear component on A, for example, can be divided into two components, a linear x linear and a linear x quadratic. The first is a linear trend, across the levels of B, on the linear trend components of A, i.e., it is a tendency for the linear trend components of A to increase or decrease over levels of B. The second is a quadratic trend, across levels of B, on the linear components of A.

Since A and B are both numerical, however, and there is no good mathematical reason why one should take precedence over the other, the linear x linear trend can also be thought of as a linear trend, across levels of A, on the linear trend components of B. Similarly, the linear x quadratic trend can be thought of as a linear trend, across the levels of A, on the quadratic components of B. These points will be clearer when the actual analyses are described.

Carrying this partitioning across the trend components in Table 11-3, we also obtain a quadratic x linear, quadratic x quadratic, cubic x linear, and cubic x quadratic trend. There are thus a total of six trend components, one for each of the six degrees of freedom in the AB interaction. Each can be thought of as a trend, across one factor, on the components of a trend on the other factor.

Table 11-12.   Trend analysis on Factor B, data from Table 11-1.

		B_1	B_2	B_3						
School size:		400	800	1200	Σc^2	C_k	$(C_k')^2$	F	α	$\hat{\omega}^2$
	Mean	45.00	52.75	57.75						
c_{ik}	Linear	-1	0	1	2	12.75	3251.2	81.3	$<.001$	.13
	Quad.	1	-2	1	6	-2.75	50.4	1.26	—	.00

Alternatively, these trends can be expressed in terms of an extension of the model equation. We have seen that the model equation with one factor numerical can be extended to include a term for each trend component. The model equation for the two-way analysis, with both factors numerical, can also be extended to represent each trend component. If we let V_i represent the numerical value of the ith level of Factor A, as before, and let W_j represent the numerical value of the jth level of Factor B, then the extended equation is

$$
\begin{aligned}
X_{ijk} = \mu + &[a_1(V_i - \overline{V}) + a_2(V_i - \overline{V})^2 + a_3(V_i - \overline{V})^3 + \cdots] + \quad \textbf{(11-1)} \\
&[b_1(W_j - \overline{W}) + b_2(W_j - \overline{W})^2 + b_3(W_j - \overline{W})^3 + \cdots] + \\
&[ab_{11}(V_i - \overline{V})(W_j - \overline{W}) + ab_{12}(V_i - \overline{V})(W_j - \overline{W})^2 + \cdots + \\
&ab_{21}(V_i - \overline{V})^2(W_j - \overline{W}) + ab_{22}(V_i - \overline{V})^2(W_j - \overline{W})^2 + \cdots] + \\
&\epsilon_{ijk}
\end{aligned}
$$

The terms in the first set of brackets represent the A main effect; each value of a_k represents one trend component, to be tested with a planned comparison with one degree of freedom. The terms in the second set of brackets represent the B main effect; each b_l represents one B main effect trend component. In the same way, the terms in the third set of brackets are those of the AB interaction; each ab_{kl} represents one trend component of the Ab interaction. The term ab_{11} represents the linear x linear component, ab_{12} represents the linear x quadratic, and so on.

Generally, Equation 11-1 is used when estimating the μ_{ij}, in much the same way that Equation 10-1 is used in the one-way trend test; those terms that are significant are kept, and those that are not significant are discarded.

Significance Tests

Each of the above interpretations of two-way trends suggests a different method of testing for them. Since the method suggested by Equation 11-1 is in some ways more basic, it will be described first.

First Method. Just as the rationale for the tests of one-way trend components is difficult to explain, so is the rationale for the two-way tests—we will simply describe the tests without attempting to give their rationale. It is well to remember, however, that all such tests are valid only if every cell contains the same number of observations and the values of both the V_i and the W_j are equally spaced.

The procedure is illustrated in Table 11-13. With this procedure, it is necessary to construct a two-way table of coefficients for each trend component. To find the linear x linear coefficients, we write the values of the coefficients (c_{i1}) that were used to test the linear component of the A main effect in one margin of the table. In Table 11-13, since the columns represent the A main effect, these coefficients are written as column headings. In the other margin we write the coefficients for the linear trend of the B main effect. The entry in each cell is then the product of that cell's row and column headings. The entry in the top left-hand cell, for example, is

Table 11-13. Trend analysis on AB interaction, data from Table 11-1, both A and B numerical.

		Linear × Linear A						Quadratic × Quadratic A		
		−3	−1	1	3		1	−1	−1	1
	−1	3	1	−1	−3	1	1	−1	−1	1
B	0	0	0	0	0	B −2	−2	2	2	−2
	1	−3	−1	1	3	1	1	−1	−1	1

$$C_{11} = 17, \ (C'_{11})^2 = 72.2$$
$$F = 1.81, \ \alpha > .10$$

$$C_{22} = 5, \ (C'_{22})^2 = 10.4$$
$$F = 0.26, \ \alpha > .10$$

		Linear × Quadratic A						Cubic × Linear A		
		−3	−1	1	3		−1	3	−3	1
	1	−3	−1	1	3	−1	1	−3	3	−1
B	−2	6	2	−2	−6	B 0	0	0	0	0
	1	−3	−1	1	3	1	−1	3	−3	1

$$C_{12} = -33, \ (C'_{12})^2 = 90.8$$
$$F = 2.27, \ \alpha > .10$$

$$C_{31} = -1, \ (C'_{31})^2 = 0.2$$
$$F = .01, \ \alpha > .10$$

		Quadratic × Linear A						Cubic × Quadratic A		
		1	−1	−1	1		−1	3	−3	1
	−1	−1	1	1	−1	1	−1	3	−3	1
B	0	0	0	0	0	B −2	2	−6	6	−2
	1	1	−1	−1	1	1	−1	3	−3	1

$$C_{21} = -17, \ (C'_{21})^2 = 361.2$$
$$F = 9.03, \ \alpha < .01$$

$$C_{32} = 29, \ (C'_{32})^2 = 70.1$$
$$F = 1.75, \ \alpha > .10$$

$(-1)(-3) = +3$. The cell entries are the coefficients, $c_{ij,11}$, for testing the linear x linear trend component. The test itself is an ordinary planned comparison, using the cell entries for the linear x linear part of Table 11-13 as coefficients for the cell means in Table 11-1. The linear x linear C value is calculated as

$$C_{11} = 3(31) + 1(43) - 1(48) - 3(58) -$$
$$3(37) - 1(59) + 1(66) + 3(69) = 17.$$

The value of $(C'_{11})^2$ (in this case, 72.25) is then found in the usual way, using n as the multiplier, and the sum of the squares of the coefficients in the cells of the linear x linear table (in this case, 40) as the divisor. The F ratio, 1.81, is not significant.

The rest of the trend components are tested in exactly the same way, but the values used as row and column headings are different. For the linear x quadratic component, for example, the column values are those for the linear component of the A main effect, as before, but the row values are those for the quadratic component of the B main effect. The cell entries are the products of the row and column values, as before, and the test is calculated as a planned comparison.

Table 11-13 shows that the only significant trend is the quadratic x linear, indicating that there is a linear trend in the quadratic components of A. This linear trend can be seen in Figure 11-1, which shows that the amount of curvature across levels of A increases from the nearly straight trend in school B_1 to the highly curved function in school B_3. Again, an alternative interpretation of the quadratic x linear trend can be seen in Figure 11-5, which plots trends across levels of B for different levels of A. Here we see a quadratic trend in the slope of the function, with the greatest slope occurring for A_3 and the next greatest for A_2. (Differences in the overall heights of the curves reflect the A main effect.)

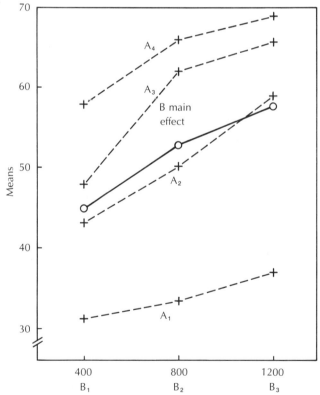

Figure 11-5. Cell means in Table 11-1 plotted as a function of Factor B.

Second Method. The other method of calculating $(C')^2$ for each trend component is generally simpler computationally. It makes use of the fact that the linear x quadratic trend, for example, is a quadratic trend, over the levels of B, for the linear trend on A. Accordingly, to test the linear x linear and linear x quadratic trends, we first find the C values for the linear trend on A, at each level of B. These were found in Table 11-3 and are reproduced in the upper table of Table 11-14. We then test for linear and quadratic trends on these C values. To find the linear x quadratic trend, for example, we do a quadratic trend test on the linear C values: C_{12} $= 1(86) - 2(111) + 1(103) = -33$. This, it can be seen from Table 11-13, is the same as the linear x quadratic C value calculated by the previous method. The values of $(C')^2$ are then found by squaring each C value, multiplying by the number of observations in each cell, and dividing by the sum of the squares of the coefficients as before. The sum of the squares of the coefficients can be found, however, without first finding every coefficient. To find the sum of the squares of the coefficients for the linear x quadratic trend, we first find the sum of the squared coefficients for the

Table 11-14. Alternative way to calculate AB interaction trends, data from Table 11-1, both A and B Numerical.

	A Linear								
	B_1	B_2	B_3	Σc^2					
School size:	400	800	1200						
C_1:	86	111	103	20	C	$(C')^2$	F	α	$\hat{\omega}^2$
Linear × Linear	−1	0	1	2	17	72.2	1.81	—	.00
Linear × Quadratic	1	−2	1	6	−33	90.8	2.27	—	.00

	A Quadratic								
	B_1	B_2	B_3	Σc^2					
School size:	400	800	1200						
C_2:	−2	−13	−19	4	C	$(C')^2$	F	α	$\hat{\omega}^2$
Quadratic × Linear	−1	0	1	2	−17	361.2	9.03	<.01	.01
Quadratic × Quadratic	1	−2	1	6	5	10.4	0.26	—	.00

	A Cubic								
	B_1	B_2	B_3	Σc^2					
School size:	400	800	1200						
C_3:	12	−3	11	20	C	$(C')^2$	F	α	$\hat{\omega}^2$
Cubic × Linear	−1	0	1	2	−1	0.2	0.01	—	.00
Cubic × Quadratic	1	−2	1	6	29	70.1	1.75	—	.00

linear trend on the A main effect (20, as can be seen from Table 11-2). We then multiply it by the sum of squared coefficients for the quadratic trend on the B main effect (6, from Table 11-12), to get $20 \times 6 = 120$. The $(C')^2$ for the linear x quadratic trend is thus $10(-33)^2/120 = 90.75$. The values for the other trends are found the same way. The complete set of tests is shown in Table 11-14. Estimates of variances and of the ω^2 follow the usual procedures for contrasts.

Estimation

The significant quadratic x linear interaction, together with the significant linear and quadratic trends in the A main effect, plus the significant linear trend in the B main effect, suggest that the data can be accounted for most economically with the equation

$$\hat{\mu}_{ij} = \hat{\mu} + \hat{a}_1^* c_{i1(a)} + \hat{a}_2^* c_{i2(a)} + \hat{b}_1^* c_{j1(b)} + \hat{ab}_{12}^* c_{ij.21(ab)}.$$

In this equation $c_{i1(a)}$ is the coefficient of level A_i in testing the linear component of the A main effect, i.e.,

$$c_{11(a)} = -3, c_{21(a)} = -1, c_{31(a)} = 1, c_{41(a)} = 3.$$

Similarly, $c_{i2(a)}$ is the coefficient of the level A_i in testing the quadratic component of the A main effect, and $c_{j1(b)}$ is the coefficient of level B_j in testing the linear component of the B main effect. The value $c_{ij.21(ab)}$ is the coefficient of level AB_{ij} in testing the quadratic x linear component of the AB interaction:

$$c_{11.21(ab)} = 3, c_{12.21(ab)} = 0, c_{13.21(ab)} = -3, c_{21.21(ab)} = 1, \cdots$$

Table 11-15. Estimated effects and cell means, data from Table 11-1, both A and B numerical.

Estimated effects:

$\hat{\mu}$	a_1^*	a_2^*	b_1^*	ab_{21}^*
51.83	5.00	-2.83	6.38	-2.12

Obtained (upper values) and estimated (lower values) cell means:

	A_1	A_2	A_3	A_4
B_1	31	43	48	58
	29.7	41.2	51.2	59.7
B_2	33	50	62	66
	34.0	49.7	59.7	64.0
B_3	37	59	66	69
	38.3	58.2	68.2	68.3

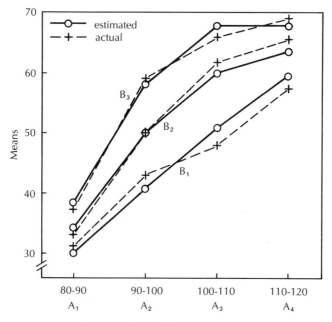

Figure 11-6. Estimated and obtained means, both A and B numerical; data from Table 11-1.

and so on. Unfortunately, the subscripts make these terms look more complicated than they are. The equation for $\hat{\mu}_{11}$, for example, would be

$$\hat{\mu}_{11} = 51.83 + \hat{a}_1^*(-3) + \hat{a}_2^*(1) + \hat{b}_1^*(-1) + \hat{ab}_{12}^*(-1).$$

Estimating the effects a_1^*, a_2^*, b_1^*, and ab_{12}^* follows the procedure described on pp. 230–231. In particular, the equation used is Equation 10-12, with the divisor being in each case the sum of the squares of the coefficients used for testing that effect. The estimates are

$$\hat{a}_1^* = 100/20 = 5$$
$$\hat{a}_2^* = -11.33/4 = -2.83$$
$$\hat{b}_1^* = 12.75/2 = 6.38$$
$$\hat{ab}_{12}^* = -17/8 = -2.12.$$

The estimate of μ_{11} is then $51.83 + 5.00(-3) - 2.83(1) + 6.38(-1) - 2.12$ (-1), or 29.7. The other estimates, along with the obtained cell means, are shown in Table 11-15 and Figure 11-6. In Figure 11-6 the increasing quadratic trend over levels of B is especially clear.

Three-way Interactions with Two Numerical Factors

Three-way and higher-way designs in which two factors are numerical require only minor extensions of the methods discussed above. The extensions will be illustrated using the data in Table 11-6, assuming that the three income levels (upper, middle, and lower), Factor B, are regarded as equally spaced on some numerical scale of social level.

The trends on the B main effect are tested in the usual way, and the AB and BC interactions are divided into trends exactly as in the two-way design discussed above. The results of these analyses are in Table 11-16. Only the linear trend on the B main effect and the linear x linear trend on the BC interaction are significant.

Significance Tests

The ABC interaction can be divided into the same eight components as the BC interaction (Table 11-16), just as the AB interaction, with B numerical, is divided into the same number of components as the B main effect (Table 11-16). To make significance tests, C and $(C')^2$ are separately calculated for each trend component at each level of A. The three $(C')^2$ for a given trend are then summed and the $(C')^2$ for the corresponding component of the BC interaction is subtracted. In this example only the BC interaction is subtracted because it is the only lower-order effect that involves both numerical factors. The result is then treated as a sum of

Table 11-16. Trend analyses on B main effect and two-way interactions, data from Table 11-6.

B Main Effect:

	C	Σc^2	$(C')^2$	F	α	$\hat{\omega}^2$
Linear	7.68	2	2211.8	10.05	$<.01$	.03
Quadratic	−1.60	6	32.0	0.15	—	.00

AB Interaction:

	SS	df	MS	F	α	$\hat{\omega}^2$
Linear	243.3	2	121.6	0.55	—	<0
Quadratic	101.0	2	50.5	0.23	—	<0

BC Interaction:

	C	Σc^2	$(C')^2$	F	α	$\hat{\omega}^2$
Linear × Linear	148.40	20	16516.9	75.1	$<.001$	.24
Linear × Quadratic	−5.20	28	14.5	0.07	—	.00
Linear × Cubic	−8.80	20	58.1	0.26	—	.00
Linear × Quartic	18.40	140	36.3	0.16	—	.00
Quadratic × Linear	14.00	60	49.0	0.22	—	.00
Quadratic × Quadratic	−30.80	84	169.4	0.77	—	.00
Quadratic × Cubic	12.00	60	36.0	0.16	—	.00
Quadratic × Quartic	0.00	420	0.0	0.00	—	.00

squares with $(I - 1)$ degrees of freedom. To illustrate, with the linear x linear component, the $(C')^2$ for level A_1 is calculated by multiplying the cell means at A_1:

	C_1	C_2	C_3	C_4	C_5
B_1	24.0	28.8	22.8	13.2	4.8
B_2	31.8	20.4	15.6	19.2	30.6
B_3	2.4	13.2	32.4	33.6	34.8

by the linear x linear coefficients

		C_1 -2	C_2 -1	C_3 0	C_4 1	C_5 2
B_1	-1	2	1	0	-1	-2
B_2	0	0	0	0	0	0
B_3	1	-2	-1	0	1	2

The resulting C value (139.2) is then squared, multiplied by 5 (the number of observations on which each cell mean is based), and divided by 20 (the sum of the squares of the coefficients). The final $(C')^2$ is 4844.2. The same procedure is followed for the other two levels of A, giving $(C')^2$ values of 3317.8 for A_2 and 9101.2 for A_3. These values are summed and the $(C')^2$ for the linear x linear component of the BC interaction is subtracted to obtain the sum of squares: $4844.2 + 3317.8 + 9101.2 - 16516.9 = 746.3$, with 2 degrees of freedom. The mean square is then $746.3/2 = 373.2$, and $F = 373.2/220 = 1.70$, which is not significant.

The remaining seven tests can be made the same way. The tests are shown in Table 11-17. The quadratic x quadratic interaction is significant at the .02 level; none of the others are significant.

The method illustrated in Table 11-14, for the data in Table 11-1, can also be used to calculate the C values in Table 11-17. We can generate the C values in Table 11-17 from those in Table 11-9 by performing linear and quadratic trend tests, over levels of B, on each level of A. The C values for the quadratic trend at A_1, for example, are -30.0, 54.0, and -37.2, in Table 11-9. To perform a linear test on these values, we use the coefficients for the linear trend on B to get $-1(-30.0) + 0(54.0) + 1(-37.2) = -7.2$, the C value for the linear x quadratic trend at A_1. The divisor used in finding $(C')^2$, i.e., the sum of the squared coefficients, is the product of the sum of the squared linear coefficients on B times the sum of the squared quadratic coefficients on C, i.e., $2 \times 14 = 28$.

Table 11-17. Trend analysis on three-way interaction, data from Table 11-6, Factors B and C numerical.

	Linear × Linear		Linear × Quadratic		Linear × Cubic		Linear × Quartic	
	C	$(C')^2$	C	$(C')^2$	C	$(C')^2$	C	$(C')^2$
A_1	139.2	4844.2	−7.2	9.3	−20.4	104.0	46.8	78.2
A_2	115.2	3317.8	−7.2	9.3	−8.4	17.6	−3.6	0.5
A_3	190.8	9101.2	−1.2	0.3	2.4	1.4	12.0	5.1
Sum		17263.2		18.9		123.0		83.8
BC Interaction		16516.9		14.5		58.1		36.3
SS		746.3		4.4		64.9		47.5
df		2		2		2		2
MS		373.2		2.2		32.4		23.8
F		1.70		0.01		0.15		0.11
α		—		—		—		—
$\hat{\omega}^2$		.01		−.01		−.01		−.01

	Quadratic × Linear		Quadratic × Quadratic		Quadratic × Cubic		Quadratic × Quartic	
	C	$(C')^2$	C	$(C')^2$	C	$(C')^2$	C	$(C')^2$
A_1	38.4	122.9	−175.2	1827.1	1.2	0.1	46.8	26.1
A_2	−24.0	48.0	127.2	963.1	18.0	27.0	−20.4	5.0
A_3	27.6	63.5	−44.4	117.3	16.8	23.5	−26.4	8.3
Sum		234.4		2907.5		50.6		39.4
BC Interaction		49.0		169.4		36.0		0.0
SS		185.4		2738.1		14.6		39.4
df		2		2		2		2
MS		92.7		1369.0		7.3		19.7
F		0.42		6.22		0.03		0.09
α		—		.003		—		—
$\hat{\omega}^2$		.00		.03		−.01		−.01

The remaining C and $(C')^2$ values can be found the same way, with a significant saving in labor, since complete tables of coefficients need not be made.

Estimation

The significant effects are now the linear component of the B main effect, the quadratic component of the AC interaction, the linear x linear component of the BC interaction, and the quadratic x quadratic component of the ABC interaction. The estimation equation is therefore

$$\hat{\mu}_{ijk} = \hat{\mu} + \hat{b}_1^* c_{j1} + \hat{c}^* (a_i)_2 c_{k2} + \hat{bc}_{11}^* c_{jk.11} + \hat{bc}^* (a_i)_{22} c_{jk.22}.$$

Here again we have run out of different symbols; a starred c refers to a level of the C main effect, and an unstarred c refers to a coefficient used in testing for trends. Each starred term is estimated separately from the related C value, by dividing that C value by the sum of its squared coefficients. The complete set of estimates is in Table 11-18, and the estimated cell means are in Figure 11-7.

Table 11-18. Estimated effects and cell means, data from Table 11-6, B and C numerical.

Estimated effects:

$\hat{\mu}$	$\hat{b}_1^*$	$\hat{c}^*(a_1)_2$	$\hat{c}^*(a_2)_2$	$\hat{c}^*(a_3)_2$	$\hat{bc}_{11}^*$
20.43	3.84	-0.31	-2.54	4.00	7.42

		$\hat{bc}^*(a_1)_{22}$	$\hat{bc}^*(a_2)_{22}$	$\hat{bc}^*(a_3)_{22}$
		-2.09	1.51	-0.53

Obtained (upper values) and estimated (lower values) cell means:

		C_1	C_2	C_3	C_4	C_5
	B_1	24.0 / 26.6	28.8 / 26.4	22.8 / 21.4	13.2 / 11.6	4.8 / -3.0
A_1	B_2	31.8 / 28.2	20.4 / 16.6	15.6 / 12.7	19.2 / 16.6	30.6 / 28.2
	B_3	2.4 / 4.6	13.2 / 19.2	32.4 / 29.1	33.6 / 34.1	34.8 / 34.3
	B_1	26.4 / 29.4	26.4 / 25.0	16.8 / 18.6	10.8 / 10.2	3.6 / -0.3
A_2	B_2	7.2 / 9.3	24.0 / 26.0	30.0 / 31.6	27.6 / 26.0	9.6 / 9.3
	B_3	10.8 / 7.4	21.6 / 17.9	25.2 / 26.3	32.4 / 32.7	32.4 / 37.0
	B_1	37.2 / 38.4	18.0 / 20.5	6.0 / 9.6	2.4 / 5.7	3.6 / 8.7
A_3	B_2	31.2 / 30.6	12.0 / 15.4	10.8 / 10.3	15.6 / 15.4	28.8 / 30.6
	B_3	9.6 / 16.4	9.6 / 13.4	18.0 / 17.3	31.2 / 28.2	52.8 / 46.0

276

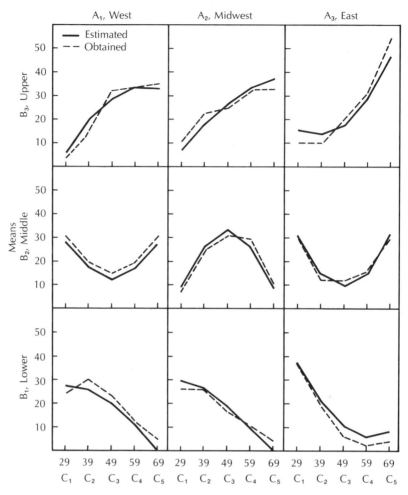

Figure 11-7. Estimated and obtained means, both B and C numerical; data from Table 11-6.

MORE THAN TWO NUMERICAL FACTORS

Experimental designs with more than two numerical factors are relatively rare, nevertheless, their analysis is a logical extension of the methods described in the previous sections. Main effects and two-way interactions are of course handled exactly as in the previous sections, no matter what the total number of numerical factors may be, as are all three-way interactions in which no more than two factors are numerical. The remaining effects are three-way interactions involving three numerical factors and interactions involving a total of more than three factors.

Three-way Interactions

Three-way interactions will be illustrated with a further analysis of the data in Table 11-6. This time we will assume that the three areas of the country (Factor A) are approximately equally spaced geographically, so that we can assign to them the numerical values -1 for the west, 0 for the midwest, and $+1$ for the east. All three factors in that design are then numerical. In this case there is no use dividing the A main effect or the AB interaction into trend components, since the sums of squares of both are so small that no single trend could be significant. (Remember, the $(C')^2$ for any trend cannot exceed the sum of squares for the effect of which it is a part.) The analysis of the AC interaction is straightforward and appears in Table 11-19. The linear x quadratic and quadratic x quadratic components are significant.

The three way interaction, with 16 degrees of freedom, can now be divided into 16 trend components. The 16 components correspond to the 16 ways of combining the two possible components (linear and quadratic) on A, the two on B, and the four on C. They are listed as the 16 rows in Table 11-20, to which we will return later.

Each trend component is tested by a planned comparison whose coefficients are the products of the corresponding coefficients for the individual factors. Table 11-21 illustrates the method of finding the coefficients for the linear x quadratic x quadratic component. Since the component is linear in Factor A, we use the coefficients for the linear trend on A, -1, 0, 1; the component is quadratic in B, so for B we use the quadratic coefficients, 1, -2, 1; and the component is quadratic in C, so we use the quadratic components, 2, -1, -2, -1, 2. The term in each cell is the product of the three corresponding coefficients for the individual factors. Thus, the coefficient in cell ABC_{111} is $(-1)(1)(2) = -2$, the term in cell ABC_{121} is $(-1)(-2)(2) = 4$, and so on. The terms in the cells are the coefficients for calculating the linear x quadratic x quadratic interaction as a planned comparison.

Table 11-19.   Trend components of AC interaction, data from Table 11-6.

	C	Σc^2	$(C')^2$	F	α	$\hat{\omega}^2$
Linear × Linear	-1.20	20	1.1	0.00	—	.00
Linear × Quadratic	60.40	28	1954.4	8.88	.004	.03
Linear × Cubic	-5.60	20	23.5	0.11	—	.00
Linear × Quartic	-7.60	140	6.2	0.03	—	.00
Quadratic × Linear	16.40	60	67.2	0.31	—	.00
Quadratic × Quadratic	122.80	84	2692.8	12.24	< .001	.04
Quadratic × Cubic	-4.80	60	5.8	0.03	—	.00
Quadratic × Quartic	51.60	420	95.1	0.43	—	.00

Table 11-20. Trend tests on three-way interaction, data from Table 11-6, all factors numerical.

			Factor A Linear							
	c_{ij}	A_1 -1	A_2 0	A_3 1	Σc^2 2	C	$(C')^2$	F	α	$\hat{\omega}^2$
B	C									
Linear	Linear	139.2	115.2	190.8	20	51.6	332.8	1.51	—	.00
	Quadratic	−7.2	−7.2	−1.2	28	6.0	3.2	0.01	—	.00
	Cubic	−20.4	−8.4	2.4	20	22.8	65.0	0.30	—	.00
	Quartic	46.8	−3.6	12.0	140	−34.8	21.6	0.10	—	.00
Quad.	Linear	38.4	−24.0	27.6	60	−10.8	4.9	0.02	—	.00
	Quadratic	−175.2	127.2	−44.4	84	130.8	509.2	2.31	—	.00
	Cubic	1.2	18.0	16.8	60	15.6	10.1	0.05	—	.00
	Quartic	46.8	−20.4	−26.4	420	−73.2	31.9	0.14	—	.00

			Factor A Quadratic							
	$c_{i.j}$	A_1 1	A_2 -2	A_3 1	Σc^2 6	C	$(C')^2$	F	α	$\hat{\omega}^2$
B	C									
Linear	Linear	139.2	115.2	190.8	20	99.6	413.3	1.88	—	.00
	Quadratic	−7.2	−7.2	−1.2	28	6.0	1.1	0.00	—	.00
	Cubic	−20.4	−8.4	2.4	20	−1.2	0.1	0.00	—	.00
	Quartic	46.8	−3.6	12.0	140	66.0	25.9	0.12	—	.00
Quad.	Linear	38.4	−24.0	27.6	60	114.0	180.5	0.82	—	.00
	Quadratic	−175.2	127.2	−44.4	84	−474.0	2228.9	10.13	.002	.03
	Cubic	1.2	18.0	16.8	60	−18.0	4.5	0.02	—	.00
	Quartic	46.8	−20.4	−26.4	420	61.2	7.4	0.03	—	.00

Calculating the full set of coefficients for all 16 trend components, however, becomes very tedious. A shorter method for testing the trends is shown in Table 11-20. There we make use of the fact that we have already analyzed the same data, with B and C numerical, in Table 11-17. To generate the 16 trend components in Table 11-20, we simply perform a linear and a quadratic trend test on each of the eight sets of C values in Table 11-17. To find the linear x quadratic x quadratic trend, for example, we first write down the three C values for the quadratic x quadratic trend, across A: −175.2 at A_1, 127.2 at A_2, and −44.4 at A_3. The C value for a linear trend on these three values is $-1(-175.2) + 0(127.2) + 1(-44.4) = 130.8$. The sum of the squared coefficients, i.e., the divisor when finding $(C')^2$, is found by multiplying the sum of the A linear trend (2) with that for the BC quadratic x quadratic trend (84), to get 168. Therefore, $(C')^2$ is $5(130.8)^2/168 = 509.2$. The remaining trend components, calculated the same way, are shown in Table 11-20; only the quadratic x quadratic x quadratic trend is significant. Estimates of proportion of variance accounted for and estimates of effects are calculated just as for lower-order trends.

Table 11-21. Coefficients for testing linear × quadratic × quadratic interaction component on data in Table 11-6.

		A Lin.	B Quad.	C Quadratic				
				C_1 2	C_2 −1	C_3 −2	C_4 −1	C_5 2
A_1	B_1	−1	1	−2	1	2	1	−2
	B_2	−1	−2	4	−2	−4	−2	4
	B_3	−1	1	−2	1	2	1	−2
A_2	B_1	0	1	0	0	0	0	0
	B_2	0	−2	0	0	0	0	0
	B_3	0	1	0	0	0	0	0
A_3	B_1	1	1	2	−1	−2	−1	2
	B_2	1	−2	−4	2	4	2	−4
	B_3	1	1	2	−1	−2	−1	2

They are shown, along with new estimates of cell means, in Table 11-22. The estimated cell means are plotted in Figure 11-8.

Higher-way Interactions

Higher-way interactions in which all factors are numerical are analyzed just like the above interactions. Each interaction can be analyzed into as many trends as there are degrees of freedom, with each trend being a kind of product of the possible trends on the individual factors. The coefficients of the trends on the interactions, moreover, are the products of the coefficients of the trends on the individual factors.

Calculating all of the products, however, is tedious — it is simpler to proceed by the methods used in Tables 11-3 and 11-14, and in Tables 11-9, 11-17, and 11-20. By this method, only one factor is singled out at first, and C values are calculated for all possible trends on that factor, at all possible combinations of levels of the other factors. Then a second factor is singled out, and, for each trend on the first factor and all possible values of the remaining factors, C values are computed for all trends on the new factor. This second set of C values is calculated, however, on the C values of the initial analysis. The number of new C values will be the product of the numbers of trends possible on each of the first two factors chosen. This procedure is continued until all of the numerical factors have been singled out and their trends have been calculated. The divisor for finding each $(C')^2$ is then found as the product of the coefficients for the corresponding trends on the individual factors.

For interactions involving a mixture of numerical and nonnumerical factors, the procedure is only a little more complicated. First, the procedure described in the last paragraph is carried out, but only the numerical

Table 11-22. Estimated effects and cell means, data from Table 11-6, all factors numerical.

Estimated effects:

$\hat{\mu}$	$\hat{b}_1^*$	ac_{12}^*	ac_{22}^*	bc_{11}^*	abc_{222}^*
20.43	3.84	2.16	1.46	7.42	−0.94

Obtained (upper values) and estimated (lower values) cell means:

		C_1	C_2	C_3	C_4	C_5
	B_1	24.0	28.8	22.8	13.2	4.8
		28.2	25.6	19.9	10.8	−1.5
A_1	B_2	31.8	20.4	15.6	19.2	30.6
		22.8	19.2	18.1	19.2	22.8
	B_3	2.4	13.2	32.4	33.6	34.8
		6.2	18.5	27.6	33.3	35.8
	B_1	26.4	26.4	16.8	10.8	3.6
		29.4	25.0	18.7	10.2	−0.3
A_2	B_2	7.2	24.0	30.0	27.6	9.6
		7.1	27.1	33.8	27.1	7.1
	B_3	10.8	21.6	25.2	32.4	32.4
		7.4	17.9	26.4	32.7	37.0
	B_1	37.2	18.0	6.0	2.4	3.6
		36.8	21.3	11.2	6.5	7.1
A_3	B_2	31.2	12.0	10.8	15.6	28.8
		31.4	14.9	9.4	14.9	31.4
	B_3	9.6	9.6	18.0	31.2	52.8
		14.8	14.2	18.9	29.0	44.5

factors are singled out for trend calculations. The result will be as many trends as there are combinations of trends on the numerical factors, and as many C values for each trend as there are combinations of levels of non-numerical factors.

Suppose, for example, that the three-way design of Table 11-6 were expanded to a four-way design by the addition of a fourth factor, D, with four levels. If A, B, and C were all numerical, as before, the four-way interaction would be divisible into 16 trends (the 16 trends found in Table 11-20), and there would be four C values (for the four levels of D) for each trend. If only B and C were numerical, there would be eight trends (See

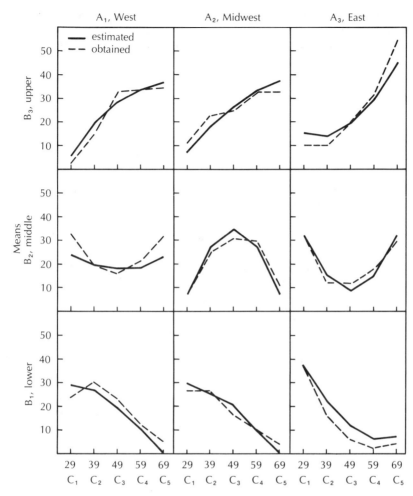

Figure 11-8. Estimated and obtained means, all three factors numerical; data from Table 11-6.

Table 11-17), and for each trend there would be 12 C values, one for each combination of the three levels of A and the four levels of D.

For each C value, then, a $(C')^2$ is calculated, and the $(C')^2$ values for a given trend are summed. From this sum is subtracted one or more other sums of squares and $(C')^2$. To determine which terms need to be subtracted, find all main effects and lower-order interactions that involve *all* of the numerical factors; then subtract the $(C')^2$ or sum of squares (whichever is appropriate) for each. In the first example above, with A, B, and C numerical, the only term that would be subtracted would be that for the corresponding trend component of the ABC interaction. With only B and C numerical, it would be necessary to subtract the corresponding trends in the ABC interaction, the BCD interaction, and BC interaction.

The degrees of freedom are the total number of $(C')^2$ that were summed, minus the sum of the degrees of freedom of the terms that were subtracted. As a check, the sum of the final degrees of freedom, over all trends, should equal the degrees of freedom for the interaction as a whole.

Estimates of effects, cell means, and proportions of variance accounted for then follow exactly the same procedures as were described earlier.

FACTORS WITH UNEQUALLY SPACED NUMERICAL VALUES

All of the above assumes that each factor has equally spaced numerical values. As we saw in Chapter 10, equally spaced numerical values are necessary to avoid excessive complication in a complete trend analysis. Linear trends on main effects and interactions, however, can be performed even when the numerical values are not equally spaced. Instead of using the linear coefficients in the appendix, it is necessary to calculate your own linear coefficients. The linear coefficients calculated are the deviations of the numerical values of the factors from their means. In the example in Table 11-1, for example, if the school sizes had been 400, 1100, and 1200, we could still have tested the linear trend on the B main effect, but the coefficients would have been -500, 200, and 300, the differences between the school sizes and their mean of 900. Only the linear trend could be tested without encountering problems too complicated for this text. In the interaction (assuming A has equally-spaced numerical values) the linear x linear, quadratic x linear, and cubic x linear trends could be tested, using the above three values for the coefficients of the B factor. Once again, however, only the linear trend in B can be calculated conveniently.

Warning. Although assumptions are not explicitly mentioned in this chapter, they cannot be forgotten. Since all of the tests described here are planned comparisons or tests based on planned comparisons, all of the assumptions necessary for planned comparisons are necessary for trend tests. These assumptions have been discussed in detail in previous chapters and will not be discussed again here. However, it is well to remember that planned comparisons are seldom robust with respect to the assumption of equality of variances and covariances; the special methods of handling this problem, discussed in Chapters 6 through 8, apply equally well to trend analyses. It should also be remembered that the rather elaborate tests described in this chapter need not necessarily be made just because one or more factors have numerical values. It is always possible to proceed with the analyses as though the factors did not have numerical values. The tests described here merely provide a way of taking what would otherwise be a few overall tests and partitioning them into a larger number of more specific tests that may give a more accurate picture of the differences in the data.

Exercises

1. Reanalyze the data in Problem 2, Chapter 5, assuming that the levels of B represent different, equally spaced academic levels, i.e., do a trend analysis on the levels of Factor B. Then do a trend analysis assuming that both A and B are numerically valued factors. What conclusions do you draw about analyzing numerically valued factors with only two levels?

2. Do a trend analysis on the data in Problem 3, Chapter 6.

3. In Problem 1, Chapter 7, the logarithms of the distances are equally spaced, making a trend test with equally spaced values appropriate, so long as it is remembered that the values on the distance factor are the *logs* of the distances rather than the distances themselves. Do the trend analysis under this assumption, testing for significance and estimating cell means. Then plot the actual and estimated cell means against the actual distances rather than plotting them against the logs of the distances. Discuss and interpret the results in terms of the graph.

4. Suppose in Table 7-8 that there was a third level of A, with values:

	C_1	C_2	A_3 C_3	C_4	C_5
B_1	11	7	10	2	14
B_2	6	−7	3	4	1
B_3	−3	−7	2	−3	1

Do complete trend tests, assuming that Factors A, B, and C, in turn, are numerically valued. Do trend tests assuming that each pair of factors, AB, AC, and BC, is numerically valued. Finally, do a trend test assuming that ABC are all numerically valued. Whenever C is assumed to be numerically valued, assume that the entire design is a fixed effects design; in that case, use the highest trend on the highest interaction as the error term.

5. Do a trend test on the data in Table 8-1, assuming that Factor A is numerically valued. Do significance tests and estimate cell means using effects significant at the .05 level.

6. Do a set of trend analyses like those in Problem 4 (i.e., on all possible combinations of numerical factors) on the data in Table 8-6, letting all factors be assumed to be fixed whenever necessary.

7. Perform a trend analysis on Factor A in the following repeated measures design; estimate and graph the cell means, using effects that are significant at the .05 level.

		20	40	60	80	ΣX	ΣX^2
	S_1	11	10	15	23	59	975
	S_2	2	5	9	13	29	279
	S_3	6	3	12	13	34	358
Subjects	S_4	7	4	13	15	39	459
	S_5	11	12	9	19	51	707
	S_6	7	5	12	20	44	618
ΣX		44	39	70	103	256	
ΣX^2		380	319	844	1853		3396

8. (a) Do a trend analysis on the latitudes main effect in Problem 2, Chapter 8.

(b) With regard to the same data, a different experimenter argues that larger rats ran more slowly than smaller ones; he felt that the difference due to latitude might be due partly to the size of the rat used. He weighed each rat to test his hypothesis. A linear correlation between weight and running time would seem to be indicated, but a simple linear correlation could not be used to test the hypothesis of no correlation between weight and running time, because mean weights of animals may have been different in different laboratories. How could the test for a linear correlation be modified so that it takes into account possible differences in mean weight among the different laboratories?

12

Multifactor Designs with Random Numerical Factors

We found that very special problems arose in the one-way design with a random numerical factor. More complicated problems arise in multifactor designs when one or more random numerical factors exist. In the one-way design there was only one observation per cell, and the values of the factor levels were not equally spaced. In the multifactor design there are the additional problems that not all cells contain observations and that the pattern of cell entries prevents us from making simple orthogonal tests.

TWO-WAY DESIGN

The data in Table 12-1 are the same data as in Table 3-1, with the addition of a second, random numerical factor. The data are scores on a certain test obtained by students taking each of three different college courses. Suppose the experimenter suspects that the differences in scores might be due more to differences in the ages of the students taking the courses than to differences among the courses they are taking. The significant effect observed in Chapter 3 might then be due to a tendency for students of different ages to take different courses rather than to a basic difference in the type of student or the course itself. The solution to this problem is to use a two-way design with Factor B being age. The A main effect would then be averaged over the levels of B, and it would therefore be independent of B.

Difficulties in selecting subjects prevent the experimenter from selecting the ages that he will study, so he does the next best thing: he ascertains the age of each subject selected and he treats these ages as a second random numerical factor.

Table 12-1. Data from Table 3-1, with the addition of a random numerical factor, age.

	Age	A$_1$	A$_2$	A$_3$	$\overline{X}$
			Course		
B$_{10}$	31	6			6.0
B$_9$	28			5	5.0
B$_8$	27	5			5.0
B$_7$	26	5		4 7	5.3
B$_6$	24			5	5.0
B$_5$	23	1	1		1.0
B$_4$	22		2 3		2.5
B$_3$	21	3			3.0
B$_2$	20		2	2	2.0
B$_1$	18		0		0.0
$\overline{X}$		4.0	1.6	4.6	3.4

Factors A and B, however, are obviously not independent in this design. There is a definite tendency for older students to take classes A$_1$ and A$_3$, the highest scoring classes. Since there is also a definite tendency for older students to obtain higher scores on the test, the differences among the levels of Factor A may well be due to differences among levels of Factor B.

These difficulties require a completely different approach to the analysis of the data. Instead of regarding the ages as the levels of a second factor, we treat them as a second set of data. This second set of data is called the *covariate,* and the analysis is called an *analysis of covariance.* The data are usually listed in the more economical way shown in Table 12-2, where the V_{ij} are the values of the covariate, i.e., the levels of Factor B. The method of analysis is also easier to describe if the random factor is regarded as a covariate rather than as a factor; in most of our discussion we will so treat it.

The Model

The model for the analysis is that for the two-way design with one factor numerical. Unless we considerably limit the model, however, there will be calculational problems. First, we will limit the trend analysis on the co-

Table 12-2. Data in Table 12-1 in analysis of covariance format.

	A$_1$ X	A$_1$ V	A$_2$ X	A$_2$ V	A$_3$ X	A$_3$ V	Overall X	Overall V
	5	27	1	23	5	28		
	3	21	2	22	4	26		
	5	26	2	20	7	26		
	6	31	0	18	5	24		
	1	23	3	22	2	20		
Sum	20	128	8	105	23	124	51	357
Mean	4.0	25.6	1.6	21.0	4.6	24.8	3.4	23.8
SS	96	3336	18	2221	119	3112	233	8669
Variance	4.0	14.8	1.3	4.0	3.3	9.2	2.87	9.33
Sum of cross products		537		173		586		1296
Covariance		6.25		1.25		3.90		3.80
$\hat{\alpha}_i$		−0.13		−0.66		0.79		
$\hat{\theta}$				.407				

variate to a test of a linear trend only (just like we did for the one-way random model in Chapter 10) and, second, we will assume that there is no interaction between the two factors. Neither of these assumptions is essential to the analysis, but each is useful for simplifying both the calculations and the interpretation of the results. Later, we will briefly discuss ways of relaxing these assumptions.

With these assumptions, the model equation is

$$X_{ij} = \mu + \alpha_i + \theta \ (V_{ij} - \bar{V}_{..}) + \epsilon_{ij}.$$

The Greek θ in this equation is the value of the linear B main effect, and V_{ij} is the level of B, i.e., the value of the covariate, associated with observation X_{ij}. Normally, if we were dealing with an ordinary two-way analysis of variance, we would use the symbol V_j, without the i subscript, to denote this value. In this case, however, the two subscripts together remind us that each observation will in general be associated with a definite value of the covariate. They also point up the similarity between this design and the correlation design discussed in Chapter 10 (pp. 238–246). If we assume that the α_i are zero, and we replace θ with a_1, the two models are identical.

Estimates of Effects

Because of the lack of independence between the levels of the two factors, special methods are needed to estimate the effects and test for them. These methods are illustrated in Table 12-3, to which we will be referring throughout this section. The first column of Table 12-3, labeled SV, shows the sums of squares for an ordinary analysis of variance on the V_{ij}—this analysis is performed exactly as though the V_{ij}, not the X_{ij}, were the data. The third

Table 12-3. Analysis of data in Table 12-2.

	SX	SV	SC	ASV	ASC	AF	ASX	df	AMX	F	α
m	173.4						173.4	1	173.4	120.5	<.001
bet	25.2	60.4	36.6	172.4	82.2	39.19	4.58	2	2.29	1.59	
w	34.4	112.0	45.6			18.57	15.83	11	1.439		
t	59.6	172.4	82.2			39.19	20.41	13			

column, labeled SC, contains a slightly different kind of calculation. The values in this column are found by taking sums of cross-products instead of sums of squares. The analysis producing these values is shown in Table 12-4. The values in the first column, labeled RC, are found in exactly the same manner as the RS in an ordinary analysis of variance, except that cross-products instead of squares are summed. RC_{bet}, for example is

$$(1/n) \ \Sigma_i \ t_i u_i = (1/5) \ [(20)(128) + (8)(105) + (23)(124)] = 1250.4,$$

where t_i is the sum of the X_{ij} and u_i is the sum of the V_{ij} in group A_i. Similarly,

$$RC_t = \Sigma_i \ X_{ij} V_{ij} = 5(27) + 3(21) + \cdots + 2(20) = 1296$$
$$SC_m = TU/N = (51)(357)/15 = 1213.8,$$

where T is the grand total of the X_{ij} and U is the grand total of the V_{ij}. These formulas are identical to those for the ordinary analysis of variance (Table 3-4) except that each squared value is replaced by a cross-product. The SC are then calculated from the RC just exactly as for an ordinary analysis of variance:

$$SC_{bet} = RC_{bet} - SC_m$$
$$SC_w = RC_t - RC_{bet}$$
$$SC_t = RC_t - SS_m.$$

We can now estimate θ from the values in Table 12-3. The estimate is $SC_w/SV_w = 45.6/112.0 = .407$. This estimate is a kind of average of the estimates obtained from each group singly. The best estimate, using only the data in one group, would be the covariance of the X_{ij} and V_{ij} in that

Table 12-4. Calculation of SC values in Table 12-3.

	RC	SC
m		1213.8
bet	1250.4	36.6
w		45.6
tot	1296.0	82.2

group, divided by the variance of the V_{ij} (see pp. 239–242). It is easy to show that SC_w is proportional to the average of the covariances within the individual groups and SV_w is proportional to the average of the variances. The estimate $(\hat{\theta})$ is the ratio of these averages. Because the values are based entirely on within-group data, the estimate is independent of the A main effect.

The estimates of the α_i are equal to the estimates from an ordinary analysis, adjusted for the linear dependence between the X_{ij} and V_{ij}:

$$\hat{\alpha}_i = (\overline{X}_{i.} - \overline{X}_{..}) - \hat{\theta}\,(\overline{V}_{i.} - \overline{V}_{..}).$$

These estimates are in Table 12-2. Although the differences between cell means are relatively large, the adjusted estimates are all very close to zero. We suspect, therefore, that the adjusted group differences will not be significant.

The best estimate of the grand mean (μ) is $\overline{X}_{..}$, just as in an analysis of variance. The estimate of μ is unchanged because in the model equation each V_{ij} has its grand mean $(\overline{V}_{..})$ subtracted from it. We can now estimate the cell means that we would have obtained if we had controlled the values of the covariate so that every V_{ij} was equal to $\overline{V}_{..}$, e.g., so that every student in the study was 23.8 years old. Each estimate is simply the grand mean $(\overline{X}_{..})$ plus $\hat{\alpha}_i$:

$$\hat{\mu}_1 = 3.40 - .13 = 3.27$$
$$\hat{\mu}_2 = 3.40 - .66 = 2.74$$
$$\hat{\mu}_3 = 3.40 + .79 = 4.19.$$

Adjusted estimates such as these must be interpreted with caution; their values depend heavily on the assumptions of linearity and equal slope.

Significance Tests

The denominator for all our F tests is the estimate of error. In an ordinary analysis of variance this would be MS_w. In analysis of covariance, however, part of MS_w is assumed to be due to a systematic relationship between the V_{ij} and X_{ij}. To obtain a pure error estimate, we must subtract that portion of MS_w that is due to the covariance between X_{ij} and V_{ij}. We adjust the error by subtracting an adjustment factor (AF in Table 12-3) from SX_w:

$$AF_w = (SC_w{}^2)/SV_w = (45.6)^2/112.0 = 18.57$$
$$ASX_w = 34.40 - 18.57 = 15.83.$$

The error mean square is

$$AMX_w = ASX_w/(N - I - 1) = 15.83/11 = 1.439.$$

The degrees of freedom are now $(N - I - 1)$ because we have subtracted one degree of freedom for the estimate of θ. If the correlation between V_{ij} and X_{ij} is large, this adjustment may substantially reduce the error estimate.

The numerator for testing the null hypothesis H_0: $\theta = 0$ is the adjustment factor, AF_w, treated as a mean square with one degree of freedom:

$$F_{(1, N-I-1)} = AF_w / AMX_w = 18.57/1.439 = 12.90,$$

which, with one and 11 degrees of freedom, is significant beyond the .01 level.

To test the more general hypothesis H_0: $\theta = \theta^*$, we would use as our mean square $(\hat{\theta} - \theta^*)^2 / SV_w$, which is equal to AF_w, when $\theta^* = 0$.

To test the A main effect, we cannot simply sum the squares of the estimated α_i as we did for the analysis of variance; since the $\hat{\alpha}_i$ and the adjusted error term both depend on $\hat{\theta}$, they are not independent. To find a numerator that is independent of the error estimate, we use a more round-about method involving the calculation of two new terms, ASV_{bet} and ASC_{bet}:

$$ASV_{bet} = SV_{bet} + SV_w$$
$$ASC_{bet} = SC_{bet} + SC_w.$$

In the one-way design we are considering, $ASV_{bet} = SV_t$ and $ASC_{bet} = SC_t$. In more complex designs these simple equalities will not hold. However, the basic principles illustrated in the above equations will hold. The ASV for any effect will be equal to the SV for that effect plus the SV for the error term, and the ASC will be the SC for the effect plus the SC for the error.

We next calculate the adjustment factor, labeled AF, for the A main effect. The adjustment factor for an effect is the square of ASC for that effect, divided by ASV:

$$AF_{bet} = (ASC)^2 / ASV = (82.2)^2 / 172.4 = 39.19.$$

Finally, the adjusted sum of squares ASX_{bet} is

$$= SX_{bet} + AF_w - AF_{bet}$$
$$= 25.2 + 18.57 - 39.19 = 4.58.$$

The degrees of freedom are $(I - 1)$, just as for an analysis of variance, so that

$$AMX_{bet} = ASX_{bet} / (I - 1) = 2.29,$$
$$F_{(I-1, N-I-1)} = AMX_{bet} / AMX_w = 1.59,$$

which is not significant.

No adjustment factor is needed to test for the grand mean in this model, because the model compensates for this by subtracting $\overline{V}_{..}$ from each V_{ij}. However, the adjusted mean square within is used as the denominator:

$$F_{(1, N-I-1)} = (T^2/N) / AMX_w.$$

INTERPRETATION OF SIGNIFICANCE TESTS

Interpretation of the results of an analysis of covariance is complicated by the fact that the V_{ij} and $\overline{X}_{i.}$ are not independent. In the example just given the analysis of variance was significant, but the analysis of covariance was not. Table 12-5 shows an example in which the analysis of covariance is not significant, but the analysis of variance is. The data in this table are the same as those in Table 12-2 except that the group means are different. In Table 12-5 the $\bar{X}_{i.}$ are about equal, but the $\bar{V}_{i.}$ are very different from each other. When the effects of these $\bar{V}_{i.}$ are factored out, a sizeable effect results.

Table 12-6 shows a result that is much more rare, but nevertheless can occur. Both analyses are significant for these data, but the effects are in different directions, with μ_2 smallest in the analysis of variance and largest in the analysis of covariance.

These differences occur because differences among the X_{ij} within a group are no longer regarded as due to random error. Instead, they are regarded as partly due to random error and partly due to a linear relationship with the V_{ij}. Basically, the analysis of covariance model assumes that the X_{ij} in each group are a linear function of the V_{ij}. It does not assume that the same linear function holds for each group, but it does assume that the linear functions all have the same slope, θ. Figures 12-1, 12-2, and 12-3 show the best estimates of the functions, along with their slopes, for each group. Since the functions are all assumed to have the same slope, they can differ only in their overall heights. The null hypothesis of no A main

Table 12-5. Sample data for which analysis of covariance is significant but analysis of variance is not.

| | A_1 | | A_2 | | A_3 | | Overall | |
	X	V	X	V	X	V	X	V
	5	27	3	23	3	28		
	3	21	4	22	2	26		
	5	26	4	20	5	26		
	6	31	2	18	3	24		
	1	23	5	22	0	20		
Sum	20	128	18	105	13	124	51	357
Mean	4.0	25.6	3.6	21.0	2.6	24.8	3.4	23.8
SS	96	3336	70	2221	47	3112	213	8669
Variance	4.0	14.8	1.3	4.0	3.3	9.2	2.87	9.33
Sum of cross products		537		383		338		1258
Covariance		6.25		1.25		3.90		3.80
$\hat{\alpha}_i$		−0.13		1.34		−1.21		
$\hat{\theta}$				.407				

	SX	SV	SC	ASV	ASC	AF	ASX	df	AMX	F	α
m	173.4							1	173.4		
bet	5.2	60.4	−1.4	172.4	44.2	11.33	12.44	2	6.220	4.32	.05
w	34.4	112.0	45.6			18.57	15.83	11	1.439		
t	39.6	172.4	44.2			11.37		13			

Table 12-6. Sample data for which analyses of variance and covariance are both significant, but in opposite directions.

| | A_1 | | A_2 | | A_3 | | Overall | |
	X	V	X	V	X	V	X	V
	5	25	1	13	5	40		
	3	19	2	12	4	38		
	5	24	2	10	7	38		
	6	29	0	8	5	36		
	1	21	3	12	2	32		
Sum	20	118	8	55	23	184	51	357
Mean	4.0	23.6	1.6	11.0	4.6	36.8	3.4	23.8
SS	96	2844	18	621	119	6808	233	10273
Variance	4.0	14.8	1.3	4.0	3.3	9.2	2.87	9.33
Sum of cross products		497		93		862		1452
Covariance		6.25		1.25		3.90		3.80
$\hat{\alpha}_i$		0.68		3.41		−4.09		
$\hat{\theta}$				.407				

	SX	SV	SC	ASV	ASC	AF	ASX	df	MX	F	α
m	173.4							1	173.4		
bet	25.2	1664.4	192.6	1776.4	238.2	31.94	11.83	2	5.915	4.11	.05
w	34.4	112.0	45.6			18.57	15.83	11	1.439		
t	59.6	1776.4	238.2			31.94	29.48	13			

effect hypothesizes that these functions all have the same height, i.e., that in fact they are all exactly the same function.

In Figure 12-1 and Table 12-2 the analysis of variance is significant because the mean of A_2 is much smaller than those of A_1 and A_3. The mean of the covariate for A_2 is also much smaller, however, so that the three regression lines very nearly coincide. Consequently, the analysis of covariance is not significant. We would conclude from these data that differences in test scores were not due directly to differences in the quality of the courses, but rather that students taking some subjects were older than those taking other subjects.

In Figure 12-2 and Table 12-5 the three group means are nearly equal, so the analysis of variance is not significant. Since the means of the covariate in the three groups are not equal, the regression lines differ from each other significantly. From these data we would conclude that the classes did differ in how much they helped the students to improve their test scores, but that these differences were obscured in the raw data because the most helpful courses tended to be taken by younger students with lower initial scoring ability.

In Figure 12-3 and Table 12-6 the differences among the group means are large, but the differences among the covariate means are even larger. As a

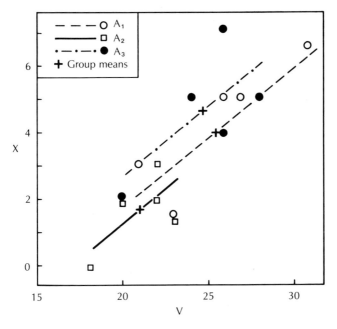

Figure 12-1. Data from Table 12-2 plotted against the covariate.

consequence, even though the mean of group A_2 is lower than those of groups A_1 and A_3, the overall height of its regression line is greater. The conclusion would probably be that the most helpful course, A_2, was so good that it more than overcame a large initial handicap on the part of those who took it.

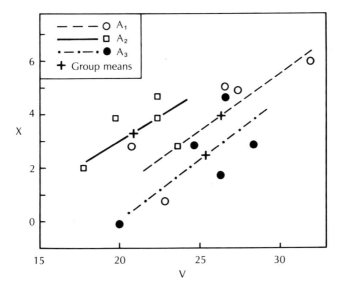

Figure 12-2. Data from Table 12-5 plotted against the covariate.

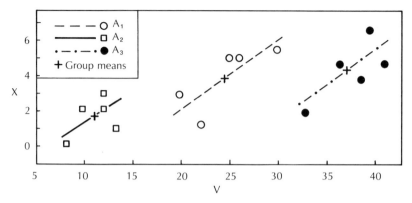

Figure 12-3. Data from Table 12-6 plotted against the covariate.

Such interpretations require extreme care, however. In Figure 12-1 and Table 12-2, for example, although the courses seem to have no direct effect on test scores, there may be an indirect effect. Courses A_1 and A_3 obviously tend to attract older students. Although there may be nothing in the course itself that would enable the average student to raise his score, nevertheless, higher-scoring students are found in those classes.

Such interpretations may also involve unwarranted extrapolations beyond the range of the data. We concluded from Figure 12-3 and Table 12-6 that A_2 is more helpful but it attracts younger students. Extending the regression line, we would be tempted to conclude that A_2 would also be most helpful for older students. However, since there are no older students in our sample from group A_2, we have no way of knowing whether such a generalization would hold. It could be, in fact, that older students avoid A_2 because it would actually be harmful to them in some way.

For some problems the analysis of covariance may be inappropriate and even misleading. Suppose, for example, we were confronted with three people each of whom had taken one of the three courses, and our job was to predict, from the data in Table 12-2, who had the lowest score. If all we knew about each was the course he had taken, the analysis of covariance would be irrelevant to our decision. No matter what the effects of age may be, it is still a fact that students in A_2 tend to have substantially lower scores than do those in A_1 and A_3, so we should choose the person from A_2. On the other hand, if we knew the ages of the three people, then the analysis of covariance would be relevant. Our best guess is to choose the youngest of the three, regardless of which course he might have taken. Similarly, if we were advising a student who wished to score highly, we should probably advise him to take whichever course he enjoys most, forgetting about trying to raise his test score by taking one course rather than another.

In a similar, classical example, suppose the data in Table 12-2 were achievement scores and the covariate was time spent studying. We would be naive to conclude from the negative results of the analysis of covariance that there were no differences among the classes. We would be wiser to try to discover why students study more in some courses than they do in others.

Some of the interpretation problems discussed here can be eliminated by assuring that the covariate is independent of the effect being treated. In a study of aggressiveness in four groups of chickens, for example, the covariate might be the weights of the chickens. If the chickens have been assigned randomly to the four groups, the values of the covariate should be independent of the groups effect, and most interpretation problems will be eliminated.

Some statisticians recommend that analysis of covariance never be used unless the covariate is independent of the groups effect. However, this would limit the applicability of the method more than necessary. There are many occasions, like the examples given above, for which the independence assumption is not met but the analysis of covariance might still apply.

ASSUMPTIONS

In addition to those of the ordinary analysis of variance, the analysis of covariance requires some special assumptions. It assumes that the relationship between the X_{ij} and V_{ij} is linear and that the slope of the function relating them is the same for all groups. The effects of departures from the assumptions have not been tested. For this and other reasons described below it is probably best to avoid an analysis of covariance whenever an alternative approach is feasible.

Linearity

The assumption that the relationship between the X_{ij} and V_{ij} is linear is less restrictive than it at first appears. If the relationship between the data and the covariate is not linear, for example, then perhaps the relationship between the data and the square of the covariate is linear. We would then simply treat the squares of the covariate values as the V_{ij}, and do the analysis as above. The same would hold true if a linear relationship held between the data and the square root, the logarithm, or some other function of the covariate. (Other, more complicated ways of relaxing this assumption are discussed briefly on p. 309.)

The assumption can never be completely eliminated, however; a transformation on the covariate simply raises the question of whether the transformed V_{ij} are linearly related to the X_{ij}. Moreover, we do not know the effects of violation of the assumptions on the analysis of covariance. We found that, usually, an analysis of variance is robust with respect to all assumptions other than random sampling—an analysis of covariance may be robust to linearity, but we cannot be sure.

Equal Slopes

It is possible to test the assumption of equal slopes. The test is straightforward but the calculations are somewhat tedious because they require calculation of variances and covariances for every group. The method divides ASX_w into two portions, one that is strictly error and another that is due to the unequal slopes.

The model equation for this test is generally assumed to be

$$X_{ij} = \mu + \alpha_i + \theta \ (V_{ij} - \bar{V}_.) + \theta_i \ (V_{ij} - \bar{V}_{i.}) + \epsilon_{ij}.$$

The added term acts like an interaction, summing to zero across the levels of A and B.

For this model the estimates of μ, θ, and α_i are the same as before, and their adjusted sums of squares are also found in the same way. To test the null hypothesis that the θ_i are all zero, we first find the quantities

$$SV_{w(i)} = \Sigma_j \ (V_{ij} - \bar{V}_{i.})^2$$
$$SC_{w(i)} = \Sigma_j \ (X_{ij} - \bar{X}_{i.})(V_{ij} - \bar{V}_{i.}).$$

These terms are, respectively, the variance of the V_{ij} and the covariance, in Group A_i, each multiplied by n_i. If they are summed over groups, they will give, respectively, SV_w, and SC_w. The best estimate of θ_i is then $[SC_{w(i)}/SV_{w(i)}] - \hat{\theta}$. These estimates are shown, for the data in Table 12-2, in Table 12-7. The sum of squares for testing the null hypothesis that the θ_i are all zero (the hypothesis of equal slopes) is

$$\begin{aligned} ASX_{ab} &= [\Sigma_i \ (SC_{w(i)})^2/SV_{w(i)}] - AF_w \\ &= (25.0)^2/59.2 + (5.0)^2/16.0 + (15.6)^2/36.8 - 18.57 \\ &= 0.16. \end{aligned}$$

The degrees of freedom are $(I - 1)$, so

$$AMX_{ab} = ASX_{ab}/(I - 1) = 0.16/2 = 0.08.$$

The error sum is

$$\begin{aligned} ASX_w^* &= ASX_w - ASX_{ab} = SX_w - \Sigma_i \ (SC_{w(i)})^2/SV_{w(i)} \\ &= 15.83 - 0.16 = 15.67. \end{aligned}$$

The degrees of freedom are the degrees of freedom of ASX_w minus $(I-1)$, which comes to

$$N - I - 1 - (I - 1) = N - 2I.$$

Table 12-7. Testing equal-slope hypotheses, data from Table 12-2 (see test for remainder of calculations).

	$SV_{w(i)}$	$SC_{w(i)}$	$\hat{\theta}_i$
A_1	59.2	25.0	.015
A_2	16.0	5.0	−.095
A_3	36.8	15.6	.017
Sum	112.0	45.6	

Thus,

$$AMX_w^* = ASX_w^*/(N - 2I) = 15.67/9 = 1.741$$
$$F_{(I-1,N-2I)} = AMX_{ab}/AMX_w^* = 0.08/1.741 = 0.05.$$

We cannot reject the hypothesis of equal slopes for the data in Table 12-2, so the analysis of covariance seems appropriate. We must be careful, however, because we are in effect using a nonsignificant statistical test to justify pooling two sums of squares, ASX_{err} and ASX_{ab}, into the ASX_w. (The problems inherent in such pooling were discussed on pp. 106–108.)

Even if the slopes are not equal, we can still conduct the analysis of covariance; the only modification in the theory given above is the replacement of ASX_w with ASX_w^* in every test. Interpretation of the results, however, becomes considerably more complicated when the slopes are not equal. The original null hypothesis that the linear functions were in fact all identical must then be changed to state that the different linear functions all cross each other at the grand mean ($\overline{V}_{..}$) of the covariate. Such a null hypothesis is probably not only uninteresting in most cases, but its meaning is probably unclear in many others. Furthermore, it may sometimes be necessary to predict the nature of the function outside the range of the data actually studied in one or more groups. As we have said before, such a prediction is risky. For these reasons, it is probably best to avoid an analysis of covariance when the slopes are not equal.

ALTERNATIVES TO THE ANALYSIS OF COVARIANCE

The analysis of covariance is often abused as a way to avoid the extra effort required for good experimental controls. Controlling for the values of the covariate frequently requires careful selection of subjects. The alternative of letting the values of the covariate select themselves is too tempting to resist, especially if a computer is available to do the work of analysis.

Experimental Control of the Covariate

Nevertheless, there are a number of advantages to controlling the covariate values experimentally. Suppose, for example, that we had determined the ages of all of our potential subjects in advance. We could have then divided our subjects into, say, three or four age groups in each course, and selected a certain number of subjects in each age group from each course. The results, if we had used four age groups, would have been a 3 x 4 analysis of variance. The more complicated analysis of covariance would not have been required, nor would the assumptions of linearity and equal slope. Furthermore, if four groups did not provide a fine enough division of ages, age could still be added as a covariate. The design would then be a two-way analysis of covariance (or, equivalently, a three-way analysis of variance with one random numerical factor). The analysis of such a design is only a little more complicated than of the one way analysis of covariance. It is discussed in the next section.

The principle advantages of controlling for the values of the covariate are the reduced number of assumptions required for the analysis and the consequent increase in freedom in analyzing the data. The interaction between the covariate and the main factor, for example, can be tested and estimated. There are also disadvantages, however, in controlling for the covariate. In the design of Table 12-2, controlling for the covariate would have hidden the tendency for older students to take certain courses. An entirely separate study, using ages as data, would have been necessary to reveal the relationship between age and the tendency to take certain courses. In the analysis of covariance both the relationship and the tendency are clear.

The example discussed on pp. 294–295, for which the covariate was time spent studying, may pose a more difficult problem. Presumably, the way to control for subjects studying would be to have some way of making sure that every student studies for a preset amount of time – perhaps by specifying certain times during which the students must study and prohibiting them from studying any other time. Such a procedure would probably be highly impractical, since both the very good and very poor students would resist such restrictions. More seriously, however, such controls might change the nature of the covariate; regulated study time might not be used as efficiently as unregulated study time by some students, whereas it may be used more efficiently by others. Thus, the variable of study time might not be the same variable in the two experiments.

Neither approach – statistical control of the covariate by analysis of covariance nor experimental control through prior selection of the covariate values – is universally the best. The experimenter needs to weigh very carefully both the advantages and disadvantages of each in light of his purposes in conducting the experiment. He also needs to consider the possibility that his particular question might be best answered by not controlling for the covariate at all, e.g., the experimenter might be interested simply in determining which course has the highest scoring students, regardless of the students' ages.

Other Kinds of Statistical Control

Even when experimental control of the covariate is not desirable, an analysis of covariance might not be the best method of statistically controlling for the covariate. Frequently, for example, the actual function relating the covariate to the data is at least approximately known. A classical case is an extinction study in which all subjects are given a standard number of trials for a learning task, after which they are divided into two or more groups in which learning is extinguished under different conditions. Frequently, an analysis of covariance is performed on the extinction data, with the level of initial learning the covariate. Such data probably violate the assumption of equal slopes, however. A reasonable assumption would be that the amount of extinction is approximately proportional to the level of initial learning, with the proportionality differing among groups. During extinction, the level of learning for one group might decrease on the average to as little as one-fourth of its original level, whereas that for another group might decrease by only one-half. The analysis of covariance assumes that the slopes are equal, and tests for differences in the intercepts. In this experiment it is probably more appropriate to test for differences in the slopes; the solution for analyzing such data is simple. The model assumes that each extinction score is approximately proportional to its corresponding learning score. The appropriate analysis would thus involve dividing each extinction score by the associated learning score and doing an ordinary analysis of variance on the ratios. Such an analysis would not only be more appropriate, it would also avoid the extra problems (in both assumptions and computations) of an analysis of covariance.

In the foregoing example the data were assumed to be proportional to the covariate; this is by no means the only relationship that may hold. The data might be proportional, for example, to the square, the square root, the logarithm, or some other function of the covariate. Or it might be that the algebraic difference between the covariate and the variate is a constant for a given group, with the value of that constant possibly different from one group to another. To test whether the value of the constant differs from group to group, an analysis of variance would be performed on the differences between the values of the variate and the covariate. In the example in Table 12-2 we found that the slope of the function relating the variate to the covariate was close to one half (actually closer to 0.4). If we had good a priori reasons for believing that the slope should be one-half, we could have subtracted one-half the value of the covariate from each X_{ij} and performed an analysis of variance on the resulting differences.

The possible functions that we might choose are endless, with each reducing the problem to a simple analysis of variance instead of the more complicated analysis of covariance. The important restriction here is that we must have good a priori reasons for believing that the assumed function actually holds (at least approximately) for the data being analyzed. In the absence of such prior knowledge, this alternative would not be available.

GENERAL ANALYSIS OF COVARIANCE—FIXED EFFECTS

The analysis of variance can be easily generalized to an analysis of covariance for any fixed-effects design. The generalization for random and mixed designs is less simple. In this section we will consider the generalization only for fixed-effects designs.

We will illustrate the generalization with two analyses: first, a planned comparison on the *adjusted means* for the data in Table 12-5, testing the null hypothesis H_0: $\mu_1 + \mu_3 - 2\mu_2 = 0$, and, second, an analysis of covariance on the two-way design in Table 12-8. The X_{ijk} in this table are identical to those in Table 5-2, giving scores of mental patients after taking drugs. To these data we have added a covariate representing ratings by the patients' doctors of the severity of each patient's illness. The ratings are assumed to be on a ten-point scale, a ten representing the most severe illness.

The Model Equation

The model equation for an analysis of covariance is identical to that for the corresponding analysis of variance, except for the addition of one term representing the covariate. In every case the additional term is θ times the difference between the observed value of V and its grand mean, $\bar{V}$. For the two-way design, for example,

$$X_{ijk} = \mu + \alpha_i + \beta_j + \alpha\beta_{ij} + \theta \ (V_{ijk} - \bar{V}...) + \epsilon_{ijk}.$$

Estimates of Effects

To obtain estimates of the effects, we first need to estimate θ. To do this, we construct a table like Table 12-3, but which contains all of the effects in the design being analyzed. The table for the experiment in Table 12-8, for example, is in Table 12-9; it includes in its row headings all of the effects found in a two-way analysis of variance.

In any fixed-effects analysis of variance there must be an error sum of squares. Ordinarily, this is SS_w, here designated by SX_w. In a two-way design with only one observation per cell it may be SX_{ab}. If there is more than one observation per cell but we have good reason to assume that

Table 12-8. Data from Table 5-2, with severity of illness as a cofactor.

		B_1		B_2		B_3		Row summary	
		X	V	X	V	X	V	X	V
A_1		8	5	8	7	5	3		
		4	3	10	7	4	6		
		0	4	6	4	8	6		
	Sum	12	12	24	18	18	15	54	45
	Mean	4.0	4.0	8.0	6.0	6.0	5.0	6.0	5.0
A_2		10	5	0	2	15	8		
		6	6	4	7	9	3		
		14	7	2	6	12	7		
	Sum	30	18	6	15	36	18	72	51
	Mean	10.0	6.0	2.0	5.0	12.0	6.0	8.0	5.7
Column	Sum	42	30	30	33	54	33	126	96
summary	Mean	7.0	5.0	5.0	5.5	9.0	5.5	7.0	5.3

Table 12-9. Analysis of data in Table 12-8.

	SX	SV	SC	ASV	ASC	AF	ASX	df	AMX	F	α
m	882							1	882	161.7	<.001
a	18	2	6	46	51	56.54	7.48	1	7.48	1.37	
b	48	1	0	45	45	45.00	49.02	2	24.51	4.49	.04
ab	144	7	30	51	75	110.29	79.73	2	39.86	7.31	.01
w	106	44	45			46.02	59.98	11	5.453		
t	316	54	81			121.50	194.50				

there is no interaction, it may be $SX_{\text{pooled}} = SX_w + SX_{ab}$. In the designs of Chapter 9 it is SX_{rem}. In this chapter, to be completely general, we will designate this error sum of squares as SX_{err}.

In every design we begin first by calculating SX, SV, and SC for each effect and for the error term. The SX and SV values are calculated exactly as in an ordinary analysis of variance, except that the calculations for SV are on the values of the covariate instead of the data.

The rule for calculating the SC values is a straightforward generalization of that for calculating the SX and SV. In the formula for calculating an SX there is always some point in the calculations at which a number or group of numbers must be squared. At this point in calculating an SC, instead of taking squares, the cross-products of the corresponding values for the Xs and Vs are taken. The formulas for calculating the SX values in a two-way design, for example, involve first finding RX values; each RX is a sum of squared values divided by some constant. The SC, therefore, are also found by first finding RC values, each RC being a sum of cross-products divided by a constant.

$$SC_m = N\bar{X}_{..}\bar{V}_{..} = TU/N$$
$$RC_{\text{bet}} = \Sigma_i\, n_i\bar{X}_{i.}\bar{V}_{i.} = \Sigma_i\, t_i u_i/n_i$$
$$RC_t = \Sigma_{ij}\, X_{ij}V_{ij}.$$

where U and u_i represent the same sums on the V_{ij} as T and t_i represent on the X_{ij}. The SC values are then found by subtraction, just like the SX and the SV.

The "SX" value for a planned comparison is $(C')^2$, the C value squared and divided by a constant. Here again, we replace the square with a cross-product. If we let

$$CX = \Sigma_i\, c_i\bar{X}_{i.} = 4.0 + 2.6 - (2)(3.6) = -0.6,$$
$$CV = \Sigma_i\, c_i\bar{V}_{i.} = 25.6 + 24.8 - (2)(21.0) = 8.4,$$

where c_i is the coefficient of the planned comparison, then

$$(CX')^2 = (CX)^2/(\Sigma_i\, c_i^2/n_i) = .30,$$
$$(CV')^2 = (CV)^2/(\Sigma_i\, c_i^2/n_i) = 58.8,$$
$$(CC')^2 = (CX)(CV)/(\Sigma_i\, c_i^2/n_i) = -4.20.$$

Note that although the values of SX and SV can never be negative, SC may frequently be negative.

The estimate of θ in any design is $\hat{\theta} = SC_{err}/SV_{err}$. For Tables 12-8 and 12-9 this is $45/44 = 1.02$. The estimates of other effects are found by first finding the estimates under an ordinary analysis of variance and then finding the same estimates, treating the covariate as the data. The best estimate of the effect is then the analysis of variance estimate, minus $\hat{\theta}$ times the corresponding estimate for the covariate.

To estimate α_1, for example in the two-way design, we first find the estimate for the ordinary analysis of variance to be

$$\overline{X}_{1..} - \overline{X}_{...} = 6.0 - 7.0 = -1.0.$$

We next find the corresponding value for the covariate to be

$$\overline{V}_{1..} - \overline{V}_{...} = 5.0 - 5.3 = -0.3.$$

The best estimate of α_1 is then

$$\hat{\alpha}_1 = -1.0 - (1.02)(-0.3) = -0.7.$$

The complete set of estimates for the data in Table 12-8 are shown in Table 12-10, along with adjusted estimates of cell, row, and column means, calculated by adding the corresponding estimated effects. Similarly, to estimate ψ for the planned comparison, we first find the estimate that would be made in the analysis of variance to be

$$\Sigma_i\, c_i \overline{X}_{i.} = 4.0 - (2)(3.6) + 2.6 = -0.6.$$

The corresponding value, calculated on the covariate, is

$$\Sigma_i\, c_i \overline{V}_{i.} = 25.6 - (2)(21.0) + 24.8 = 8.4.$$

The best estimate of ψ is then

$$\hat{\psi} = -0.6 - (1.02)(8.4) = -9.2.$$

Table 12-10. Estimates of effects on data in Table 12-8.

	B_1	B_2	B_3	Rows
A_1	$\alpha\hat{\beta} = -1.3$ $\hat{\mu} = 5.3$	3.1 7.2	-1.8 6.3	$\hat{\alpha} = -.7$ $\hat{\mu} = 6.3$
A_2	$\alpha\hat{\beta} = 1.3$ $\hat{\mu} = 9.3$	-3.1 2.4	1.8 11.3	$\hat{\alpha} = .7$ $\hat{\mu} = 7.7$
Columns	$\hat{\beta} = 0.3$ $\hat{\mu} = 7.3$	-2.2 4.8	1.8 8.8	$\hat{\mu} = 7.0$

Significance Tests

The adjustment factor (AF) for the error sum of squares is always

$$AF_{err} = (SC_{err})^2/SV_{err}.$$

Similarly, the adjustment factor for sum of squares total is

$$AF_t = (SC_t)^2/SV_t.$$

The adjustment factor for any effect is first found by calculating ASV and ASC for that effect:

$$ASV_{eff} = SV_{eff} + SV_{err}$$
$$ASC_{eff} = SC_{eff} + SC_{err}$$
$$AF_{eff} = (ASC_{eff})^2/ASV_{eff}.$$

These values are shown, for the two-way design, in Table 12-9. For the planned comparison the $(C')^2$ take the place of the sums of squares for the effect:

$$ASV_{eff} = (CV')^2 + SV_{err} = 58.8 + 112.0 = 170.8$$
$$ASC_{eff} = (CC')^2 + SC_{err} = -4.20 + 45.6 = 41.4$$
$$AF_{eff} = (41.4)^2/170.8 = 10.03.$$

Occasionally, both the numerator and the denominator in the ratio for calculating an AF will be zero (the denominator will never be zero unless the numerator is also). In that case, the adjustment factor is assumed to be zero.

The adjusted sum of squares for error and sum of squares total are the SX minus the AF:

$$ASX_{err} = SX_{err} - AF_{err}$$
$$ASX_t = SX_t - AF_t.$$

The calculations are shown in Table 12-9, the sum of squares within being the error term.

The adjusted sum of squares for each effect is

$$ASX_{eff} = SX_{eff} - AF_{eff} + AF_{err}.$$

Note that the formulas for specific effects differ from those for error and total in that AF_{err} is added. For the planned comparison,

$$ASX_{eff} = .30 - 10.03 + 18.57 = 8.84.$$

The adjusted sums of squares for the two-way design are in Table 12-9.

The degrees of freedom for ASX_{err} are equal to the degrees of freedom for the corresponding analysis of variance, minus one. In Table 12-9, for example, the degrees of freedom for the analysis of variance would be $IJ(n-1) = 12$. Thus, in the analysis of covariance ASX_w has 11 degrees of freedom. Similarly, the degrees of freedom for ASX_t is equal to the degrees of freedom for SX_t minus one. For the design in Tables 12-8 and

12-9 this is $17 - 1 = 16$. The degrees of freedom for the adjusted sums of squares of effects are all equal to their degrees of freedom in the ordinary one way analysis of variance, as is shown in Table 12-9.

Finally, each adjusted mean square (AMX) is equal to ASX divided by its degrees of freedom. The denominator for all significance tests is AMX_{err}, and the test is an ordinary F test. The test of the planned comparison, for example, is $F_{(1,11)} = 8.84/1.439 = 6.14$. Table 12-9 has F ratios for the two-way design.

It should be noted that the adjusted sums of squares in Table 12-9 do not add up to ASX_t. In the analysis of covariance these sums of squares are not orthogonal because each depends on the estimate of θ. For this reason estimates of variances of effects and of proportions of variance accounted for are difficult to make, and not very meaningful. The same problem arises with planned comparisons. Planned comparisons that would be orthogonal in an ordinary analysis of variance are not generally orthogonal in an analysis of covariance.

No adjusted sum of squares need be calculated to test for the grand mean. The grand mean is always tested by

$$F = N\bar{X}^2/AMX_{err}.$$

For the example in Tables 12-8 and 12-9, $F_{(1,11)} = 882/5.453 = 161.7$.

The null hypothesis

$$H_0: \theta = \theta^*$$

is always tested by the ratio

$$F = (\theta - \theta^*)SC_{err}/AMX_{err}.$$

For the null hypothesis

$$H_0: \theta = 0$$

this reduces to

$$F = AF_{err}/AMX_{err}.$$

For the data in Tables 12-8 and 12-9,

$$F_{(1,11)} = 46.02/5.453 = 8.44, \alpha = .02$$

Testing for Equal Slopes

The equal slope assumption assumes that the same value of θ is appropriate within each cell. The test of that assumption requires that the error sum of squares be SX_w. We can then test the assumption of equal slopes by treating the design as a one-way design with as many groups as there are cells. The design in Table 12-8 would be treated as a one-way design with six groups, just as it was in Table 5-1. Estimates of θ and the test of equal slopes both proceed as described on pp. 295–297. For the data in Table 12-8, the F ratio turns out to be $F_{(5,6)} = 1.038/9.13 = 0.11$. If the assumption of equal slopes were rejected, AMX_w^* could be used as the denominator for testing the remaining effects. Interpretation would be almost impossibly complicated, however. Note again that this test is appropriate only if ASX_w $= ASX_{err}$. If some other error term is used, then this test for equal slopes cannot be made.

RANDOM AND MIXED DESIGNS

The procedure for random and mixed designs is basically the same as for fixed-effects designs, except that careful attention must be paid to the difference between the test of an effect and the use of the mean square for an effect to test another effect. We will illustrate the general principles with the data in Table 12-8, assuming this time that Factor B is random. The necessary calculations, to be described below, are summarized in Table 12-11.

Estimates of effects in random and mixed designs are found exactly as in the fixed-effects design. In addition, tests on θ are identical to those in fixed-effects designs.

Finally, any effect in a mixed or random design that would normally be tested with MX_{err} in the denominator is tested in exactly the same way as for a fixed-effects design. For the data in Table 12-8 with Factor B random, the B main effect and AB interaction would normally be tested in an analysis of variance by an F ratio with MS_w in the denominator. In the analysis of covariance they are therefore tested with AMX_w in the denominator, just as they would be in a fixed-effects design.

Effects that would not ordinarily be tested against MX_{err}, however, require special treatment. In our example the grand mean and the A main effect fit this category. To test such an effect, we first find the appropriate

Table 12-11. Analysis of data in Table 12-8, with Factor B random.

	SX	SV	SC	DAF	DSX	df	DMX
m	882						
a	18	2	6				
b	48	1	0	0	48.00	1	48.00
ab	144	7	30	128.57	15.43	1	15.43
w	106	44	45	46.02	59.98	11	5.453
t	316	54	81	121.50	194.50		

	ASV	ASC	AF	ASX	df	AMX	F	α
m				882	1	882	18.38	—
a	9	36	144.00	2.57	1	2.57	0.17	—
b	45	45	45.00	49.02	2	24.51	4.49	.04
ab	51	75	110.29	79.73	2	39.86	7.31	.01

effect for the denominator of the F ratio, using the expected mean squares for an ordinary analysis of variance. For the A main effect in our example, the denominator is the AB interaction, and for the grand mean it is the B main effect. We then follow the procedure for a fixed-effects design, except that we treat the denominator sums of squares as though they were error. This requires that we devise entirely new adjusted sums of squares and adjustment factors for the effects that will appear in the denominators of our F ratios. To test the A main effect, for example, we treat the AB interaction as if it were error. Since the adjustment factor for error is $AF_{err} = (SC_{err})^2/SV_{err}$, the adjustment factor for the AB interaction is

$$DAF_{ab} = (SC_{ab})^2/SV_{ab} = (30^2)/7 = 128.57.$$

The term DAF, instead of AF, is used here to indicate that this adjustment factor applies only when the AB interaction appears in the denominator of the F ratio. For actually testing the AB interaction, the value of AF in Table 12-9 still applies. The appropriate denominator for testing the grand mean is the B main effect, so for the B main effect we calculate the adjustment factor:

$$DAF_b = (SC_b)^2/SV_b = (0)^2/1 = 0.$$

Sometimes the numerator and denominator of this ratio are both zero, in which case DAF should be set to zero. Such a case may arise, for example, in a one-way design with repeated measures (i.e., a two-way design with subjects as one factor) if only one measure on the covariance is taken for each subject. The experiment may perhaps consist of measuring amount of learning on different trials, with scores on an initial intelligence test being the covariate. In such a case the same value of the covariate applies to all treatments given to the same subject. SV_a and SV_b will both be zero; whenever an SV is zero, ASX is equal to its SX. Whenever the denominator for calculating DAF is set to zero, the AF of the effect being tested should then also be set to zero. In the two way design we are now considering, for example, if SV_{ab} were zero, then DAF_{ab} and AF_a (the effect tested using the AB interaction in the denominator) are both set to zero. The result with both adjustment factors set to zero is an ordinary analysis of variance, instead of an analysis of covariance, on the effect being tested. Whenever the SV for the denominator effect is zero, the appropriate test is an ordinary analysis of variance.

To find the ASX for the effect being tested, we first find ASV_{eff} and ASC_{eff}, again using the denominator effect as error. In testing the A main effect in our example,

$$ASV_a = SV_a + SV_{ab} = 2 + 7 = 9$$
$$ASC_a = SC_a + SC_{ab} = 6 + 30 = 36$$
$$AF_a = (ASC_a)^2/ASV_a = (36)^2/9 = 144.$$

As usual, no adjustment is made in testing the grand mean.

The adjusted sum of squares for the denominator of the F ratio is, as in the fixed effects model;

$$DSX_{err} = SX_{err} - DAF_{err},$$

using whatever effect is to serve as the denominator for the test being made:

$$DSX_{ab} = SX_{ab} - DAF_{ab} = 144 - 128.57 = 15.43$$
$$DSX_b = SX_b - DAF_b = 48 - 0 = 48.00.$$

The adjusted sum of squares for the effect being tested is

$$ASX_{eff} = SX_{eff} - AF_{eff} + DAF_{err},$$

which, for the A main effect in our example, is

$$ASX_a = SX_a - AF_a + DAF_{ab} = 18 - 144 + 128.57 = 2.57.$$

Finally, the degrees of freedom for the numerator of the F ratio are the same as they would be for an ordinary analysis of variance, whereas the degrees of freedom for the denominator are one less than they would be.

$$AMX_a = 2.57/1 = 2.57$$
$$DMX_{ab} = 15.43/1 = 15.43.$$

In our example it just happens that the degrees of freedom for each effect and its denominator are all equal to one, so that the F ratios for testing the A main effect and the grand mean both have just one degree of freedom in the numerator and one degree of freedom in the denominator. The F ratios and their significance levels are shown in Table 12-11.

The same procedure applies for quasi-F ratios. The quasi-F and its degrees of freedom are found as described in earlier chapters, except that for all effects except the one being tested the DSX are treated as sums of squares and the degrees of freedom are one less than usual.

It should be pointed out that the procedures described in this section are to a degree compromise methods. They do not make use of all of the information available for estimating θ. Since estimates of θ play a part in all of the tests described, the tests do not use all of the data that they might, and thus they are not as powerful as they potentially might be. However, they have the virtue that they do use statistically correct F ratios (except for the quasi-F, which are of course only approximately distributed as F). Any more powerful tests would have to be based on much more complicated statistics.

ADJUSTED GRAND MEAN

Very occasionally, the user might choose to regard the grand mean itself as a function of the values of the covariate. He might then wish to test the null hypothesis that the grand mean would have a particular value if the influence of the covariate were subtracted out.

The model equation normally used in analysis of covariance assumes that the grand mean is unaffected by the values of the covariate. The equation can be modified, however, so that the grand mean is affected. The new equation for the one-way design is

$$X_{ij} = \mu + \alpha_i + \theta V_{ij} + \epsilon_{ij}.$$

With this model the estimates of α_i and θ are the same as before, as are the tests of θ and the A main effect. The best estimate of μ, however, is now

$$\hat{\mu} = \overline{X}_{..} - \theta \overline{V}_{..},$$

and the test of μ requires calculation of ASV, ASC, AF, and ASX values, just as for the test of any other effect. To test the null hypothesis $\mu = \mu^*$;

$$SX_m = N(\overline{X}_{..} - \mu^*)^2, \quad SV_m = N\overline{V}_{..}^2, \quad SC_m = N\overline{V}_{..}(\overline{X}_{..} - \mu^*)$$

and adjustments factors and adjusted sums of squares are calculated from these.

The result is a different test than that described in the previous sections; it is a test of the value the grand mean would take if all V_{ij} had been held constant at a value of zero.

MULTIPLE COVARIATES

It is possible to have more than one covariate in the model. For the one-way analysis of variance, for example, the model might be

$$X_{ij} = \mu + \alpha_i + \theta(V_{ij} - \overline{V}_{..}) + \lambda(W_{ij} - \overline{W}_{..}) + \epsilon_{ii}.$$

In this model both V_{ij} and W_{ij} are covariates. In a study like that in Table 12-2, the V_{ij} might be the chronological ages of the students, as before, with the W_{ij} perhaps their overall grade-point averages.

Conceptually, the problem with two or more covariates is not much more complicated than with one. The assumptions and calculations required, however, are much more complicated. Each covariate must be linearly related to the data, and the slope of the function relating each covariate to the data must be the same for all groups (though it may be different for different covariates). Finally, no provision is made in the model for any interaction among the covariates.

Computationally, the problem involves calculating SV values for every covariate, and SC values not only for every covariate but for the covariance between every pair of covariates as well. For the error term in the above model, for example, we would have to calculate three different SC values: one for X and V taken together, one for X and W taken together, and one for V and W taken together. If there were three covariates, four values of SV and six values of SC (four things taken two at a time) would have to be calculated. The same number of SV and SC would also have to be calculated for each effect. Finally, each AF and DAF would require the solution of a set of linear equations, the number of equations and the number of unknowns both being equal to the number of covariates. Estimates and tests of the slope values, θ, λ, etc., are even more complex. The specific calculations for an analysis this complex are usually best done by computer. For the interested reader, the basic calculations for a fixed-effects model are given in Scheffe (1959); the generalization to mixed and random models is straightforward.

RELAXING THE ASSUMPTIONS

Despite the problems involved, multiple covariates can occasionally be useful not only when different kinds of covariates are used but also when the assumptions for a single covariate are not met. Suppose, for example, that we suspected that the relationship between the X and V in Table 12-2, had both a linear and quadratic component. One solution to such a problem would be to let $W_{ij} = V_{ij}^2$ and do an analysis of covariance with two covariates. If we suspected a cubic component as well, we could add a third covariate, $U_{ij} = V_{ij}^3$.

This relatively simple device provides a potentially endless variety of ways to relax the assumptions required in an analysis of covariance. However, the relaxation is only partial in each case — in the example above the data must still be linearly related to the V_{ij} and W_{ij}. The problems attendant on an analysis of covariance are reduced, but definitely not eliminated by a device such as this.

Exercises

1. In Problem 2, Chapter 3, suppose the experimenter had also recorded the age, in days, of each rat. The complete data are as follows:

| A_1 | | A_2 | | A_3 | | A_4 | | A_5 | |
X	V	X	V	X	V	X	V	X	V
7	14	5	16	9	14	7	13	10	14
8	13	4	12	11	16	12	16	7	12
5	12	4	13	6	15	8	14	3	13
9	16	6	15	8	13	5	12	12	15
10	15	3	14	7	12	11	16	13	15

(a) Do an analysis of covariance on these data, estimating θ, estimating α_i, and testing the null hypotheses

$$H_0(1): \alpha_i = 0, \text{ all } \alpha_i$$
$$H_0(2): \theta = 0.$$

(b) Test the two contrasts in Problem 7a, Chapter 4, using an analysis of covariance.

2. The following data are identical to those of Problem 6, Chapter 8, with the addition of a covariate. The covariate is performance in a maze-learning task. Do a complete analysis of covariance on the data.

| | | Task a first | | | | | Task b first | | | |
| | | Task a | | Task b | | | Task a | | Task b | |
		X	V	X	V		X	V	X	V
	Subject					Subject				
Enriched	E1	45	18	62	18	E5	50	17	68	17
	E2	20	14	39	14	E6	43	14	63	14
	E3	20	12	39	12	E7	66	19	84	19
	E4	33	15	55	15	E8	59	20	82	20
	N1	28	12	44	12	N5	50	18	71	18
Normal	N2	53	16	69	16	N6	43	15	59	15
	N3	36	13	53	13	N7	42	17	66	17
	N4	52	21	75	21	N8	50	15	62	15
	R1	69	20	86	20	R5	63	20	78	20
Restricted	R2	59	15	77	15	R6	48	14	60	14
	R3	81	21	107	21	R7	61	20	82	20
	R4	83	22	101	22	R8	37	15	63	15

3. The following data are the same as in Problem 4, Chapter 8, with the addition of a covariate. The covariate is the size of the hospital (in hundreds of patients). Do a complete analysis of covariance on these data.

| | Hospital 1 | | | | Hospital 2 | | | |
| | Ward 1A | | Ward 1B | | Ward 2A | | Ward 2B | |
	X	V	X	V	X	V	X	V
	3	5	2	5	3	2	3	2
Ward A	3	5	3	5	2	2	3	2
First	1	5	2	5	0	2	1	2
	3	5	1	5	3	2	2	2
	1	5	0	5	3	2	1	2
Ward B	2	5	2	5	1	2	2	2
First	1	5	0	5	2	2	0	2
	0	5	0	5	1	2	1	2

Appendix

Table A-1. Upper-tail significance levels of the standard normal distribution.

	0	1	2	3	4	5	6	7	8	9
0.00	.5000	.4960	.4920	.4880	.4840	.4801	.4761	.4721	.4681	.4641
.10	.4602	.4562	.4522	.4483	.4443	.4404	.4364	.4325	.4286	.4247
.20	.4207	.4168	.4129	.4090	.4052	.4013	.3974	.3936	.3897	.3859
.30	.3821	.3783	.3745	.3707	.3669	.3632	.3594	.3557	.3520	.3483
.40	.3446	.3409	.3372	.3336	.3300	.3264	.3228	.3192	.3156	.3121
.50	.3085	.3050	.3015	.2981	.2946	.2912	.2877	.2843	.2810	.2776
.60	.2743	.2709	.2676	.2643	.2611	.2578	.2546	.2514	.2483	.2451
.70	.2420	.2389	.2358	.2327	.2296	.2266	.2236	.2206	.2177	.2148
.80	.2119	.2090	.2061	.2033	.2005	.1977	.1949	.1922	.1894	.1867
.90	.1841	.1814	.1788	.1762	.1736	.1711	.1685	.1660	.1635	.1611
1.00	.1587	.1562	.1539	.1515	.1492	.1469	.1446	.1423	.1401	.1379
1.10	.1357	.1335	.1314	.1292	.1271	.1251	.1230	.1210	.1190	.1170
1.20	.1151	.1131	.1112	.1093	.1075	.1056	.1038	.1020	.1003	.0985
1.30	.0968	.0951	.0934	.0918	.0901	.0885	.0869	.0853	.0838	.0823
1.40	.0808	.0793	.0778	.0764	.0749	.0735	.0721	.0708	.0694	.0681
1.50	.0668	.0655	.0643	.0630	.0618	.0606	.0594	.0582	.0571	.0559
1.60	.0548	.0537	.0526	.0516	.0505	.0495	.0485	.0475	.0465	.0455
1.70	.0446	.0436	.0427	.0418	.0409	.0401	.0392	.0384	.0375	.0367
1.80	.0359	.0351	.0344	.0336	.0329	.0322	.0314	.0307	.0301	.0294
1.90	.0287	.0281	.0274	.0268	.0262	.0256	.0250	.0244	.0239	.0233
2.00	.0228	.0222	.0217	.0212	.0207	.0202	.0197	.0192	.0188	.0183
2.10	.0179	.0174	.0170	.0166	.0162	.0158	.0154	.0150	.0146	.0143
2.20	.0139	.0136	.0132	.0129	.0125	.0122	.0119	.0116	.0113	.0110
2.30	.0107	.0104	.0102	.0099	.0096	.0094	.0091	.0089	.0087	.0084
2.40	.0082	.0080	.0078	.0075	.0073	.0071	.0069	.0068	.0066	.0064
2.50	.0062	.0060	.0059	.0057	.0055	.0054	.0052	.0051	.0049	.0048
2.60	.0047	.0045	.0044	.0043	.0041	.0040	.0039	.0038	.0037	.0036
2.70	.0035	.0034	.0033	.0032	.0031	.0030	.0029	.0028	.0027	.0026
2.80	.0026	.0025	.0024	.0023	.0023	.0022	.0021	.0021	.0020	.0019
2.90	.0019	.0018	.0018	.0017	.0016	.0016	.0015	.0015	.0014	.0014
3.00	.0013	.0013	.0013	.0012	.0012	.0011	.0011	.0011	.0010	.0010
3.10	.0010	.0009	.0009	.0009	.0008	.0008	.0008	.0008	.0007	.0007
3.20	.0007	.0007	.0006	.0006	.0006	.0006	.0006	.0005	.0005	.0005
3.30	.0005	.0005	.0005	.0004	.0004	.0004	.0004	.0004	.0004	.0003
3.40	.0003	.0003	.0003	.0003	.0003	.0003	.0003	.0003	.0003	.0002
3.50	.0002	.0002	.0002	.0002	.0002	.0002	.0002	.0002	.0002	.0002
3.60	.0002	.0002	.0001	.0001	.0001	.0001	.0001	.0001	.0001	.0001
3.70	.0001	.0001	.0001	.0001	.0001	.0001	.0001	.0001	.0001	.0001
3.80	.0001	.0001	.0001	.0001	.0001	.0001	.0001	.0001	.0001	.0001

Table A-2. Values of the standard normal distribution for selected two-tailed significance levels.

	SIGNIFICANCE LEVEL									
	0	1	2	3	4	5	6	7	8	9
0.000		3.891	3.719	3.615	3.540	3.481	3.432	3.390	3.353	3.320
.001	3.291	3.264	3.239	3.216	3.195	3.175	3.156	3.138	3.121	3.105
.002	3.090	3.076	3.062	3.048	3.036	3.023	3.011	3.000	2.989	2.978
.003	2.968	2.958	2.948	2.938	2.929	2.920	2.911	2.903	2.894	2.886
.004	2.878	2.870	2.863	2.855	2.848	2.841	2.834	2.827	2.820	2.814
.005	2.807	2.801	2.794	2.788	2.782	2.776	2.770	2.765	2.759	2.753
.006	2.748	2.742	2.737	2.732	2.727	2.721	2.716	2.711	2.706	2.702
.007	2.697	2.692	2.687	2.683	2.678	2.674	2.669	2.665	2.661	2.656
.008	2.652	2.648	2.644	2.640	2.636	2.632	2.628	2.624	2.620	2.616
.009	2.612	2.608	2.605	2.601	2.597	2.594	2.590	2.586	2.583	2.579
.01	2.576	2.543	2.512	2.484	2.457	2.432	2.409	2.387	2.366	2.346
.02	2.326	2.308	2.290	2.273	2.257	2.241	2.226	2.212	2.197	2.183
.03	2.170	2.157	2.144	2.132	2.120	2.108	2.097	2.086	2.075	2.064
.04	2.054	2.044	2.034	2.024	2.014	2.005	1.995	1.986	1.977	1.969
.05	1.960	1.951	1.943	1.935	1.927	1.919	1.911	1.903	1.896	1.888
.06	1.881	1.873	1.866	1.859	1.852	1.845	1.838	1.832	1.825	1.818
.07	1.812	1.805	1.799	1.793	1.787	1.780	1.774	1.768	1.762	1.757
.08	1.751	1.745	1.739	1.734	1.728	1.722	1.717	1.711	1.706	1.701
.09	1.695	1.690	1.685	1.680	1.675	1.670	1.665	1.660	1.655	1.650
.1	1.645	1.598	1.555	1.514	1.476	1.440	1.405	1.372	1.341	1.311
.2	1.282	1.254	1.227	1.200	1.175	1.150	1.126	1.103	1.080	1.058
.3	1.036	1.015	.994	.974	.954	.935	.915	.896	.878	.860
.4	.842	.824	.806	.789	.772	.755	.739	.722	.706	.690
.5	.674	.659	.643	.628	.613	.598	.583	.568	.553	.539
.6	.524	.510	.496	.482	.468	.454	.440	.426	.412	.399
.7	.385	.372	.358	.345	.332	.319	.305	.292	.279	.266
.8	.253	.240	.228	.215	.202	.189	.176	.164	.151	.138
.9	.126	.113	.100	.088	.075	.063	.050	.038	.025	.013

Table A-3. Values of the chi-square distribution for selected one-tailed significance levels.

DF	SIGNIFICANCE LEVEL								
	.5000	.2500	.1000	.0500	.0250	.0100	.0050	.0025	.0010
1	.455	1.323	2.706	3.841	5.024	6.635	7.879	9.141	10.83
2	1.386	2.773	4.605	5.991	7.378	9.210	10.60	11.98	13.82
3	2.366	4.108	6.251	7.815	9.348	11.34	12.84	14.32	16.27
4	3.357	5.385	7.779	9.488	11.14	13.28	14.86	16.42	18.47
5	4.351	6.626	9.236	11.07	12.83	15.09	16.75	18.39	20.51
6	5.348	7.841	10.64	12.59	14.45	16.81	18.55	20.25	22.46
7	6.346	9.037	12.02	14.07	16.01	18.48	20.28	22.04	24.32
8	7.344	10.22	13.36	15.51	17.53	20.09	21.95	23.77	26.12
9	8.343	11.39	14.68	16.92	19.02	21.67	23.59	25.46	27.88
10	9.342	12.55	15.99	18.31	20.48	23.21	25.19	27.11	29.59
11	10.34	13.70	17.28	19.68	21.92	24.72	26.76	28.73	31.26
12	11.34	14.85	18.55	21.03	23.34	26.22	28.30	30.32	32.91
13	12.34	15.98	19.81	22.36	24.74	27.69	29.82	31.88	34.53
14	13.34	17.12	21.06	23.68	26.12	29.14	31.32	33.43	36.12
15	14.34	18.25	22.31	25.00	27.49	30.58	32.80	34.95	37.70
16	15.34	19.37	23.54	26.30	28.85	32.00	34.27	36.46	39.25
17	16.34	20.49	24.77	27.59	30.19	33.41	35.72	37.95	40.79
18	17.34	21.60	25.99	28.87	31.53	34.81	37.16	39.42	42.31
19	18.34	22.72	27.20	30.14	32.85	36.19	38.58	40.88	43.82
20	19.34	23.83	28.41	31.41	34.17	37.57	40.00	42.34	45.31
21	20.34	24.93	29.62	32.67	35.48	38.93	41.40	43.78	46.79
22	21.34	26.04	30.81	33.92	36.78	40.29	42.80	45.20	48.27
23	22.34	27.14	32.01	35.17	38.08	41.64	44.18	46.62	49.73
24	23.34	28.24	33.20	36.42	39.36	42.98	45.56	48.03	51.18
25	24.34	29.34	34.38	37.65	40.65	44.31	46.93	49.44	52.62
26	25.34	30.43	35.56	38.89	41.92	45.64	48.29	50.83	54.05
27	26.34	31.53	36.74	40.11	43.19	46.96	49.64	52.22	55.47
28	27.34	32.62	37.92	41.34	44.46	48.28	50.99	53.59	56.89
29	28.34	33.71	39.09	42.56	45.72	49.59	52.34	54.97	58.30
30	29.34	34.80	40.26	43.77	46.98	50.89	53.67	56.33	59.70
40	39.34	45.62	51.81	55.76	59.34	63.69	66.77	69.70	73.40
60	59.33	66.98	74.40	79.08	83.30	88.38	91.95	95.34	99.61
120	119.3	130.1	140.2	146.6	152.2	158.9	163.6	168.1	173.6

Table A-4. Values of the t distribution for selected two-tailed significance levels.

DF	SIGNIFICANCE LEVEL								
	.500	.200	.100	.050	.020	.010	.005	.002	.001
1	1.000	3.078	6.314	12.71	31.82	63.66	127.3	318.3	636.6
2	.816	1.886	2.920	4.303	6.965	9.925	14.09	22.33	31.60
3	.765	1.638	2.353	3.182	4.541	5.841	7.453	10.21	12.92
4	.741	1.533	2.132	2.776	3.747	4.604	5.597	7.173	8.610
5	.727	1.476	2.015	2.571	3.365	4.032	4.773	5.893	6.869
6	.718	1.440	1.943	2.447	3.143	3.707	4.317	5.207	5.959
7	.711	1.415	1.895	2.365	2.998	3.499	4.029	4.785	5.408
8	.706	1.397	1.860	2.306	2.896	3.355	3.833	4.501	5.041
9	.703	1.383	1.833	2.262	2.821	3.250	3.690	4.297	4.781
10	.700	1.372	1.812	2.228	2.764	3.169	3.581	4.144	4.587
11	.697	1.363	1.796	2.201	2.718	3.106	3.497	4.025	4.437
12	.695	1.356	1.782	2.179	2.681	3.055	3.428	3.930	4.317
13	.694	1.350	1.771	2.160	2.650	3.012	3.372	3.852	4.220
14	.692	1.345	1.761	2.145	2.624	2.977	3.326	3.787	4.140
15	.691	1.341	1.753	2.131	2.602	2.947	3.286	3.733	4.072
16	.690	1.337	1.746	2.120	2.583	2.921	3.252	3.686	4.015
17	.689	1.333	1.740	2.110	2.567	2.898	3.222	3.646	3.965
18	.688	1.330	1.734	2.101	2.552	2.878	3.197	3.610	3.922
19	.688	1.328	1.729	2.093	2.539	2.861	3.174	3.579	3.883
20	.687	1.325	1.725	2.086	2.528	2.845	3.153	3.552	3.850
21	.686	1.323	1.721	2.080	2.518	2.831	3.135	3.527	3.819
22	.686	1.321	1.717	2.074	2.508	2.819	3.119	3.505	3.792
23	.685	1.319	1.714	2.069	2.500	2.807	3.104	3.485	3.768
24	.685	1.318	1.711	2.064	2.492	2.797	3.090	3.467	3.745
25	.684	1.316	1.708	2.060	2.485	2.787	3.078	3.450	3.725
26	.684	1.315	1.706	2.056	2.479	2.779	3.067	3.435	3.707
27	.684	1.314	1.703	2.052	2.473	2.771	3.056	3.421	3.690
28	.683	1.313	1.701	2.048	2.467	2.763	3.047	3.408	3.674
29	.683	1.311	1.699	2.045	2.462	2.756	3.038	3.396	3.659
30	.683	1.310	1.697	2.042	2.457	2.750	3.030	3.385	3.646
40	.681	1.303	1.684	2.021	2.423	2.704	2.971	3.307	3.551
60	.679	1.296	1.671	2.000	2.390	2.660	2.915	3.232	3.460
80	.678	1.292	1.664	1.990	2.374	2.639	2.887	3.195	3.416
100	.677	1.290	1.660	1.984	2.364	2.626	2.871	3.174	3.390
INF	.674	1.282	1.645	1.960	2.326	2.576	2.807	3.090	3.291

Table A-5. Values of the F distribution for selected one-tailed significance levels.

DF NUM	DEN	.5000	.2500	.1000	.0500	.0250	.0100	.0050	.0025	.0010
					SIGNIFICANCE LEVEL					
1	1	1.000	5.828	39.86	161.4	647.8	4052.			
1	2	.667	2.571	8.526	18.51	38.51	98.50	198.5	398.5	998.5
1	3	.585	2.024	5.538	10.13	17.44	34.11	55.55	89.58	166.9
1	4	.549	1.807	4.545	7.709	12.22	21.20	31.33	45.67	74.13
1	5	.528	1.692	4.060	6.608	10.01	16.26	22.78	31.41	47.18
1	6	.515	1.621	3.776	5.987	8.813	13.74	18.63	24.81	35.51
1	8	.499	1.538	3.458	5.318	7.571	11.26	14.69	18.78	25.41
1	10	.490	1.491	3.285	4.965	6.937	10.04	12.83	16.04	21.04
1	12	.484	1.461	3.177	4.747	6.554	9.330	11.75	14.49	18.64
1	15	.478	1.432	3.073	4.543	6.200	8.683	10.80	13.13	16.58
1	20	.472	1.404	2.975	4.351	5.871	8.096	9.944	11.94	14.82
1	24	.469	1.390	2.927	4.260	5.717	7.823	9.551	11.40	14.03
1	30	.466	1.376	2.881	4.171	5.568	7.562	9.179	10.89	13.29
1	40	.463	1.363	2.835	4.085	5.424	7.314	8.828	10.41	12.61
1	60	.460	1.349	2.791	4.001	5.286	7.077	8.494	9.962	11.97
1	120	.458	1.336	2.748	3.920	5.152	6.851	8.179	9.539	11.38
1	INF	.455	1.323	2.706	3.841	5.024	6.635	7.879	9.141	10.83
2	1	1.500	7.500	49.50	199.5	799.5	5000.			
2	2	1.000	3.000	9.000	19.00	39.00	99.00	199.0	399.0	999.0
2	3	.881	2.280	5.462	9.552	16.04	30.82	49.80	79.93	148.5
2	4	.828	2.000	4.325	6.944	10.65	18.00	26.28	38.00	61.25
2	5	.799	1.853	3.780	5.786	8.434	13.27	18.31	24.96	37.12
2	6	.780	1.762	3.463	5.143	7.260	10.92	14.54	19.10	27.00
2	8	.757	1.657	3.113	4.459	6.059	8.649	11.04	13.89	18.49
2	10	.743	1.598	2.924	4.103	5.456	7.559	9.427	11.57	14.91
2	12	.735	1.560	2.807	3.885	5.096	6.927	8.510	10.29	12.97
2	15	.726	1.523	2.695	3.682	4.765	6.359	7.701	9.173	11.34
2	20	.718	1.487	2.589	3.493	4.461	5.849	6.986	8.206	9.953
2	24	.714	1.470	2.538	3.403	4.319	5.614	6.661	7.771	9.339
2	30	.709	1.452	2.489	3.316	4.182	5.390	6.355	7.365	8.773
2	40	.705	1.435	2.440	3.232	4.051	5.179	6.066	6.986	8.251
2	60	.701	1.419	2.393	3.150	3.925	4.977	5.795	6.632	7.768
2	120	.697	1.402	2.347	3.072	3.805	4.787	5.539	6.301	7.321
2	INF	.693	1.386	2.303	2.997	3.690	4.607	5.301	5.995	6.913
3	1	1.709	8.200	53.59	215.7	864.2	5404.			
3	2	1.135	3.153	9.162	19.16	39.17	99.17	199.2	399.2	999.2
3	3	1.000	2.356	5.391	9.277	15.44	29.46	47.47	76.05	141.1
3	4	.941	2.047	4.191	6.591	9.979	16.69	24.26	34.95	56.18
3	5	.907	1.884	3.619	5.409	7.764	12.06	16.53	22.43	33.20
3	6	.886	1.784	3.289	4.757	6.599	9.780	12.92	16.87	23.70
3	8	.860	1.668	2.924	4.066	5.416	7.591	9.596	11.98	15.83
3	10	.845	1.603	2.728	3.708	4.826	6.552	8.081	9.833	12.55
3	12	.835	1.561	2.606	3.490	4.474	5.952	7.226	8.651	10.80
3	15	.826	1.520	2.490	3.287	4.153	5.417	6.476	7.634	9.335
3	20	.816	1.481	2.380	3.098	3.859	4.938	5.818	6.757	8.098
3	24	.812	1.462	2.327	3.009	3.721	4.718	5.519	6.364	7.553
3	30	.807	1.443	2.276	2.922	3.589	4.510	5.239	5.999	7.053
3	40	.802	1.424	2.226	2.839	3.463	4.313	4.976	5.659	6.595
3	60	.798	1.405	2.177	2.758	3.343	4.126	4.729	5.343	6.171
3	120	.793	1.387	2.130	2.680	3.227	3.949	4.497	5.048	5.781
3	INF	.789	1.369	2.084	2.605	3.116	3.782	4.279	4.773	5.422

(Table continued)

Table A-5, cont.

DF NUM	DEN	.5000	.2500	.1000	.0500	.0250	.0100	.0050	.0025	.0010
4	1	1.823	8.581	55.83	224.6	899.6	5625.			
4	2	1.207	3.232	9.243	19.25	39.25	99.25	199.2	399.2	999.2
4	3	1.063	2.390	5.343	9.117	15.10	28.71	46.19	73.95	137.0
4	4	1.000	2.064	4.107	6.388	9.605	15.98	23.15	33.30	53.44
4	5	.965	1.893	3.520	5.192	7.388	11.39	15.56	21.05	31.08
4	6	.942	1.787	3.181	4.534	6.227	9.148	12.03	15.65	21.92
4	8	.915	1.664	2.806	3.838	5.053	7.006	8.805	10.94	14.39
4	10	.899	1.595	2.605	3.478	4.468	5.994	7.343	8.887	11.28
4	12	.888	1.550	2.480	3.259	4.121	5.412	6.521	7.761	9.633
4	15	.878	1.507	2.361	3.056	3.804	4.893	5.803	6.796	8.252
4	20	.868	1.465	2.249	2.866	3.515	4.431	5.174	5.967	7.096
4	24	.863	1.445	2.195	2.776	3.379	4.218	4.890	5.596	6.589
4	30	.858	1.424	2.142	2.690	3.250	4.018	4.623	5.253	6.124
4	40	.854	1.404	2.091	2.606	3.126	3.828	4.374	4.934	5.698
4	60	.849	1.385	2.041	2.525	3.008	3.649	4.140	4.637	5.307
4	120	.844	1.365	1.992	2.447	2.894	3.480	3.921	4.362	4.947
4	INF	.839	1.346	1.945	2.372	2.786	3.319	3.715	4.106	4.617
5	1	1.894	8.820	57.24	230.2	921.8	5764.			
5	2	1.252	3.280	9.293	19.30	39.30	99.30	199.3	399.3	999.3
5	3	1.102	2.409	5.309	9.013	14.88	28.24	45.39	72.62	134.6
5	4	1.037	2.072	4.051	6.256	9.364	15.52	22.45	32.26	51.70
5	5	1.000	1.895	3.453	5.050	7.146	10.97	14.94	20.18	29.75
5	6	.977	1.785	3.108	4.387	5.988	8.746	11.46	14.88	20.80
5	8	.948	1.658	2.726	3.687	4.817	6.632	8.302	10.28	13.48
5	10	.932	1.585	2.522	3.326	4.236	5.636	6.872	8.287	10.48
5	12	.921	1.539	2.394	3.106	3.891	5.064	6.071	7.196	8.892
5	15	.911	1.494	2.273	2.901	3.576	4.556	5.372	6.262	7.567
5	20	.900	1.450	2.158	2.711	3.289	4.103	4.762	5.463	6.460
5	24	.895	1.428	2.103	2.621	3.155	3.895	4.486	5.106	5.976
5	30	.890	1.407	2.049	2.534	3.026	3.699	4.228	4.776	5.533
5	40	.885	1.386	1.997	2.449	2.904	3.514	3.986	4.469	5.128
5	60	.880	1.366	1.946	2.368	2.786	3.339	3.760	4.185	4.757
5	120	.875	1.345	1.896	2.290	2.674	3.174	3.548	3.922	4.416
5	INF	.870	1.325	1.847	2.214	2.567	3.017	3.350	3.677	4.103
6	1	1.942	8.983	58.20	234.0	937.1	5859.			
6	2	1.282	3.312	9.326	19.33	39.33	99.33	199.3	399.3	999.3
6	3	1.129	2.422	5.285	8.941	14.73	27.91	44.84	71.71	132.8
6	4	1.062	2.077	4.010	6.163	9.197	15.21	21.97	31.54	50.52
6	5	1.024	1.894	3.405	4.950	6.978	10.67	14.51	19.58	28.83
6	6	1.000	1.782	3.055	4.284	5.820	8.466	11.07	14.35	20.03
6	8	.971	1.651	2.668	3.581	4.652	6.371	7.952	9.828	12.86
6	10	.954	1.576	2.461	3.217	4.072	5.386	6.545	7.871	9.926
6	12	.943	1.529	2.331	2.996	3.728	4.821	5.757	6.803	8.379
6	15	.933	1.482	2.208	2.790	3.415	4.318	5.071	5.891	7.092
6	20	.922	1.437	2.091	2.599	3.128	3.871	4.472	5.111	6.018
6	24	.917	1.414	2.035	2.508	2.995	3.667	4.202	4.763	5.550
6	30	.912	1.392	1.980	2.421	2.867	3.473	3.949	4.442	5.122
6	40	.907	1.371	1.927	2.336	2.744	3.291	3.713	4.144	4.730
6	60	.901	1.349	1.875	2.254	2.627	3.119	3.492	3.868	4.372
6	120	.896	1.328	1.824	2.175	2.515	2.956	3.285	3.612	4.044
6	INF	.891	1.307	1.774	2.099	2.408	2.802	3.091	3.375	3.743

DF NUM	DEN	SIGNIFICANCE LEVEL								
		.5000	.2500	.1000	.0500	.0250	.0100	.0050	.0025	.0010
10	1	2.042	9.320	60.19	241.9	968.7	6056.			
10	2	1.345	3.377	9.392	19.40	39.40	99.40	199.4	399.4	999.4
10	3	1.183	2.445	5.230	8.786	14.42	27.23	43.69	69.81	129.2
10	4	1.113	2.082	3.920	5.964	8.844	14.55	20.97	30.04	48.05
10	5	1.073	1.890	3.297	4.735	6.619	10.05	13.62	18.32	26.92
10	6	1.048	1.771	2.937	4.060	5.461	7.874	10.25	13.24	18.41
10	8	1.018	1.631	2.538	3.347	4.295	5.814	7.211	8.866	11.54
10	10	1.000	1.551	2.323	2.978	3.717	4.849	5.846	6.987	8.754
10	12	.989	1.500	2.188	2.753	3.374	4.296	5.085	5.966	7.292
10	15	.977	1.449	2.059	2.544	3.060	3.805	4.424	5.097	6.081
10	20	.966	1.399	1.937	2.348	2.774	3.368	3.847	4.354	5.075
10	24	.961	1.375	1.877	2.255	2.640	3.168	3.587	4.025	4.638
10	30	.955	1.351	1.819	2.165	2.511	2.979	3.344	3.720	4.239
10	40	.950	1.327	1.763	2.077	2.388	2.801	3.117	3.438	3.874
10	60	.945	1.303	1.707	1.993	2.270	2.632	2.904	3.177	3.541
10	120	.939	1.279	1.652	1.910	2.157	2.472	2.705	2.935	3.237
10	INF	.934	1.255	1.599	1.831	2.048	2.321	2.519	2.711	2.959
12	1	2.067	9.406	60.71	243.9	976.7	6106.			
12	2	1.361	3.393	9.408	19.41	39.41	99.42	199.4	399.4	999.4
12	3	1.197	2.450	5.216	8.745	14.34	27.05	43.39	69.32	128.3
12	4	1.126	2.083	3.896	5.912	8.751	14.37	20.70	29.66	47.40
12	5	1.085	1.888	3.268	4.678	6.525	9.888	13.38	17.99	26.41
12	6	1.060	1.767	2.905	4.000	5.366	7.718	10.03	12.95	17.99
12	8	1.029	1.624	2.502	3.284	4.200	5.667	7.015	8.613	11.19
12	10	1.012	1.543	2.284	2.913	3.621	4.706	5.661	6.754	8.445
12	12	1.000	1.490	2.147	2.687	3.277	4.155	4.906	5.744	7.005
12	15	.989	1.438	2.017	2.475	2.963	3.666	4.250	4.884	5.812
12	20	.977	1.387	1.892	2.278	2.676	3.231	3.678	4.151	4.823
12	24	.972	1.362	1.832	2.183	2.541	3.032	3.420	3.826	4.393
12	30	.966	1.337	1.773	2.092	2.412	2.843	3.179	3.525	4.001
12	40	.961	1.312	1.715	2.003	2.288	2.665	2.953	3.246	3.642
12	60	.956	1.287	1.657	1.917	2.169	2.496	2.742	2.988	3.315
12	120	.950	1.262	1.601	1.834	2.055	2.336	2.544	2.749	3.016
12	INF	.945	1.237	1.546	1.752	1.945	2.185	2.358	2.527	2.742
15	1	2.093	9.493	61.22	245.9	984.9	6157.			
15	2	1.377	3.410	9.425	19.43	39.43	99.43	199.4	399.4	999.4
15	3	1.211	2.455	5.200	8.703	14.25	26.87	43.08	68.82	127.4
15	4	1.139	2.083	3.870	5.858	8.657	14.20	20.44	29.26	46.76
15	5	1.098	1.885	3.238	4.619	6.428	9.722	13.15	17.66	25.91
15	6	1.072	1.762	2.871	3.938	5.269	7.559	9.814	12.65	17.56
15	8	1.041	1.617	2.464	3.218	4.101	5.515	6.814	8.355	10.84
15	10	1.023	1.534	2.244	2.845	3.522	4.558	5.471	6.514	8.127
15	12	1.012	1.480	2.105	2.617	3.177	4.010	4.721	5.515	6.709
15	15	1.000	1.426	1.972	2.403	2.862	3.522	4.070	4.665	5.535
15	20	.989	1.374	1.845	2.203	2.573	3.088	3.502	3.940	4.562
15	24	.983	1.347	1.783	2.108	2.437	2.889	3.246	3.618	4.139
15	30	.978	1.321	1.722	2.015	2.307	2.700	3.006	3.320	3.753
15	40	.972	1.295	1.662	1.924	2.182	2.522	2.781	3.044	3.400
15	60	.967	1.269	1.603	1.836	2.061	2.352	2.570	2.788	3.078
15	120	.961	1.243	1.545	1.750	1.945	2.192	2.373	2.551	2.783
15	INF	.956	1.216	1.487	1.666	1.833	2.039	2.187	2.330	2.513

(Table continued)

Table A-5, cont.

DF NUM	DEN	.5000	.2500	.1000	.0500	.0250	.0100	.0050	.0025	.0010
					SIGNIFICANCE LEVEL					
20	1	2.119	9.581	61.74	248.0	993.1	6209.			
20	2	1.393	3.426	9.441	19.45	39.45	99.45	199.4	399.4	999.4
20	3	1.225	2.460	5.184	8.660	14.17	26.69	42.77	68.31	126.4
20	4	1.152	2.083	3.844	5.803	8.560	14.02	20.17	28.86	46.10
20	5	1.111	1.882	3.207	4.558	6.329	9.552	12.90	17.32	25.39
20	6	1.084	1.757	2.836	3.874	5.168	7.396	9.588	12.35	17.12
20	8	1.053	1.609	2.425	3.150	3.999	5.359	6.608	8.088	10.48
20	10	1.035	1.523	2.201	2.774	3.419	4.405	5.274	6.267	7.803
20	12	1.023	1.468	2.060	2.544	3.073	3.858	4.530	5.279	6.405
20	15	1.011	1.413	1.924	2.328	2.756	3.372	3.882	4.438	5.248
20	20	1.000	1.358	1.794	2.124	2.464	2.938	3.318	3.720	4.290
20	24	.994	1.331	1.730	2.027	2.327	2.738	3.062	3.401	3.873
20	30	.989	1.303	1.667	1.932	2.195	2.549	2.823	3.105	3.493
20	40	.983	1.276	1.605	1.839	2.068	2.369	2.598	2.831	3.145
20	60	.978	1.248	1.543	1.748	1.944	2.198	2.387	2.576	2.826
20	120	.972	1.220	1.482	1.659	1.825	2.035	2.188	2.339	2.534
20	INF	.967	1.191	1.421	1.571	1.708	1.878	2.000	2.117	2.266
30	1	2.145	9.670	62.27	250.1	1001.	6261.			
30	2	1.410	3.443	9.458	19.46	39.46	99.47	199.5	399.5	999.5
30	3	1.239	2.465	5.168	8.617	14.08	26.50	42.46	67.80	125.4
30	4	1.165	2.082	3.817	5.746	8.461	13.84	19.89	28.45	45.43
30	5	1.123	1.878	3.174	4.496	6.227	9.379	12.66	16.98	24.87
30	6	1.097	1.751	2.800	3.808	5.065	7.229	9.358	12.04	16.67
30	8	1.065	1.600	2.383	3.079	3.894	5.198	6.396	7.816	10.11
30	10	1.047	1.512	2.155	2.700	3.311	4.247	5.071	6.012	7.469
30	12	1.035	1.454	2.011	2.466	2.963	3.701	4.331	5.033	6.089
30	15	1.023	1.397	1.873	2.247	2.644	3.214	3.687	4.200	4.950
30	20	1.011	1.340	1.738	2.039	2.349	2.778	3.123	3.488	4.005
30	24	1.006	1.311	1.672	1.939	2.209	2.577	2.868	3.171	3.593
30	30	1.000	1.282	1.606	1.841	2.074	2.386	2.628	2.876	3.217
30	40	.994	1.253	1.541	1.744	1.943	2.203	2.401	2.602	2.872
30	60	.989	1.223	1.476	1.649	1.815	2.028	2.187	2.346	2.555
30	120	.983	1.192	1.409	1.554	1.690	1.860	1.984	2.105	2.262
30	INF	.978	1.160	1.342	1.459	1.566	1.696	1.789	1.878	1.990
60	1	2.172	9.759	62.79	252.2	1010.	6313.			
60	2	1.426	3.459	9.475	19.48	39.48	99.48	199.5	399.5	999.5
60	3	1.254	2.470	5.151	8.572	13.99	26.32	42.15	67.28	124.5
60	4	1.178	2.082	3.790	5.688	8.360	13.65	19.61	28.03	44.75
60	5	1.136	1.874	3.140	4.431	6.123	9.202	12.40	16.62	24.33
60	6	1.109	1.744	2.762	3.740	4.959	7.057	9.122	11.72	16.21
60	8	1.077	1.589	2.339	3.005	3.784	5.032	6.177	7.535	9.727
60	10	1.059	1.499	2.107	2.621	3.198	4.082	4.859	5.747	7.122
60	12	1.046	1.439	1.960	2.384	2.848	3.535	4.123	4.778	5.762
60	15	1.034	1.380	1.817	2.160	2.524	3.047	3.480	3.951	4.637
60	20	1.023	1.319	1.677	1.946	2.223	2.608	2.916	3.242	3.703
60	24	1.017	1.289	1.607	1.842	2.080	2.403	2.658	2.924	3.295
60	30	1.011	1.257	1.538	1.740	1.940	2.208	2.415	2.628	2.920
60	40	1.006	1.225	1.467	1.637	1.803	2.019	2.184	2.350	2.574
60	60	1.000	1.191	1.395	1.534	1.667	1.836	1.962	2.087	2.252
60	120	.994	1.156	1.320	1.429	1.530	1.656	1.747	1.836	1.950
60	INF	.989	1.116	1.240	1.318	1.388	1.473	1.533	1.589	1.660

DF NUM DEN		SIGNIFICANCE LEVEL								
		.5000	.2500	.1000	.0500	.0250	.0100	.0050	.0025	.0010
INF	1	2.198	9.849	63.33	254.3	1018.	6366.			
INF	2	1.443	3.476	9.491	19.50	39.50	99.50	199.5	399.5	999.5
INF	3	1.268	2.474	5.134	8.526	13.90	26.13	41.83	66.75	123.5
INF	4	1.192	2.081	3.761	5.628	8.257	13.46	19.32	27.61	44.05
INF	5	1.149	1.869	3.105	4.365	6.015	9.020	12.14	16.26	23.79
INF	6	1.122	1.737	2.722	3.669	4.849	6.880	8.879	11.39	15.74
INF	8	1.089	1.578	2.293	2.928	3.670	4.859	5.951	7.245	9.333
INF	10	1.070	1.484	2.055	2.538	3.080	3.909	4.639	5.472	6.762
INF	12	1.058	1.422	1.904	2.296	2.725	3.361	3.904	4.509	5.420
INF	15	1.046	1.359	1.755	2.066	2.395	2.868	3.260	3.686	4.307
INF	20	1.034	1.294	1.607	1.843	2.085	2.421	2.690	2.975	3.378
INF	24	1.028	1.261	1.533	1.733	1.935	2.211	2.428	2.654	2.968
INF	30	1.023	1.226	1.456	1.622	1.787	2.006	2.176	2.350	2.589
INF	40	1.017	1.188	1.377	1.509	1.637	1.805	1.932	2.060	2.233
INF	60	1.011	1.147	1.291	1.389	1.482	1.601	1.689	1.776	1.890
INF	120	1.006	1.099	1.193	1.254	1.310	1.381	1.431	1.480	1.543
INF	INF	1.000	1.000	1.000	1.000	1.000	1.000	1.000	1.000	1.000

Table A-6. Upper α point of Studentized range, α = 0.01.

v \ I	2	3	4	5	6	7	8	9	10	11	12	13	14	15	16	17	18	19	20
1	8.93	13.4	16.4	18.5	20.2	21.5	22.6	23.6	24.5	25.2	25.9	26.5	27.1	27.6	28.1	28.5	29.0	29.3	29.7
2	4.13	5.73	6.77	7.54	8.14	8.63	9.05	9.41	9.72	10.0	10.3	10.5	10.7	10.9	11.1	11.2	11.4	11.5	11.7
3	3.33	4.47	5.20	5.74	6.16	6.51	6.81	7.06	7.29	7.49	7.67	7.83	7.98	8.12	8.25	8.37	8.48	8.58	8.68
4	3.01	3.98	4.59	5.03	5.39	5.68	5.93	6.14	6.33	6.49	6.65	6.78	6.91	7.02	7.13	7.23	7.33	7.41	7.50
5	2.85	3.72	4.26	4.66	4.98	5.24	5.46	5.65	5.82	5.97	6.10	6.22	6.34	6.44	6.54	6.63	6.71	6.79	6.86
6	2.75	3.56	4.07	4.44	4.73	4.97	5.17	5.34	5.50	5.64	5.76	5.87	5.98	6.07	6.16	6.25	6.32	6.40	6.47
7	2.68	3.45	3.93	4.28	4.55	4.78	4.97	5.14	5.28	5.41	5.53	5.64	5.74	5.83	5.91	5.99	6.06	6.13	6.19
8	2.63	3.37	3.83	4.17	4.43	4.65	4.83	4.99	5.13	5.25	5.36	5.46	5.56	5.64	5.72	5.80	5.87	5.93	6.00
9	2.59	3.32	3.76	4.08	4.34	4.54	4.72	4.87	5.01	5.13	5.23	5.33	5.42	5.51	5.58	5.66	5.72	5.79	5.85
10	2.56	3.27	3.70	4.02	4.26	4.47	4.64	4.78	4.91	5.03	5.13	5.23	5.32	5.40	5.47	5.54	5.61	5.67	5.73
11	2.54	3.23	3.66	3.96	4.20	4.40	4.57	4.71	4.84	4.95	5.05	5.15	5.23	5.31	5.38	5.45	5.51	5.57	5.63
12	2.52	3.20	3.62	3.92	4.16	4.35	4.51	4.65	4.78	4.89	4.99	5.08	5.16	5.24	5.31	5.37	5.44	5.49	5.55
13	2.50	3.18	3.59	3.88	4.12	4.30	4.46	4.60	4.72	4.83	4.93	5.02	5.10	5.18	5.25	5.31	5.37	5.43	5.48
14	2.49	3.16	3.56	3.85	4.08	4.27	4.42	4.56	4.68	4.79	4.88	4.97	5.05	5.12	5.19	5.26	5.32	5.37	5.43
15	2.48	3.14	3.54	3.83	4.05	4.23	4.39	4.52	4.64	4.75	4.84	4.93	5.01	5.08	5.15	5.21	5.27	5.32	5.38
16	2.47	3.12	3.52	3.80	4.03	4.21	4.36	4.49	4.61	4.71	4.81	4.89	4.97	5.04	5.11	5.17	5.23	5.28	5.33
17	2.46	3.11	3.50	3.78	4.00	4.18	4.33	4.46	4.58	4.68	4.77	4.86	4.93	5.01	5.07	5.13	5.19	5.24	5.30
18	2.45	3.10	3.49	3.77	3.98	4.16	4.31	4.44	4.55	4.65	4.75	4.83	4.90	4.98	5.04	5.10	5.16	5.21	5.26
19	2.45	3.09	3.47	3.75	3.97	4.14	4.29	4.42	4.53	4.63	4.72	4.80	4.88	4.95	5.01	5.07	5.13	5.18	5.23
20	2.44	3.08	3.46	3.74	3.95	4.12	4.27	4.40	4.51	4.61	4.70	4.78	4.85	4.92	4.99	5.05	5.10	5.16	5.20
24	2.42	3.05	3.42	3.69	3.90	4.07	4.21	4.34	4.44	4.54	4.63	4.71	4.78	4.85	4.91	4.97	5.02	5.07	5.12
30	2.40	3.02	3.39	3.65	3.85	4.02	4.16	4.28	4.38	4.47	4.56	4.64	4.71	4.77	4.83	4.89	4.94	4.99	5.03
40	2.38	2.99	3.35	3.60	3.80	3.96	4.10	4.21	4.32	4.41	4.49	4.56	4.63	4.69	4.75	4.81	4.86	4.90	4.95
60	2.36	2.96	3.31	3.56	3.75	3.91	4.04	4.16	4.25	4.34	4.42	4.49	4.56	4.62	4.67	4.73	4.78	4.82	4.86
120	2.34	2.93	3.28	3.52	3.71	3.86	3.99	4.10	4.19	4.28	4.35	4.42	4.48	4.54	4.60	4.65	4.69	4.74	4.78
∞	2.33	2.90	3.24	3.48	3.66	3.81	3.93	4.04	4.13	4.21	4.28	4.35	4.41	4.47	4.52	4.57	4.61	4.65	4.69

Table A-6, Cont. $\alpha = 0.05$.

v \ I	2	3	4	5	6	7	8	9	10	11	12	13	14	15	16	17	18	19	20
1	18.0	27.0	32.8	37.1	40.4	43.1	45.4	47.4	49.1	50.6	52.0	53.2	54.3	55.4	56.3	57.2	58.0	58.8	59.6
2	6.08	8.33	9.80	10.9	11.7	12.4	13.0	13.5	14.0	14.4	14.7	15.1	15.4	15.7	15.9	16.1	16.4	16.6	16.8
3	4.50	5.91	6.82	7.50	8.04	8.48	8.85	9.18	9.46	9.72	9.95	10.2	10.3	10.5	10.7	10.8	11.0	11.1	11.2
4	3.93	5.04	5.76	6.29	6.71	7.05	7.35	7.60	7.83	8.03	8.21	8.37	8.52	8.66	8.79	8.91	9.03	9.13	9.23
5	3.64	4.60	5.22	5.67	6.03	6.33	6.58	6.80	6.99	7.17	7.32	7.47	7.60	7.72	7.83	7.93	8.03	8.12	8.21
6	3.46	4.34	4.90	5.30	5.63	5.90	6.12	6.32	6.49	6.65	6.79	6.92	7.03	7.14	7.24	7.34	7.43	7.51	7.59
7	3.34	4.16	4.68	5.06	5.36	5.61	5.82	6.00	6.16	6.30	6.43	6.55	6.66	6.76	6.85	6.94	7.02	7.10	7.17
8	3.26	4.04	4.53	4.89	5.17	5.40	5.60	5.77	5.92	6.05	6.18	6.29	6.39	6.48	6.57	6.65	6.73	6.80	6.87
9	3.20	3.95	4.41	4.76	5.02	5.24	5.43	5.59	5.74	5.87	5.98	6.09	6.19	6.28	6.36	6.44	6.51	6.58	6.64
10	3.15	3.88	4.33	4.65	4.91	5.12	5.30	5.46	5.60	5.72	5.83	5.93	6.03	6.11	6.19	6.27	6.34	6.40	6.47
11	3.11	3.82	4.26	4.57	4.82	5.03	5.20	5.35	5.49	5.61	5.71	5.81	5.90	5.98	6.06	6.13	6.20	6.27	6.33
12	3.08	3.77	4.20	4.51	4.75	4.95	5.12	5.27	5.39	5.51	5.61	5.71	5.80	5.88	5.95	6.02	6.09	6.15	6.21
13	3.06	3.73	4.15	4.45	4.69	4.88	5.05	5.19	5.32	5.43	5.53	5.63	5.71	5.79	5.86	5.93	5.99	6.05	6.11
14	3.03	3.70	4.11	4.41	4.64	4.83	4.99	5.13	5.25	5.36	5.46	5.55	5.64	5.71	5.79	5.85	5.91	5.97	6.03
15	3.01	3.67	4.08	4.37	4.59	4.78	4.94	5.08	5.20	5.31	5.40	5.49	5.57	5.65	5.72	5.78	5.85	5.90	5.96
16	3.00	3.65	4.05	4.33	4.56	4.74	4.90	5.03	5.15	5.26	5.35	5.44	5.52	5.59	5.66	5.73	5.79	5.84	5.90
17	2.98	3.63	4.02	4.30	4.52	4.70	4.86	4.99	5.11	5.21	5.31	5.39	5.47	5.54	5.61	5.67	5.73	5.79	5.84
18	2.97	3.61	4.00	4.28	4.49	4.67	4.82	4.96	5.07	5.17	5.27	5.35	5.43	5.50	5.57	5.63	5.69	5.74	5.79
19	2.96	3.59	3.98	4.25	4.47	4.65	4.79	4.92	5.04	5.14	5.23	5.31	5.39	5.46	5.53	5.59	5.65	5.70	5.75
20	2.95	3.58	3.96	4.23	4.45	4.62	4.77	4.90	5.01	5.11	5.20	5.28	5.36	5.43	5.49	5.55	5.61	5.66	5.71
24	2.92	3.53	3.90	4.17	4.37	4.54	4.68	4.81	4.92	5.01	5.10	5.18	5.25	5.32	5.38	5.44	5.49	5.55	5.59
30	2.89	3.49	3.85	4.10	4.30	4.46	4.60	4.72	4.82	4.92	5.00	5.08	5.15	5.21	5.27	5.33	5.38	5.43	5.47
40	2.86	3.44	3.79	4.04	4.23	4.39	4.52	4.63	4.73	4.82	4.90	4.98	5.04	5.11	5.16	5.22	5.27	5.31	5.36
60	2.83	3.40	3.74	3.98	4.16	4.31	4.44	4.55	4.65	4.73	4.81	4.88	4.94	5.00	5.06	5.11	5.15	5.20	5.24
120	2.80	3.36	3.68	3.92	4.10	4.24	4.36	4.47	4.56	4.64	4.71	4.78	4.84	4.90	4.95	5.00	5.04	5.09	5.13
∞	2.77	3.31	3.63	3.86	4.03	4.17	4.29	4.39	4.47	4.55	4.62	4.68	4.74	4.80	4.85	4.89	4.93	4.97	5.01

(Table continued)

Table A-6, Cont.   α = 0.10.

v \ l	2	3	4	5	6	7	8	9	10	11	12	13	14	15	16	17	18	19	20
1	90.0	135	164	186	202	216	227	237	246	253	260	266	272	277	282	286	290	294	298
2	14.0	19.0	22.3	24.7	26.6	28.2	29.5	30.7	31.7	32.6	33.4	34.1	34.8	35.4	36.0	36.5	37.0	37.5	37.9
3	8.26	10.6	12.2	13.3	14.2	15.0	15.6	16.2	16.7	17.1	17.5	17.9	18.2	18.5	18.8	19.1	19.3	19.5	19.8
4	6.51	8.12	9.17	9.96	10.6	11.1	11.5	11.9	12.3	12.6	12.8	13.1	13.3	13.5	13.7	13.9	14.1	14.2	14.4
5	5.70	6.97	7.80	8.42	8.91	9.32	9.67	9.97	10.2	10.5	10.7	10.9	11.1	11.2	11.4	11.6	11.7	11.8	11.9
6	5.24	6.33	7.03	7.56	7.97	8.32	8.61	8.87	9.10	9.30	9.49	9.65	9.81	9.95	10.1	10.2	10.3	10.4	10.5
7	4.95	5.92	6.54	7.01	7.37	7.68	7.94	8.17	8.37	8.55	8.71	8.86	9.00	9.12	9.24	9.35	9.46	9.55	9.65
8	4.74	5.63	6.20	6.63	6.96	7.24	7.47	7.68	7.87	8.03	8.18	8.31	8.44	8.55	8.66	8.76	8.85	8.94	9.03
9	4.60	5.43	5.96	6.35	6.66	6.91	7.13	7.32	7.49	7.65	7.78	7.91	8.03	8.13	8.23	8.32	8.41	8.49	8.57
10	4.48	5.27	5.77	6.14	6.43	6.67	6.87	7.05	7.21	7.36	7.48	7.60	7.71	7.81	7.91	7.99	8.07	8.15	8.22
11	4.39	5.14	5.62	5.97	6.25	6.48	6.67	6.84	6.99	7.13	7.25	7.36	7.46	7.56	7.65	7.73	7.81	7.88	7.95
12	4.32	5.04	5.50	5.84	6.10	6.32	6.51	6.67	6.81	6.94	7.06	7.17	7.26	7.36	7.44	7.52	7.59	7.66	7.73
13	4.26	4.96	5.40	5.73	5.98	6.19	6.37	6.53	6.67	6.79	6.90	7.01	7.10	7.19	7.27	7.34	7.42	7.48	7.55
14	4.21	4.89	5.32	5.63	5.88	6.08	6.26	6.41	6.54	6.66	6.77	6.87	6.96	7.05	7.12	7.20	7.27	7.33	7.39
15	4.17	4.83	5.25	5.56	5.80	5.99	6.16	6.31	6.44	6.55	6.66	6.76	6.84	6.93	7.00	7.07	7.14	7.20	7.26
16	4.13	4.78	5.19	5.49	5.72	5.92	6.08	6.22	6.35	6.46	6.56	6.66	6.74	6.82	6.90	6.97	7.03	7.09	7.15
17	4.10	4.74	5.14	5.43	5.66	5.85	6.01	6.15	6.27	6.38	6.48	6.57	6.66	6.73	6.80	6.87	6.94	7.00	7.05
18	4.07	4.70	5.09	5.38	5.60	5.79	5.94	6.08	6.20	6.31	6.41	6.50	6.58	6.65	6.72	6.79	6.85	6.91	6.96
19	4.05	4.67	5.05	5.33	5.55	5.73	5.89	6.02	6.14	6.25	6.34	6.43	6.51	6.58	6.65	6.72	6.78	6.84	6.89
20	4.02	4.64	5.02	5.29	5.51	5.69	5.84	5.97	6.09	6.19	6.29	6.37	6.45	6.52	6.59	6.65	6.71	6.76	6.82
24	3.96	4.54	4.91	5.17	5.37	5.54	5.69	5.81	5.92	6.02	6.11	6.19	6.26	6.33	6.39	6.45	6.51	6.56	6.61
30	3.89	4.45	4.80	5.05	5.24	5.40	5.54	5.65	5.76	5.85	5.93	6.01	6.08	6.14	6.20	6.26	6.31	6.36	6.41
40	3.82	4.37	4.70	4.93	5.11	5.27	5.39	5.50	5.60	5.69	5.77	5.84	5.90	5.96	6.02	6.07	6.12	6.17	6.21
60	3.76	4.28	4.60	4.82	4.99	5.13	5.25	5.36	5.45	5.53	5.60	5.67	5.73	5.79	5.84	5.89	5.93	5.98	6.02
120	3.70	4.20	4.50	4.71	4.87	5.01	5.12	5.21	5.30	5.38	5.44	5.51	5.56	5.61	5.66	5.71	5.75	5.79	5.83
∞	3.64	4.12	4.40	4.60	4.76	4.88	4.99	5.08	5.16	5.23	5.29	5.35	5.40	5.45	5.49	5.54	5.57	5.61	5.65

SOURCE: *The Analysis of Variance*, by Henry Scheffé, published by Wiley, 1959, with corrections from the original tables published in *Biometrika Tables for Statisticians*, vol. 1, by E. S. Pearson and H. O. Hartley, Cambridge University Press for the Biometrika Trustees, 1954.

Table A-7. Coefficients of orthogonal polynomials.

I	Polynomial	$V = 1$	2	3	4	5	6	7	8	9	10	Σc^2
3	Linear	−1	0	1								2
	Quadratic	1	−2	1								6
4	Linear	−3	−1	1	3							20
	Quadratic	1	−1	−1	1							4
	Cubic	−1	3	−3	1							20
5	Linear	−2	−1	0	1	2						10
	Quadratic	2	−1	−2	−1	2						14
	Cubic	−1	2	0	−2	1						10
	Quartic	1	−4	6	−4	1						70
6	Linear	−5	−3	−1	1	3	5					70
	Quadratic	5	−1	−4	−4	−1	5					84
	Cubic	−5	7	4	−4	−7	5					180
	Quartic	1	−3	2	2	−3	1					28
7	Linear	−3	−2	−1	0	1	2	3				28
	Quadratic	5	0	−3	−4	−3	0	5				84
	Cubic	−1	1	1	0	−1	−1	1				6
	Quartic	3	−7	1	6	1	−7	3				154
8	Linear	−7	−5	−3	−1	1	3	5	7			168
	Quadratic	7	1	−3	−5	−5	−3	1	7			168
	Cubic	−7	5	7	3	−3	−7	−5	7			264
	Quartic	7	−13	−3	9	9	−3	−13	7			616
	Quintic	−7	23	−17	−15	15	17	−23	7			2184
9	Linear	−4	−3	−2	−1	0	1	2	3	4		60
	Quadratic	28	7	−8	−17	−20	−17	−8	7	28		2772
	Cubic	−14	7	13	9	0	−9	−13	−7	14		990
	Quartic	14	−21	−11	9	18	9	−11	−21	14		2002
	Quintic	−4	11	−4	−9	0	9	4	−11	4		468
10	Linear	−9	−7	−5	−3	−1	1	3	5	7	9	330
	Quadratic	6	2	−1	−3	−4	−4	−3	−1	2	6	132
	Cubic	−42	14	35	31	12	−12	−31	−35	−14	42	8580
	Quartic	18	−22	−17	3	18	18	3	−17	−22	18	2860
	Quintic	−6	14	−1	−11	−6	6	11	1	−14	6	780

SOURCE: B. J. Winer, *Statistical Principles in Experimental Design*, 2d ed., McGraw-Hill, 1971.

Table A-8. Arcsin transformation ($Z = 2$ arcsin R).

	R									
	0	1	2	3	4	5	6	7	8	9
0.00	0.000	.063	.089	.110	.127	.142	.155	.168	.179	.190
.01	.200	.210	.220	.229	.237	.246	.254	.262	.269	.277
.02	.284	.291	.298	.304	.311	.318	.324	.330	.336	.342
.03	.348	.354	.360	.365	.371	.376	.382	.387	.392	.398
.04	.403	.408	.413	.418	.423	.428	.432	.437	.442	.446
.05	.451	.456	.460	.465	.469	.473	.478	.482	.486	.491
.06	.495	.499	.503	.507	.512	.516	.520	.524	.528	.532
.07	.536	.539	.543	.547	.551	.555	.559	.562	.566	.570
.08	.574	.577	.581	.584	.588	.592	.595	.599	.602	.606
.09	.609	.613	.616	.620	.623	.627	.630	.633	.637	.640
.1	.644	.676	.707	.738	.767	.795	.823	.850	.876	.902
.2	.927	.952	.976	1.000	1.024	1.047	1.070	1.093	1.115	1.137
.3	1.159	1.181	1.203	1.224	1.245	1.266	1.287	1.308	1.328	1.349
.4	1.369	1.390	1.410	1.430	1.451	1.471	1.491	1.511	1.531	1.551
.5	1.571	1.591	1.611	1.631	1.651	1.671	1.691	1.711	1.731	1.752
.6	1.772	1.793	1.813	1.834	1.855	1.875	1.897	1.918	1.939	1.961
.7	1.982	2.004	2.026	2.049	2.071	2.094	2.118	2.141	2.165	2.190
.8	2.214	2.240	2.265	2.292	2.319	2.346	2.375	2.404	2.434	2.465
.90	2.498	2.501	2.505	2.508	2.512	2.515	2.518	2.522	2.525	2.529
.91	2.532	2.536	2.539	2.543	2.546	2.550	2.553	2.557	2.561	2.564
.92	2.568	2.572	2.575	2.579	2.583	2.587	2.591	2.594	2.598	2.602
.93	2.606	2.610	2.614	2.618	2.622	2.626	2.630	2.634	2.638	2.642
.94	2.647	2.651	2.655	2.659	2.664	2.668	2.673	2.677	2.681	2.686
.95	2.691	2.695	2.700	2.705	2.709	2.714	2.719	2.724	2.729	2.734
.96	2.739	2.744	2.749	2.754	2.760	2.765	2.771	2.776	2.782	2.788
.97	2.793	2.799	2.805	2.811	2.818	2.824	2.831	2.837	2.844	2.851
.98	2.858	2.865	2.872	2.880	2.888	2.896	2.904	2.913	2.922	2.931
.99	2.941	2.952	2.962	2.974	2.987	3.000	3.015	3.032	3.052	3.078

Table A-9. Inverse arcsin transformation (Z to R).

	Z									
	0	1	2	3	4	5	6	7	8	9
0.0	0.000	.000	.000	.000	.000	.001	.001	.001	.002	.002
.1	.002	.003	.004	.004	.005	.006	.006	.007	.008	.009
.2	.010	.011	.012	.013	.014	.016	.017	.018	.019	.021
.3	.022	.024	.025	.027	.029	.030	.032	.034	.036	.038
.4	.039	.041	.043	.046	.048	.050	.052	.054	.057	.059
.5	.061	.064	.066	.069	.071	.074	.076	.079	.082	.085
.6	.087	.090	.093	.096	.099	.102	.105	.108	.111	.114
.7	.118	.121	.124	.127	.131	.134	.138	.141	.145	.148
.8	.152	.155	.159	.163	.166	.170	.174	.178	.181	.185
.9	.189	.193	.197	.201	.205	.209	.213	.217	.221	.226
1.0	.230	.234	.238	.243	.247	.251	.256	.260	.264	.269
1.1	.273	.278	.282	.287	.291	.296	.300	.305	.310	.314
1.2	.319	.323	.328	.333	.338	.342	.347	.352	.357	.361
1.3	.366	.371	.376	.381	.386	.390	.395	.400	.405	.410
1.4	.415	.420	.425	.430	.435	.440	.445	.450	.455	.460
1.5	.465	.470	.475	.480	.485	.490	.495	.500	.505	.510
1.6	.515	.520	.525	.530	.535	.540	.545	.550	.554	.559
1.7	.564	.569	.574	.579	.584	.589	.594	.599	.604	.609
1.8	.614	.618	.623	.628	.633	.638	.643	.647	.652	.657
1.9	.662	.666	.671	.676	.680	.685	.690	.694	.699	.704
2.0	.708	.713	.717	.722	.726	.731	.735	.739	.744	.748
2.1	.752	.757	.761	.765	.769	.774	.778	.782	.786	.790
2.2	.794	.798	.802	.806	.810	.814	.818	.822	.826	.829
2.3	.833	.837	.841	.844	.848	.851	.855	.858	.862	.865
2.4	.869	.872	.875	.879	.882	.885	.888	.891	.895	.898
2.5	.901	.904	.906	.909	.912	.915	.918	.921	.923	.926
2.6	.928	.931	.934	.936	.938	.941	.943	.945	.948	.950
2.7	.952	.954	.956	.958	.960	.962	.964	.966	.968	.969
2.8	.971	.973	.974	.976	.977	.979	.980	.982	.983	.984
2.9	.985	.987	.988	.989	.990	.991	.992	.993	.993	.994
3.0	.995	.996	.996	.997	.997	.998	.998	.999	.999	.999
3.1	1.000									

328

Table A-10. Curves of constant power for selected numerator degrees of freedom. ν.

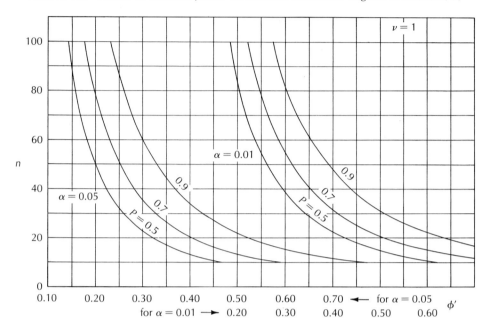

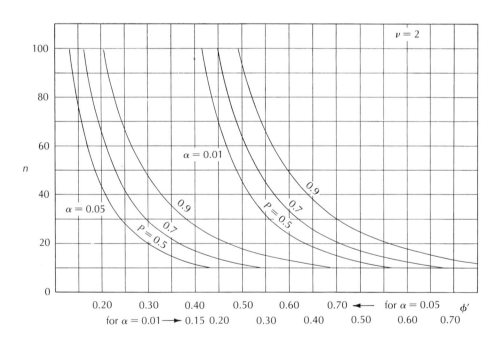

Table A-10, cont.

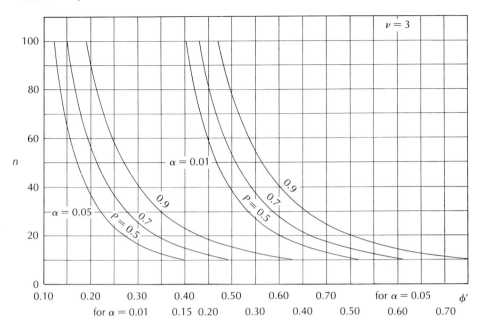

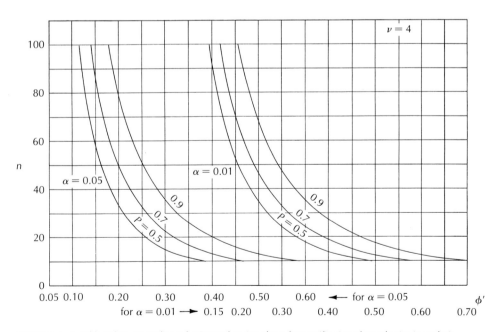

SOURCE: L. S. Feldt and M. W. Mahmoud, "Power function charts for specification of sample size in analysis of variance." *Psychometrika,* **23**:201–210.

Symbols

a, b, c, . . .   Subscripts indicating effects of Factors A, B, C, . . . , respectively; e.g., MS_{ac} is the mean square for the AC interaction. Use of parentheses indicates nesting, e.g., b(a) means A is nested in B.

a*, b*, c*   With subscripts $(k, . . .)$, symbolizes values used to estimate cell means in trend analyses.

A, B, C, . . .   Factors of an experiment; sometimes have subscripts $(i, j, k, . . .)$ indicating specific factor levels. May also indicate effects, e.g., an AC interaction.

AF Adjustment factor in analysis of covariance; usually has subscripts (a, b, . . .).

ASC Adjusted sum of cross-products in analysis of covariance; usually has subscripts (a, b, . . .).

ASX Adjusted sum of squares on X in analysis of covariance; usually has subscripts (a, b, . . .).

bet   Between; used as subscript.

c Coefficient of a planned comparison (linear combination of sample means); usually has identifying subscripts $(i, j, . . .)$.

C Sample value of a planned comparison (linear combination of sample means); may have subscripts $(k, . . .)$ indicating a particular planned comparison.

$C(X, Y)$ Covariance between X and Y.

(C') Normalized value of a planned comparison; under H_0, (C') has the standard normal distribution.

C''	Value of a planned comparison normalized for Tukey method of post hoc comparison.
d	Approximate degrees of freedom for approximate planned comparison (see pp. 61–62).
D	Denominator for approximate planned comparison (see pp. 61–62).
df	Degrees of freedom; also written DF.
e	Error variance; used as subscript.
$E(X)$	Expected value of X.
$F_{(\nu_1, \nu_2)}$	F-distributed random variable with ν_1 degrees of freedom in the numerator and ν_2 degrees of freedom in the denominator.
F'	F ratio for Scheffé method of testing post hoc comparison.
F^*	Quasi-F ratio, for which the denominator is *not* a chi square divided by its degrees of freedom.
g	Subscript indicating any general effect, e.g., $\phi_g{}^2$ (p. 104). Also, in a balanced incomplete blocks design, the number of levels of Factor B that are paired with each level of Factor A.
G	Value used in calculating confidence intervals for ω^2 with random factors.
h	In a balanced incomplete blocks design, the number of levels of Factor A that are paired with each level of Factor B.
H	Value used in calculating confidence intervals for ω^2 with random factors.
H	The efficiency factor in a balanced incomplete blocks design.
$i, j, k, \ldots$	Subscripts indicating specific levels of Factors A, B, C, $\ldots$, respectively; use of parentheses indicates nesting, e.g., $i(j)$ means J is nested in I. Used as *second order subscripts* to indicate trend coefficients of various orders.
$I, J, K, \ldots$	Number of levels of Factors A, B, C, $\ldots$, respectively.
k	When not a subscript, symbolizes a coefficient in a linear combination of mean squares.
Ku	Kurtosis.
m	Grand mean, used as subscript e.g., MS_m is the mean square for the grand mean.
MX	Mean square of X in analysis of covariance; usually has subscripts (a, b, $\ldots$).
MS	Mean square; always has subscripts (a, b, $\ldots$) indicating the specific effect.
n	Number of subjects in a group; sometimes has subscripts $(i, j, \ldots)$ indicating a specific group.

N	Total number of observations in experiment.
$N_{(\mu, \sigma^2)}$	Normally distributed random variable with mean (μ) and variance (σ^2).
Q	In a balanced incomplete blocks design Q_{ij} is one if A_i is paired with B_j, and zero otherwise.
r	Pearson product-moment correlation; r_{XY} is the correlation between X and Y.
rem	Remainder, used as subscript, e.g., SS_{rem} is remaining sum of squares after specific effects have been removed from total sum of squares.
RS	Raw sum of squares; intermediate value obtained when calculating SS; always has subscripts (a, b, . . .) indicating the specific effect.
s^2	Sample variance; sometimes has subscripts ($i, j, \ldots$) indicating subgroup on which variance is calculated.
S	Quantity used for finding confidence interval for ψ, the population value of a planned comparison; S^2 is occasionally used, symbolizing a variance calculated as an intermediate step in a derivation (see p. 60).
S^2	See S.
S'	Quantity used for finding confidence interval for the population value (ψ) in a Scheffé method of post hoc comparison.
SC	Sum of cross-products of X and V in analysis of covariance; usually has subscripts (a, b, . . .).
Sk	Skewness.
SS	Sum of squares; always has subscripts (a, b, c, . . .) indicating the specific effect.
SX	Sum of squares on X in analysis of covariance; usually has subscripts, (a, b, . . .).
t	Total; used as subscript.
t	Sample total; always appears with subscripts ($i, j, k, \ldots$) indicating a specific subgroup.
$t_{(\nu)}$	t-distributed random variable with ν degrees of freedom.
t^*	An intermediate value in calculating estimates of effects in the balanced incomplete blocks design.
$t'_{(k, \nu)}$	Studentized-range-distributed random variable with k groups and ν degrees of freedom.
T	Grand total of all observations in the entire experiment, or, when used as a column label in a table, total number of values squared and summed to obtain RS for a given effect.
T'	Quantity used for finding confidence interval for the population value (ψ) in a Tukey method of post hoc comparison.

u	Subscript indicating an unweighted average of group variances (p. 44).
V	Numerical value of factor levels used in trend tests; usually has subscripts $(i, j, \ldots)$.
$V(X)$	Variance of X.
w	Within; used as subscript.
W	Numerical value of factor levels used in trend tests; usually has subscripts $(i, j, \ldots)$.
X	Random variable; usually has subscripts $(i, j, \ldots)$ indicating a specific value or set of values.
Y	Random variable, usually some tranformation on X (see p. 137); usually has subscripts $(i, j, \ldots)$ indicating specific value or set of values.
Z	Fisher's Z transformation on the Pearson product-moment correlation (r).

α	Without subscripts, indicates α level of significance.
$\alpha, \beta, \gamma, \ldots$	With subscripts $(i, j, \ldots)$, indicate population values for specific effects.
ϵ	Error; usually has subscripts $(i, j, \ldots)$ indicating a specific subgroup.
θ	Slope coefficient in analysis of covariance.
λ	In a balanced incomplete blocks design, the number of times that any two levels of Factor A are both paired with the same level of Factor B.
μ	Population mean; usually has subscripts $(i, j, \ldots)$ indicating the subgroup over which a mean has been taken. If no subscripts are present, μ is the population grand mean.
μ^*	Population mean value specified by a null hypothesis.
μ_r^*	rth moment about the mean.
ν	Degrees of freedom. See $\chi_{(\nu)}^2$, $F_{(\nu_1, \nu_2)}$, and $t_{(\nu)}$.
ξ	Population value of Fisher's Z transformation on ρ.
ρ	Population value of Pearson product-moment correlation; ρ_{XY} is the correlation between x and y.
σ^2	Population variance; usually has subscripts (a, b, $\ldots$) indicating the type of variance.
τ^2	Quantity similar to σ^2 and having same subscripts; more convenient to use than σ^2 for some purposes (see p. 103).
ϕ	Noncentrality parameter of chi square, F, and t distributions.

ϕ'	Noncentrality parameter, similar to ϕ but unaffected by sample size; used in conjunction with Table A-10 to calculate power; $\phi' = \phi/\sqrt{n}$.
$\chi^2_{(\nu)}$	Chi-square distributed random variable with ν degrees of freedom.
ψ	Population value of planned comparison (linear combination of population means); usually has identifying subscripts $(k, \ldots)$.
ω^2	Proportion of variance accounted for; may have subscripts (a, b, . . .) indicating specific effect.
$\cdot$	Subscript, replaces subscript over which a sum or a mean was taken.
$\sim$	"Is distributed as"
$-$	When placed over a symbol, indicates a sample mean.
$\char94$	When placed over a Greek symbol, indicates an estimate of that parameter.
Σ	Summation

Bibliography

Aitkin, M. A. Multiple comparisons in psychological experiments. *British Journal of Mathematical and Statistical Psychology*, 1969, **22:** 193–198.

Alexander, H. W. A general test for trend. *Psychological Bulletin*, 1946, **43:** 533–557.

Anderson, R. L., and T. A. Bancroft. *Statistical Theory in Research.* New York: McGraw-Hill, 1952.

Anderson, R. L., and E. E. Houseman. Tables of orthogonal polynomial values extended to $N = 104$. *Iowa Agricultural Experimental Station Research Bulletin*, #297, 1942.

Anderson, T. W. *An Introduction to Multivariate Statistical Analysis.* New York: Wiley, 1958.

Bartlett, M. S. The use of transformations. *Biometrics*, 1947, **3:** 39–52.

Benjamin, L. S. Facts and artifacts in using analysis of covariance to "undo" the law of initial values. *Psychophysiology*, 1967, **4:** 187–206.

Box, G. E. P. Non-normality and tests on variances. *Biometrika*, 1953, **40:** 318–335.

——— Some theorems on quadratic forms applied in the study of analysis of variance problems: I. Effect of inequality of variance in the one-way classification. *Annals of Mathematical Statistics*, 1954, 25: 290–302.

——— Some theorems on quadratic forms applied in the study of analysis of variance problems: II. Effect of inequality of variance and of correlation of errors in the two-way classification. *Annals of Mathematical Statistics*, 1954, **25:** 484–498.

Box, G. E. P., and S. L. Anderson. Permutation theory in the derivation of robust criteria and the study of departures from assumptions. *Journal of the Royal Statistical Society,* Series B, 1955, **17:** 1–34.

Bulmer, M. G. Approximate confidence limits for components of variance. *Biometrika,* 1957, **44:** 159–167.

Cicchetti, D. V. Extension of multiple-range tests to interaction tables in the analysis of variance: A rapid approximate solution. *Psychological Bulletin,* 1972, **77:** 405–408.

Cochran, W. G. Some consequences when the assumptions for the analysis of variance are not satisfied. *Biometrics,* 1974, **3:** 39–52.

——— The distribution of the largest of a set of estimated variances as a fraction of their total. *Annals of Eugenics,* 1956, **11:** 47–52.

Cochran, W. G., and G. M. Cox. *Experimental Designs.* New York: Wiley, 1957.

Cohen, A. A note on the admissibility of pooling in the analysis of variance. *Annals of Mathematical Statistics,* 1968, **39:** 1744–1746.

Cornfield, J., and J. W. Tukey. Average values of mean squares in factorials. *Annals of Mathematical Statistics,* 1956, **27:** 907–949.

Cox, D. R. *Planning of Experiments.* New York: Wiley, 1958.

David, F. N., and N. L. Johnson. The effects of non-normality on the power function of the F-test in the analysis of variance. *Biometrika,* 1951, 38: 43–47.

Davidson, M. L. Univariate vs. multivariate tests in repeated-measures experiments. *Psychological Bulletin,* 1972, **77:** 446–452.

Davis, D. J. Flexibility and power in comparisons among means. *Psychological Bulletin,* 1969, **71:** 441–444.

DeLury, D. B. The analysis of covariance. *Biometrics,* 1948, **4:** 153–170.

Donaldson, T. S. Robustness of the F-test to errors of both kinds and the correlation between the numerator and denominator of the F-ratio. *Journal of the American Statistical Association,* 1968, **63:** 660–676.

Duncan, D. B. On the properties of the multiple comparison test. *Virginia Journal of Science,* 1952, **3:** 49–67.

——— Multiple range and multiple F-tests. *Biometrics,* 1955, 11: 1–42.

——— Multiple range tests for correlated and heteroscedastic means. *Biometrics,* 1957, **13:** 164–176.

——— A Bayesian approach to multiple comparisons. *Technometrics,* 1965, **7:** 171–222.

Duncan, O. J. Multiple comparisons among means. *Journal of the American Statistical Association,* 1961, **56:** 52–64.

Dunnett, C. W. A multiple comparisons procedure for comparing several treatments with a control. *Journal of the American Statistical Association,* 1955, **50:** 1096–1121.

——— New tables for multiple comparisons with a control. *Biometrics,* 1964, **20:** 482–491.

Eisenhart, C. The assumptions underlying the analysis of variance. *Biometrics*, 1947, **3:** 1–21.

Elashoff, J. D. Analysis of covariance: A delicate instrument. *American Educational Research Journal*, 1969, **6:** 383–401.

Evans, S. H., and E. J. Anastasio. Misuse of analysis of covariance when treatment effect and covariate are confounded. *Psychological Bulletin*, 1968, 69: 225–234.

Federer, W. T. *Experimental Design*. New York: MacMillan Company, 1955.

Federer, W. T., and M. Zelen. Analysis of multifactor classifications with unequal numbers of observations. *Biometrics*, 1966, **22:** 525–552.

Feldt, L. S., and M. W. Mahmoud. Power function charts for specification of sample size in analysis of variance. *Psychometrika*, 1958, **23:** 201–210.

Finney, D. J. *Experimental Design and Its Statistical Basis*. Chicago: University of Chicago Press, 1955.

Fisher, R. A. *The design of Experiments*. Edinburgh: Oliver and Boyd, 1947.

Fleiss, J. L. Estimating the magnitude of experimental effects. *Psychological Bulletin*, 1969, **72:** 273–276.

Fox, M. Charts of the power of the F-test. *Annals of Mathematical Statistics*, 1956, **27:** 484–497.

Games, P. A. Inverse relation between the risks of Type I and Type II errors and suggestions for the unequal n case in multiple comparisons. *Psychological Bulletin*, 1971, **75:** 97–102.

—— Multiple comparisons of means. *American Educational Research Journal*, 1971, **8:** 531–565.

Gayen, A. K. The distribution of the variance ratio in random samples of any size drawn from non-normal universes. *Biometrika*, 1950, **37:** 236–255.

Gaylon, D. W., and F. N. Hopper. Estimating the degrees of freedom for linear combinations of mean squares by Satterthwaite's formula. *Technometrics*, 1969, **11:** 691–706.

Grant, D. A. Analysis of variance tests in the analysis and comparison of curves. *Psychological Bulletin*, 1956, **53:** 141–154.

Graybill, F. A. On quadratic estimation of variance components. *Annals of Mathematical Statistics*, 1954, **25:** 367–372.

—— *An Introduction to Linear Statistical Models*. New York: McGraw-Hill, 1961.

Harris, M., D. G. Howitz, and A. M. Mood. On the determination of sample sizes in designing experiments. *Journal of the American Statistical Association*, 1948, **43:** 391–402.

Harter, H. L. Error rates and sample sizes for range tests in multiple comparisons. *Biometrics*, 1957, **13:** 511–536.

Hartley, H. O. Corrigenda: Tables of percentage points of the "Student-ized" range. *Biometrika*, 1958, **40**: 236.

—— Some recent developments in analysis of variance. *Comments on Pure and Applied Math*, 1955, **8**: 47–72.

Heilizer, F. A note on variance heterogeneity in the analysis of variance. *Psychological Reports*, 1964, **14**: 532–534.

Horsnell, G. The effect of unequal group variances on the F-test for the homogeneity of group means. *Biometrika*, 1953, **40**: 128–136.

Hsu, P. L. Contributions to the theory of student's t-test as applied to the problem of two samples. *Statistical Research Memoirs*, 1938, **2**: 1–24.

Hse, T., and Feldt, L. S. The effect of limitations on the number of criterion score values on the significance level of the F-test. *American Educational Research Journal*, 1969, **6**: 515–527.

Hummel, T. J., and J. R. Sligo. Empirical comparison of univariate and multivariate analysis of variance procedures. *Psychological Bulletin*, 1971, **76**: 49–57.

James, G. S. The comparison of several groups of observations when the ratios of the population variances are unknown. *Biometrika*, 1951, **38**: 324–329.

Johnson, R. H., and L. Jones. Multiple comparisons and error rate. *Journal of College Student Personnel*, 1972, **13**: 154–158.

Kempthorne, O. *The Design and Analysis of Experimenter*. New York: Wiley, 1952.

Kesselman, H. J. The statistic with the smaller critical value. *Psychological Bulletin*, 1974, **81**: 130–131.

Kesselman, H. J., and L. E. Toothaker. Error rates for multiple comparison methods: Some evidence concerning the misleading conclusions of Petrinovich and Hardyck. *Psychological Bulletin*, 1973, **80**: 31–32.

Kramer, C. Y. Extension of multiple range tests to group means with unequal numbers of replications. *Biometrics*, 1956, **12**: 307–310.

—— Extension of multiple range tests to group correlated adjusted means. *Biometrics*, 1975, **13**: 13–18.

Lehmann, E. L. Robust estimation in analysis of variance. *Annals of Mathematical Statistics*, 1963, **34**: 957–966.

Lippman, L. G., and C. J. Taylor. Multiple comparisons in complex ANOVA designs. *Journal of General Psychology*, 1972, **86**: 221–223.

Lord, F. M. Statistical adjustments when comparing preexisting groups. *Psychological Bulletin*, 1969, **72**: 336–337.

Lunney, G. H. Using analysis of variance with a dichotomous dependent variable: An empirical study. *Journal of Educational Measurement*, 1970, **7**: 263–269.

May, J. M. Extended and corrected tables of the upper percentage points of the "Studentized" range. *Biometrika, 1952,* **39:** 192–193.

Mazuy, K. K., and W. S. Connor. Student's *t* in a two-way classification with unequal variances. *Annals of Mathematical Statistics, 1965,* **36:** 1248–1255.

Miller, R. G. *Simultaneous Statistical Inference.* New York: McGraw-Hill, 1966.

Namboodiri, N. K. Experimental designs in which each subject is used repeatedly. *Psychological Bulletin, 1972,* **77:** 54–64.

Neufeld, R. J. Generalization of results beyond the experimental setting: Statistical versus logical considerations. *Perceptual and Motor Skills,* 1970, **31:** 443–446.

Olds, E. G., J. B. Mattson, and R. E. Odeh. Notes on the Use of Transformations in the Analysis of Variance. W.A.D.C. Tech. Rep. 56-308, Wright Air Development Center, Ohio, 1956.

O'Neill, R., and G. B. Wetherill. The present state of multiple comparison methods. *Journal of the Royal Statistical Society,* Series B, 1971, **33:** 218–240.

Overall, J. E., and S. N. Dalal. Empirical formulae for estimating appropriate sample sizes for analysis of variance designs. *Perceptual and Motor Skills,* 1968, **27:** 363–367.

Owen, D. B. The power of Student's *t* test. *Journal of the American Statistical Association,* 1965, **60:** 320–333.

Pachares, J. Table of the upper 10% points of the Studentized range. *Biometrika,* 1959, **46:** 461–466.

Patnaik, P. B. The noncentral X^2 and *F*-distributions and their approximations. *Biometrika,* 1949, **36:** 202–232.

Paull, A. E. On preliminary tests for pooling mean squares in the analysis of variance. *Annals of Mathematical Statistics,* 1950, **21:** 539–556.

Pearson, E. S. The analysis of variance in cases of non-normal variation. *Biometrika,* 1931, **23:** 114–133.

Pearson, E. S., and H. O. Hartley. Tables of the probability integral of the Studentized range. *Biometrika,* 1943, **33:** 89–99.

——— Charts of the power function of the analysis of variance tests, derived from the non-central *F*-distribution. *Biometrika,* 1951, **38:** 112–130.

——— *Biometrika Tables for Statisticians.* London: Cambridge University Press, 1962.

Petrinovich, L. F., and C. D. Hardyck, Error rates for multiple comparison methods: Some evidence concerning the frequency of erroneous conclusions. *Psychological Bulletin,* 1969, **71:** 43–54.

Plackett, R. L., and J. P. Burman. The design of optimum multifactorial experiments. *Biometrika,* 1946, **33:** 305–325.

Rawlings, R. R. Note on orthogonal analysis of variance. *Psychological Bulletin*, 1972, **77**: 373–374.

Robson, D. S. A simple method for construction of orthogonal polynomials when the independent variable is unequally spaced. *Biometrics*, 1959, **15**: 187–191.

Satterthwaite, F. E. An approximate distribution of estimates of variance components. *Biometrics Bulletin*, 1946, **2**: 110–114.

Scheffe, H. A method for judging all contrasts in the analysis of variance. *Biometrika*, 1953, **40**: 87–104.

——— A "mixed model" for the analysis of variance. *Annals of Mathematical Statistics*, 1953, **27**: 23–36.

——— Alternative models for the analysis of variance. *Annals of Mathematical Statistics*, 1956, **27**: 251–271.

——— *The Analysis of Variance*, New York: Wiley, 1959.

Smith, R. A. The effect of unequal group size on Tukey's HSD procedure. *Psychometrika*, 1971, 36: 31–34.

Spjotvoll, E. On the optimality of some multiple comparison procedures. *Annals of Mathematical Statistics*, 1972, **43**: 398–411.

Sprott, D. A. A note on combined interblock and intrablock estimation in incomplete block designs. *Annals of Mathematical Statistics*, 1956, **27**: 633–641. Correction, **28**: p. 269.

Stoloff, P. H. Correcting for heterogeneity of covariance for repeated measures designs of the analysis of variance. *Educational and Psychological Measurement*, 1970, **30**: 909–924.

Tiku, M. L. Tables of the power of the F-test. *Journal of the American Statistical Association*, 1967, **62**: 525–539.

Tukey, J. W. Comparing individual means in the analysis of variance. *Biometrics*, 1949, **5**: 99–114.

——— One degree of freedom for non-additivity. *Biometrics*, 1949, **5**: 232–242.

——— Variances of variance components: I. Balanced designs. *Annals of Mathematical Statistics*, 1956, **27**: 722–736.

——— Variances of variance components: II. The unbalanced single classification. *Annals of Mathematical Statistics*, 1957, **28**: 43–56.

——— Approximations to the upper 5% point of Fisher's B distribution and noncentral χ^2. *Biometrika*, 1957, **44**: 528–530.

Wald, A. On the power function of the analysis of variance test. *Annals of Mathematical Statistics*, 1942, **13**: 434–439.

Walsh, J. F. Use of a general analysis of variance program in missing data situations. *Behavior Research Methods and Instrumentation*, 1971, **3**: 202–203.

——— Using a general analysis of variance algorithm for covariance designs. *Behavior Research Methods and Instrumentation*, 1971, **3**: 203–204.

Welsh, B. L. The generalization of Student's problem when several different population variances are involved. *Biometrika,* 1947, **34:** 28–35.

—— On the comparison of several mean values: An alternative approach. *Biometrika,* 1951, **38:** 330–336.

Werts, C. E., and R. L. Linn, Lord's paradox: A generic problem. *Psychological Bulletin,* 1969, **72:** 423–425.

—— Problems with inferring treatment effects from repeated measures. *Educational and Psychological Measurement,* 1971, **31:** 857–866.

Wilk, M. B., and O. Kempthorne. Fixed, mixed, and random models. *Journal of the American Statistical Association,* 1955, **50:** 1144–1167.

—— Nonaddativities in a Latin square design. *Journal of the American Statistical Association,* 1957, **52:** 218–236.

Williams, J. D. Two way fixed effects analysis of variance with disproportionate cell frequencies. *Multivariate Behavioral Research,* 1972, **7:** 67–83.

Winer, B. J. *Statistical Principles in Experimental Design.* New York: McGraw-Hill, 1971.

Yates, F. Incomplete randomized blocks. *Annals of Eugenics,* 1936, **7:** 121–140.

—— The recovery of inter-block information in balanced incomplete block designs. *Annals of Eugenics,* 1940, **10:** 137–325.

Index